Synthetic Pesticide Use in Africa

World Food Preservation Center Book Series

Series Editor:
Charles L. Wilson

Postharvest Extension and Capacity Building for the Developing World
Majeed Mohammed and Vijay Yadav Tokala

Animal Sourced Foods for Developing Economies: Preservation, Nutrition, and Safety
Edited by Muhammad Issa Khan and Aysha Sameen

Bio-management of Postharvest Diseases and Mycotoxigenic Fungi
Edited by Neeta Sharma and Avantina S. Bhandari

Bio-management of Postharvest Diseases and Mycotoxigenic Fungi
Edited by Charles L. Wilson and Don M. Huber

Synthetic Pesticide Use in Africa: Impact on People, Animals and the Environment
Edited by Charles L. Wilson and Don M. Huber

For more information about this series please visit: http://worldfoodpreservationcenter.com/crc-press.html

Synthetic Pesticide Use in Africa
Impact on People, Animals, and the Environment

Edited by
Charles L. Wilson and Don M. Huber

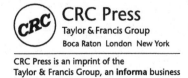

CRC Press
Taylor & Francis Group
Boca Raton London New York

CRC Press is an imprint of the
Taylor & Francis Group, an **informa** business

First edition published 2021
by CRC Press
6000 Broken Sound Parkway NW, Suite 300, Boca Raton, FL 33487-2742

and by CRC Press
2 Park Square, Milton Park, Abingdon, Oxon, OX14 4RN

© 2021 Taylor & Francis Group, LLC

CRC Press is an imprint of Taylor & Francis Group, LLC

The right of Charles L. Wilson and Don M. Huber to be identified as the authors of the editorial material, and of the authors for their individual chapters, has been asserted in accordance with sections 77 and 78 of the Copyright, Designs and Patents Act 1988.

Reasonable efforts have been made to publish reliable data and information, but the author and publisher cannot assume responsibility for the validity of all materials or the consequences of their use. The authors and publishers have attempted to trace the copyright holders of all material reproduced in this publication and apologize to copyright holders if permission to publish in this form has not been obtained. If any copyright material has not been acknowledged please write and let us know so we may rectify in any future reprint.

Library of Congress Cataloging-in-Publication Data
Names: Wilson, Charles L., editor. | Huber, D. M., editor.
Title: Synthetic pesticide use in Africa : impact on people, animals, and the environment /
edited by Charles L. Wilson and Don M. Huber.
Other titles: World Food Preservation Center book series.
Description: First edition. | Boca Raton : CRC Press, 2021. |
Series: World Food Preservation Center book series | Includes bibliographical references and index.
Identifiers: LCCN 2020057802 (print) | LCCN 2020057803 (ebook) |
ISBN 9780367436773 (hardback) | ISBN 9781003007036 (ebook)
Subjects: LCSH: Pesticides—Environmental aspects—Africa. |
Pesticides—Health aspects—Africa. | Pesticides—Government policy—Africa.
Classification: LCC SB950.3.A35 S96 2021 (print) |
LCC SB950.3.A35 (ebook) | DDC 632/.6096—dc23
LC record available at https://lccn.loc.gov/2020057802
LC ebook record available at https://lccn.loc.gov/2020057803

ISBN: 978-0-367-43677-3 (hbk)
ISBN: 978-1-032-00282-8 (pbk)
ISBN: 978-1-003-00703-6 (ebk)

Typeset in Times
by codeMantra

Contents

Preface

Plants are at the heart of human/animal health and the health of the environment. When food production is poorly managed, the result is nutrient-deficient plants that become the source of poor human/animal health and environmental degradation. Today's synthetic-chemical-dependent "Conventional Agriculture" and synthetic-chemical-free "Regenerative Agriculture" result in diametrically opposite results pertaining to crop, human/animal, and environmental health.

The world's population is estimated to reach near 10 billion people by 2050, with 70% of this increased population occurring in developing countries such as those in Africa. By some estimates, food production will need to double in order to avoid starvation in African countries. The "Second Green Revolution," with its increased use of synthetic pesticides/fertilizers and GMOs mounted by Conventional Agriculture to meet this population challenge, is not sustainable.

A UN report presented to the UN Human Rights Council recognizes that "although pesticide use has been correlated with a rise in food production, it has had *catastrophic impacts*" on human health and the environment. An average of about 200,000 people across the world die from the toxic exposure to pesticides each year according to the United Nations in their call for tighter global regulation of substances meant to control pests or weeds for plant cultivation. The report acknowledges that "increased food production has not succeeded in eliminating hunger worldwide." And it continues, "reliance on hazardous pesticides is a short-term solution that undermines the rights to adequate food and health for present and future generations." This UN report lists an array of serious illnesses and health issues with suspected links to pesticides, including cancer, Alzheimer's and Parkinson's diseases, hormone disruption, birth defects, sterility, and neurological effects.

Due to serious underreporting, the number of deaths from pesticide poisoning worldwide can only be estimated, but 99% of these fatalities are believed to be in developing countries. Pesticides banned in the developed world are often exported and sold in developing countries. This, coupled with the poor labeling of pesticides (in 'unfamiliar' languages), poor application training, and poor sanitation, leads to a large number of avoidable deaths of pesticide applicators as well as food consumers.

Tragically, pesticides are the most commonly used substances among farmers in developing countries to commit suicide. In the early 1990s, Sri Lanka had one of the world's highest suicide rates, at around 52 per 100,000 per year, up from around 8 per 100,000 in 1955. Much of this rise was traced to the introduction of pesticides into poor rural homes from the late 1960s, making them accessible for self-poisoning. Fortunately, programs have been introduced more recently to reduce these deaths.

The developed world is also paying a health price for exporting banned pesticides to developing countries, since the exported pesticides used for growing coffee, fruit, tea, and other commodities are then imported back into the United States, the EU, and other developed countries as contaminants on these imports. Although there are techniques to detect these chemical residues, only about 2% of imported produce is inspected by the US Food and Drug Administration in the United States or the European Food Safety Authority (EFSA) in the EU.

It is hoped that the knowledge available in the following chapters in "Synthetic Pesticide Use in Africa: Impact on People, Animals, and the Environment" will both enlighten you to the present serious concerns and motivate you to make the changes necessary for the sustainable production of safe, nutritious, and affordable food for the anticipated 50 billion inhabitants of this Earth in 2050.

Preface

Editors

Don M. Huber, Professor Emeritus of Plant Pathology at Purdue University, holds B.S. and M.S. degrees from the University of Idaho (1957, 1959), a Ph.D. from Michigan State University (1963), and is a graduate of the US Army Command & General Staff College, Industrial College of the Armed Forces, and National Security Program. He was Cereal Pathologist at the University of Idaho for 8 years before joining the Department of Botany & Plant Pathology at Purdue University in 1971. His agricultural research in the past 55 years has focused on the epidemiology and control of soilborne plant pathogens with emphasis on microbial ecology, cultural and biological controls, nutrient–disease interactions, pesticide–disease interactions, physiology of host–parasite relationships, and techniques for rapid microbial identification. He is a past Chairman of the USDA-APS National Plant Disease Recovery System; a member of the US Threat Pathogens Committee; former member of the Advisory Board for the Office of Technology Assessment, US Congress (now Congressional Research Service); and Global Epidemiology and Quadripartite Medical Working Groups of the Office of the US Surgeon General (OTSG). He is the author or coauthor of over 300 journal articles, Experiment Station Bulletins, book chapters and review articles; 3 books; and 84 special invited publications. He is also an active scientific reviewer; speaker; consultant to academia, industry, and government; and international research cooperator.

Charles L. Wilson's career spans 50+ years (15 years in academe at West Virginia University, University of Arkansas, and Ohio State University; 37 years with the USDA). In the last 6 years, he has been involved in the establishment of the World Food Preservation Center® LLC. Dr. Wilson has authored over 250 scientific publications, 20 patents, and 6 books. He has been invited to organize and chair numerous international symposia, to present keynote addresses at numerous additional symposia, to deliver over 40 lectures in 22 states in the United States and 15 countries, and to lead seminars and discussions with students at 10 major universities. He has organized three international BARD-sponsored workshops on food preservation by biological means. He was instrumental in the development of the first EPA-registered "biofungicide" for the control of postharvest diseases. Dr. Wilson was honored in 2010 by the USDA/BARD/ISHS at an international workshop with an award that read: "In recognition for over twenty-years of pioneering research on the Biological Control of Postharvest Diseases. Research that came from a simple idea and grew into a new field of science fostering research and the careers of young scientist throughout the world."

Contributors

Arden Andersen
Independent Family Practice and
    Occupational Medicine Physician
    and Consultant
Kansas, Missouri

Mohamed Besri
Hassan II Institute of Agronomy and
    Veterinary Medicine
Rabat, Morocco

P. Deonikar
Systems Biology Group
Massachusetts Institute of Technology
Cambridge, Massachusetts

Arthur Dunham
Department of Biology
University of Pennsylvania
Philadelphia, Pennsylvania

Ted Dupmeier
Ryan Veterinary Clinic
Ryan, Iowa

Carey Gillam
U.S. Right to Know
Oakland, California

Don M. Huber
Botany and Plant Pathology Department
Purdue University
West Lafayette, Indiana

Vinay K. Khanna
CSIR-Indian Institute of Toxicology Research
Lucknow, India

Robert J. Kremer
School of Natural Resources, University of
    Missouri
Columbia, Missouri

Andre Leu
Regeneration International
Ambassador
IFOAM - Organics International
Author, Poisoning our Children, The Myths of
    Safe Pesticides
Bonn, Germany

Gustav Komla Mahunu
Department of Food Science & Technology,
    Faculty of Agriculture, University for
    Development Studies
Tamale, Ghana

Peter Mokaya
Managing Director
Organic Consumers Alliance (OCA)
Nairobi, Kenya

Michelle Perro
Executive Director
Gordon Medical Associates
San Rafael, California

Stephanie Seneff
Computer Science and Artificial Intelligence
    Laboratory
Massachusetts Institute of Technology
Cambridge, Massachusetts

Prahlad K. Seth
Biotech Park
Lucknow, India

V.A. Shiva Ayyadurai
Systems Biology Group
Massachusetts Institute of Technology
Cambridge, Massachusetts

Qiya Yang
School of Food and Biological Engineering,
    Jiangsu University
Zhenjiang, People's Republic of China

Hongyin Zhang
School of Food and Biological Engineering,
    Jiangsu University
Zhenjiang, People's Republic of China

1 Impact of Synthetic Pesticides on the Health of the African People, Animals and Environment

Peter Mokaya

CONTENTS

1.1 INTRODUCTION

Writing the introductory chapter "The Impact of Synthetic Pesticides on the Health of the African People, Animals and Environment" is a rare honor and a herculean task. I see it as an opportunity to "swim with the dolphins" and also contribute a "drop in the ocean" to stop the raging fire in the forest. In the words of the Peace Prize Award winner Professor Wangari Mathai, the first African Woman to win the coveted Nobel Prize. This introductory book chapter is written with her famous plea to present and future generations to protect and preserve the environment from the raging fire of synthetic pesticides. We must realize that if the current generation destroys the environment, humanity, by extension, will be hurting themselves and destroy the heritage of future generations. We are "stewards" of this earth's resources and are entrusted to protect and preserve all that is within it, especially in Africa. It is to the Mother Africa that this introductory chapter is dedicated and to all its diverse peoples. Information in subsequent chapters will hopefully awaken and protect the continent from the "push to flood it" with toxic synthetic pesticides and other harmful agrochemicals.

1.2 A PUBLIC HEALTH PHYSICIAN'S PERSPECTIVE

It is against this backdrop and from a public health physician's perspective that this introduction to the harm done by synthetic pesticides, the world over and especially Africa, is presented. This information is shared with the hope and belief that humanity can find viable and sustainable solutions to the escalating use of synthetic pesticides. Synthetic pesticides that are an increasing threat to the health of soils and microorganisms while degrading the environment as well as worsening the health of animals and humanity. This present state of affairs poses an existential threat to humanity, unless we put a stop to "digging the hole" and embrace more sustainable and environmentally friendly food systems such as those promoted through regenerative agriculture.

Regenerative food systems free of synthetic pesticides are in alignment with the UN Sustainable Development Goals (SDGs) and objectives that are aimed toward reducing poverty (SDG no. 1) and hunger (SDG no. 2), as well as improving health (SDG no. 3). The SDGs also target climate change mitigation and the promotion of global partnerships (SDG no. 17). In addressing the broken food system, one should listen to the early admonition of Hippocrates, *the father of modern medicine*. In 431 B.C., Hippocrates famously said, "Let food be thy medicine and medicine be thy food".

Unfortunately, current farming and food systems promote and encourage the use of synthetic pesticides, driven by profit interests that result in more use of herbicides and other pesticides (Charles Benbrooke et al. September 2012, https://enveurope.springeropen.com/articles/10.1186/2190-4715-24-24).

Consequentially, these farming practices have resulted in an exponential increase of noncommunicable diseases (NCDs) as demonstrated by Dr. Nancy Swanson et al. (2014).

1.3 "THE BIG PICTURE" OF IMPACT OF SYNTHETIC PESTICIDES ON AFRICA

This introductory chapter seeks to provide the "big picture" on why and how synthetic pesticides are one of the leading public health concerns in the 21st century. They pose an existential threat to humans and the environment globally, and specifically, in Africa. The current chemical industrial model of farming and food systems is failing. This is supported by evidence, some of which is presented in the Appendix that includes up to ten of the latest blog posts on our website, the latest of which can be accessed on this link.

The current food system is anchored on the use of synthetic pesticides and other agrochemicals that, unfortunately, are harmful to the entire ecosystem. Evidence provided in subsequent chapters of this book demonstrates that Africa remains the last frontier of hope. Going forward, there is need to apply a multifaceted and integrated systems thinking approach as described by Peter Senge, as the Fifth Discipline. The systems thinking approach must replace the current silos-based compartmentalized and reductionist approach to agricultural food production which fails to see and address the "big picture". As a result of this current antagonistic approach to various components of the ecosystem, there is an increase in the negative impact of synthetic pesticides in Africa.

Going forward, there is an urgent need to provide transdisciplinary evidence for the need to transform the food system from the current unsustainable food systems, which overuse synthetic pesticides and other agrochemicals, to sustainable food systems, which don't use synthetic pesticides and other agrochemicals but rather adopt regenerative farming and food production systems. These alternative food systems conserve and preserve the ecosystem's biodiversity, which, in turn, improves the health of the soils, animals, the environment, and ultimately, people's health, to include the African people.

This book is, indeed, dedicated to the health of African people, their animals, and their environment. Toward that end, subsequent chapters of this book, as outlined in this introductory chapter, speak to this African audience.

1.4 EVIDENCE OF HOW SYNTHETIC PESTICIDES NEGATIVELY IMPACT THE HEALTH, NUTRITION AND SUSTAINABILITY OF THE ECOSYSTEM

The increasing use of pesticides in Africa has negatively impacted the entire ecosystem, from damaging the health of the environment to that of people as is elaborated in several chapters of this book. In Chapter 1, Dr. Arden Anderson elaborates and provides evidence on how synthetic pesticides have negatively impacted on the health, nutrition, and sustainability of the ecosystem, titled "Health, Nutrition and Sustainability…", while in Chapter 2, Prof. Don Huber explains how glyphosate negatively impacts the health of humans, animals, and the environment entitled "Glyphosate's Impact on Man, His Animals, and Environment" with Prof. Tyrone Hayes, in Chapter 3, concluding that the general negative impact of pesticides on what he has titled "From Silent Spring to Silent Night". This chapter provides evidence of how the birds are being poisoned and dying from the overuse of synthetic pesticides, as cautioned by Rachel Carson in her famous book, *The Silent Spring*, and how waterways are poisoned which leads to the death of frogs and other amphibians. Their evidence-based "expose" is an indictment of the current food systems that are deliberately "poisoning" the entire ecosystem for short-term profit interests without considering the good and long-term viability and health of ecosystems.

1.5 EVIDENCE OF HOW SYNTHETIC PESTICIDES IMPACT HUMAN HEALTH

The correlation between the increasing use of synthetic pesticides and deteriorating human health is elaborated in Chapters 4–9. Dr. Stephanie Seneff, Massachusetts Institute of Technology, discusses the links and provides evidence of "Agricultural Pesticides and the Deterioration of Human Health" in Chapter 4. Prof. Prahlad Seth discusses how synthetic pesticides harm the brain in Chapter 6 titled "Synthetic Pesticides and the Brain". These subthemes conclude with Chapter 7 with Dr. Yemisi Jeff-Agboola dissecting and presenting "Gender and Synthetic Pesticides in Africa".

1.6 EVIDENCE OF HOW SYNTHETIC PESTICIDES IMPACT CHILDREN

Children, including African children, are not spared the scourge of synthetic pesticides. None other than the passionate regenerative organic farming advocate, Andrew Leu, could do more justice to this topic. He shares damning evidence of how pesticides harm the health of children in Chapter 8 titled "Poisoning Our Children and the Myths of Safe Pesticides", which is complemented by Dr. Michelle Perro, who is the author of "What Is Making Our Children Sick?" She discusses this subject under the title "Pesticides and the Crisis in Children's Health" in Chapter 9.

1.7 THE "POISONOUS COMBO" OF SYNTHETIC PESTICIDES AND GMOs

Chapters 10 and 11 are star-studded and packed with evidence presented by Dr. Judy Carman and Prof. Gilles Eric Seralini, who tackle the evidence of the link between GMOs and increased use of synthetic pesticides, and how these two work synergistically to produce a "poisonous combo" which harms the entire ecosystem, especially humans, in Chapter 10 titled "The Safety of Crops That Are Genetically Modified to Produce Insecticides" and Chapter 11 titled "Roundup Toxicity Above Glyphosate and GMOs Tolerance to It", respectively.

1.8 NEGATIVE IMPACT OF SYNTHETIC PESTICIDES ON ANIMALS

As discussed at the onset of this introductory chapter, a systems thinking approach is adopted to present the harms and damages caused by synthetic pesticides. This is not exclusive to humans; the negative effects also affect animals and birds, in some cases leading to the extinction of some species and the ensuing drastic reduction of biodiversity. These topics are covered in Chapters 12, 13, and 14. Insect species have been reduced drastically during our lifetime, in less than 30–50 years, which is correlated to the increased use of synthetic pesticides. For example, some pollinator butterfly populations have decreased by over 30% within only the last 20 years of synthetic pesticides use.

Dr. Darcy Ogada, in Chapter 12, discusses the topic titled, "African Animals Under Synthetic Pesticide Attack", while Dr. Arthur Dunham discusses how veterinary animals have been negatively impacted by synthetic pesticides in Chapter 13, titled "A Private Veterinarian's 40 Years' Experience with Pesticide Toxicity to Animals". The section on animal health concludes with Dr. Ted Dupmeier, in Chapter 14, discussing how increased pesticide use has become a threat to animal production and their sustainability under the title "Agricultural Pesticide Threat to Animal Production and Sustainability".

1.9   NEGATIVE IMPACT OF SYNTHETIC PESTICIDES ON SOIL HEALTH

A systems thinking approach to food production and consumption policies and strategies cannot be complete without discussing the impact of synthetic pesticides on soil health as a critical part of the environment. This subtheme is covered in two chapters, Chapters 15 and 16, by Prof. Robert Kremer, under the title "Disruption of the Soil Microbiota by Agricultural Pesticides", and Dr. Michael McNeal, under the title, "Remediation of Agricultural Pesticides for Production and Sustainability".

Dr. Elaine Ingham, a soil scientist, has written extensively on the harmful effects of synthetic pesticides on soil microorganisms and how soil can regenerate and nourish biodiversity of soil microorganisms and to increase organic matter, resulting in "living soils" which sustain a rich and biodiverse flora and fauna.

1.10   CONTROL STRATEGIES FOR COMBATING SYNTHETIC PESTICIDES USE

The remaining chapters, unlike the previous chapters that have addressed the problem posed by synthetic pesticides address the subtheme of control strategies, Chapters 17 and 18, titled "Synthetic Pesticides and Postharvest Diseases" by Prof. Hongyin Zhang and "Montreal Protocol and the Methyl Bromide Phase out in Africa" by Prof. Mohammed Bersi, respectively. This is followed by Chapter 19 which discusses "Regulatory Collusion and the Illusion of Safety" by Carey Gillam, a renowned journalist and the author of the famous book *White Wash, the Story of a Weed Killer, Cancer and the Corruption of Science*. This book concludes with Chapter 19, with Dr. Shiva Ayyadurai, MIT, discussing and laying bare "The Myth of Substantial Equivalence and the Safety Evaluation of Genetically Engineered Crops".

1.11   THE WAY FORWARD

The way forward for Africa is captured in the appendix with a series of blogs written by Dr. Peter Mokaya, the Founding Director of Organic Consumers Alliance (OCA), board member of Noncommunicable Diseases Alliance and member of the pan-African Alliance for Food Sovereignty in Africa (AFSA).

Dr. Zach Bush, a triple board-certified physician, believes that use of agrochemicals, which include synthetic pesticides, is a bigger threat to humankind than terrorism. He cautions that unless humans change their direction and adopt more sustainable and synthetic chemicals-free farming systems, they will have just less than 90 years for their possible extinction. Dr. Bush pleads with Africa not to allow agrochemical corporations and Big Agriculture to "invade" Africa and "flood" it with synthetic pesticides and other synthetic inputs. He states that "we have already destroyed natural soils in the USA" and that this has consequently resulted in an exponential increase in NCDs.

Weakened immune systems from synthetic pesticides use have resulted in increased susceptibility and vulnerability to infections and more severe disease manifestation, leading to premature deaths, including those from COVID-19. At the time of writing this book, the United States is the leading country in COVID-19 incidence and prevalence rates. This epidemiological pattern is correlated with the severity of comorbidities that include diabetes, hypertension, and other noncommunicable diseases.

Africa, with its weak healthcare systems, cannot afford preventable diseases, including pandemics like COVID-19, but it can avoid and prevent such disease outbreaks by shifting and transitioning its food systems from synthetic pesticide–based farming to the more sustainable agroecological food systems that are free from synthetic chemicals.

Agroecology presents a sustainable agriculture solution. Agroecology is a food system that is natural and affordable, and it also have positive impacts on the "Quadruple Crises" which include increasing NCDs, soil degeneration, loss of biodiversity, climate crisis, and an emerging socioeconomic crisis. The agroecological food system provides a sustainable and economically viable solution to the "Quadruple Crises" all in one swoop. This is the future that Africa and the African people must embrace, and they should take the leadership in saving the planet Earth through working in partnership with like-minded leaders and other agroecological practitioners, including the authors of this book, in response to the UN's clarion call that echoed in the UNCTAD report "Wake Up Before It Is Too Late" (https://unctad.org/en/PublicationsLibrary/ditcted2012d3_en.pdf).

2 Glyphosate's Impact on Humans, Animals, and the Environment

Don M. Huber
Purdue University

CONTENTS

2.1 INTRODUCTION

Innovations for understanding new chemical syntheses and searches for new markets of residual war materials all introduced chemical farming to agriculture after World War II (WWII). The commercialization of glyphosate as an herbicide (Roundup®) in 1974 started the massive conversion of agriculture to a nearly monochemical herbicide program. The simultaneous shift from conventional tillage to no-till or minimum tillage for soil erosion control stimulated this conversion. The development and introduction of genetically engineered (GE) crops tolerant to glyphosate (Roundup Ready®) in 1996 greatly increased the volume and scope of glyphosate usage, and furthered the almost complete conversion of major segments of crop production to a monochemical herbicide strategy. With the regulatory approval of glyphosate as a crop desiccant in 2005 for preharvest application to cereals, pulses, and other crops, glyphosate fulfilled a monochemical role, with extensive chemical residues occurring in many food and feed products in the food chain. The numerous interactions of glyphosate with plant nutrition, increased disease and deterioration of nutritional density of food products, have been largely overlooked, but they appear more obvious each year as glyphosate's residual effects on mineral nutrition, microbial balance, and environmental impact become more apparent.

With the extensive use of glyphosate for general weed control and crop desiccation, the development of weeds resistant to herbicides and the rapid adoption of genetically modified (GMO) crops (such as soybean, corn (maize), cotton, canola, sugar beets, and alfalfa) for tolerance to herbicides (glyphosate, 2,4-D, dicamba, glufosinate, etc.) have greatly increased the volume of synthetic chemicals used in agriculture and their subsequent impacts on nutrient availability and health. The additional nutrient inefficiency of GMO plants has further reduced the density and availability of essential mineral nutrients for crops, animals, and humans.

Minerals are constituents of plant and animal cells and tissues, acting as activators, inhibitors, and regulators of physiological processes necessary for cell and bodily functions. Mineral deficiencies from herbicide immobilization or microbial dysbiosis are now common conditions in plants, animals, people, and the environment (Johal and Huber, 2009; Hoy, Swanson, and Seneff, 2015; Swanson, Leu, Abrahamson, et al., 2014). The ready access to chemical tools after WWII simplified some of the management decisions for weed, insect, and disease control, but it created a dependence on the "silver bullet" approach of industrial agriculture at the expense of ecological management necessary for sustainability. The focus of this chapter is primarily on glyphosate, just one of the several hundred synthetic chemicals used in modern agriculture, and its impact on soil, crop, and environmental health. Glyphosate, in its use for weed control, crop desiccation, and forest management, is the most extensively and indiscriminately used chemical "pesticide" in the world, with about half of it used in agriculture (almost ¼ billion pounds in the United States annually = 250,000,000 lbs = 113,398,093 kg) (USDA-NASS, 2015) and a similar additional amount used in forestry, parks and recreation areas, school grounds, home gardens and surroundings, utility right-of-ways, etc.

2.2   CHARACTERISTICS OF GLYPHOSATE AND OTHER HERBICIDES

Although herbicides are grouped based on the physiological pathways disrupted (ALS, auxin, photosynthesis, lipid synthesis, 5-enolpyruvulshikimate-3-phosphate synthase (EPSPS), glutamine, fatty acid, cell wall inhibiting, etc.) (Mallory-Smith and Retzinger Jr., 2003), the chelation of mineral nutrients, as critical enzyme cofactors for these physiological pathways, is a primary mode of action of herbicides and other pesticides (Table 2.1). Some metal chelates are used extensively in agriculture to increase solubility or uptake of essential micronutrients (citric acid, EDTA, amino

TABLE 2.1

Some Minerals Chelated by Various Herbicides

Chemical Family	Active Ingredient	Commercial Name	Elements Affected
Sulfonylureas	Trifloxysulfuron metal	Krismat	P
	Halosulfuron	Sempra	P
Triazines	Ametrine	Gesapax	Mn, K, Zn, Mg
	Atrazine	Gesaprin	Mn, K, Zn, Mg
	Hexazine	Velpar	Mn, K, Zn, Mg
	Terbutrine	Igran	Mn, K, Zn, Mg
Ureas	Diuron	Karmex	Mn, K
Isoxasoles	Isoxaflutole	Merlin	Mg
Glycines	Glyphosate	Roundup, Touchdown	Mn, Co, Cu, B, Fe, Zn, K
Phosphonic acids	Glufosinate	Finale, Liberty	Mn, Co, Cu, B, Fe, Zn, K
Dinitroanalines	Pendimethalins	Prowl	N/A
Chloroacetamides	Acetochlor	Harness	Mg, K
Phenoxy-carboxylic acids	2.4-D	2,5-D	Cu
Phenoxaprop		Puma, Puma gold	Cu
Dicamba	Dimethylamine salt	Banvel	Cu
Tordon	Picloram	Tordon	Cu

acids, etc.) that are critical for physiological processes; however, chelators that immobilize specific metal cofactors essential for enzyme activity (nitrification inhibitors, fungicides, antibiotics, plant growth regulators, etc.) are also used as herbicides (Table 2.1) and other biocides.

2.2.1 GLYPHOSATE AS A MINERAL CHELATOR

Glyphosate, N-(phosphonomethyl) glycine, is a synthetic amino acid of glycine. In contrast to most other herbicides that chelate with a single or few mineral species, glyphosate is a broad-spectrum chelator that attaches to and ties up both macro- and micro-essential mineral nutrients (Ca, Co, CU, Fe, K, Mg, Mn, Mo, Ni, Zn) that are critical for physiological processes in soil, microbes, plants, and animals (U.S. Patent No. 3,160,632; Bernards, Thelin, Penner, et al. 2005; Glass, 1984; Kobylecka, Ptaszyniski, and Zwolinnska, 2000; Lundager and Madsen, et al., 1978; Martell and Smith, 1974; Motekaitis and Martell, 1985; Pandy, Pandy, and Misra, 2000; Subramaniam and Hoggard, 1988). Its broad-spectrum chelating activity makes glyphosate a broad-spectrum herbicide and a potent antimicrobial agent since the functions of numerous essential enzymes are affected (Ganson and Jensen, 1988; Mucha, Drag, Dalton, et al., 2010; Nilsson, 1985; Ozturk, Yazici, Eker, et al., 2008).

Glyphosate was first patented by Stauffer Chemical Company in 1964 as a mineral chelator (U.S. patent number 3,160,632) and used to clean steam pipes and boilers. It is also a powerful antibiotic (U.S. Patent No. 7,771,736 B2; U.S. Patent No. 5,902,804; Clair, Linn, Travert, et al., 2012) that is toxic to beneficial organisms in soil (Dominguez, Brown, Sautter, de Oliveira, et al., 2016; Gaupp-Berghausen et al., 2015; Kremer and Means, 2009) and the gastrointestinal (GI) track of animals (Ackermann, Coenen, Schrodl, et al., 2015; Krüger, Shehata, Schrödl, et al., 2013; Schrödl, et al., 2014; Shehata, Schrodl, Aldin, et al., 2012; Shehata, Kühnert, Haufe, 2013) essential for nutrient availability and absorption, vitamin production, growth, immunity, and defense against stress and disease. Primary emphasis in understanding glyphosate's herbicidal activity has been on its inhibition of EPSPS at the start of the shikimate physiological pathway for secondary metabolism (Marschner, 2011. This enzyme requires reduced flavin mononucleotide (FMN) as a cofactor (catalyst) for the EPSPS enzyme whose reduction requires manganese (Mn). Thus, by immobilizing Mn by chelation, glyphosate denies the availability of reduced FMN for the EPSPS enzyme. It can also affect up to 25 other plant enzymes that require Mn as a cofactor and numerous other enzymes in both primary and secondary metabolisms that require other metal cofactors. There are actually various enzymes requiring mineral cofactors for function in the shikimate and other metabolic pathways that are inhibited by glyphosate chelation of Ca, Co, Cu, Fe, Mg, Mn, Ni, etc. that could account for impairment of this pathway exclusive of EPSPS (Abraham, 2010). The shikimate pathway is responsible for plant responses to stress and defense against pathogens through the production of amino acids, hormones, lignin, phytoalexins, flavonoids, phenols, etc. (Marchner, 2011). By inhibiting various enzymes in the shikimate pathway, plants become highly susceptible to various ubiquitous soilborne pathogens such as *Fusarium, Gaeumannomyces, Pythium, Phytophthora*, and *Rhizoctonia* (Huber, 2010; Huber and McKay-Buis, 1993; Johal and Huber, 2009). It is this pathogenic activity that actually kills the plant as "the herbicidal mode of action" since the shikimate pathway is for the production of secondary metabolites (Johal and Rahe, 1984; Levesque and Rahe, 1992, Johal and Huber, 2009). This predisposition to disease was clearly shown by Schafer, Hallett, and Johnson (2010, 2012, 2013) in their study of glyphosate-resistant weed species. Resistance to glyphosate is manifest as increased resistance to the soilborne pathogens, rather than resistance to the chemical. If glyphosate is not translocated to plant roots because of stem-boring insects or other disruption of the vascular system, aerial parts of the plant may be stunted, but the plant is not generally killed.

Recognizing glyphosate as a strong chelator to immobilize essential plant nutrients provides an understanding of the various non-herbicidal effects of glyphosate (Table 2.2). Glyphosate is a

TABLE 2.2

Some General Effects of Glyphosate

1. It is a strong mineral chelator in water, soil, plants, and animals.
2. It is rapidly absorbed by plant roots, stems, and leaves; and it moves systemically throughout the plant.
3. It accumulates in meristematic tissues of plants and in most tissues and organs of animals and man.
4. It decreases availability of minerals in soil and water.
5. It increases susceptibility of plants to draught and other stresses.
6. It increases susceptibility of plants, animals, and man to diseases.
7. It is a persistent toxicant in many soils.
8. It causes fruit and bud drop and other hormonal changes in plants.
9. It is toxic to beneficial soil and gut microorganisms that facilitate nutrient access, availability, and absorption.
10. It can be desorbed as an active toxicant in soil by phosphorus.
11. It stimulates soilborne diseases and pathogens.
12. It inhibits nitrogen fixation in soil and legume nodules.
13. It accumulates in food and feed products to impact food safety.
14. It is toxic to bees and other environmental entities.
15. It is a significant contributor to water pollution.

phloem mobile, systemic chemical that accumulates in meristematic plant tissues (root, shoot tip, reproductive structures, and legume nodules) and is released into the soil rhizosphere through root exudation or mineralization of plant residues. Degradation of glyphosate in many soils is slow or nonexistent since, as a synthetic aminophosphonic acid analogue of the natural amino acid glycine, it is not readily biodegradable and is broken down primarily by microbial co-metabolism. Although glyphosate can be rapidly immobilized in soil and plants through chelation with various cations (Ca, Mg, Cu, Fe, Mn, Ni, Zn), it is not necessarily degraded and can accumulate for years in soils and perennial plants (Bott, Tesfamariam, Kania et al., 2011; Caetano, Ramalho, Botrel, et al., 2012; Farenhorst, McQueen, Saiyed et al., 2009; Kanissery, Welsh, and Sims, 2014). Phosphorus fertilizers can desorb accumulated glyphosate that is immobilized in soil, whereupon it becomes an active toxin to damage and reduce the physiological efficiency of subsequent crops grown in that soil.

2.2.2 ANTIBIOTIC ACTIVITY OF GLYPHOSATE

Glyphosate is a patented antibiotic against many taxonomic groups of bacteria, fungi, and parasites (U.S. Patent No. 7,771,736 B2). Most organisms that utilize the shikimate pathway are especially sensitive to glyphosate. This includes beneficial soil organisms and the GI microbiota. Mammals are the only group of organisms that don't have the shikimate pathway directly; however, many of the essential gut microorganisms in animals and humans are very sensitive to glyphosate at very low levels. Intake of glyphosate in food and water at levels below 0.1 ppm can cause gut dysbiosis and lead to cancer, weak immunity, poor mineral absorption, autism, chronic botulism, inflammatory bowel diseases, and numerous other diseases (Krüger, Shehata, Schrödl, et al., 2013; Kurenbach et al., 2015; ScienceDaily, 2016; Swanson et al., 2014). There are some organisms that have an alternative shikimate pathway that is not sensitive to glyphosate. Glyphosate-tolerant organisms include many plant and animal pathogens. In fact, a very serious plant pathogen, *Agrobacterium tumefaciens*, that causes the crown gall disease in many plants was the source of the genetic resistance to glyphosate that was engineered into the Roundup Ready® crops.

In our research at Purdue University on mineral nutrition related to plant diseases (Thompson and Huber, 2007; Cheng, 2005), we verified that the biological availability of manganese (Mn) is a critical component for resistance to many plant, human, and animal diseases. Since only the reduced form of Mn ($Mn2^+$) can be taken up by plants for those reactions, the physiological efficiency for Mn is determined by the activity of Mn-reducing (Mn^{2+}) organisms, rather than Mn-oxidizing (Mn^{4+}=non-physiologically plant available) organisms. Manganese-reducing organisms were reduced 10-fold, while manganese-oxidizing organisms were increased 12-fold in the rhizospheres of wheat plants growing in a silt loam soil after a single preplant application of glyphosate, even though microbial net soil respiration remained similar to plants in the untreated soil. An application of glyphosate several years prior to planting cereal crops can predispose those plants to severe *Fusarium* head scab and the take-all disease caused by the soilborne fungal pathogen, *Gaeumannomyces graminis*. Glyphosate is toxic to Mn-reducing organisms and stimulates Mn-oxidizing organisms in the plant rhizosphere to greatly reduce soil Mn availability (Huber and McKay-Buis, 1993). Oxidation of Mn from the reduced Mn^{2+} form to the nonavailable Mn^{4+} form is a common mode of action (virulence mechanism) for many plant pathogens, and it causes various root rot, vascular wilt, and foliar diseases. Within a few hours after spore germination of the fungal pathogen *Magnaporthe grisea* on rice leaves, the fungus' hyphae oxidize Mn in leaf tissues to suppress defensive mechanisms that otherwise might prevent penetration, infection, and development of the blast disease that has caused millions of people to starve (Cheng, 2005; Thompson and Huber, 2007) prior to the development of blast-resistant varieties. The bacterial plant pathogen *Xylella fastidiosa* initiates Mn oxidation in xylem tissues that precipitate in the biofilm that forms to plug these tissues and prevent movement of water and nutrients throughout the plant (Cobine, Cruz, Navarrete, et al., 2013). More than 40 plant diseases are reported to be increased by glyphosate (Johal and Huber, 2009; Yamada, Kremer, Camargo, et al., 2009). The copper-chelating phenoxyprop group of herbicides (Puma and Puma Gold) increase the incidence and severity of ergot in cereal grains (Evans, Solberg, and Huber, 2007). The induced Cu deficiency causes male sterility so that the protective glumes around the pistil open to permit the fungal spores to infect the pistil. Numerous other plant diseases are increased by other agricultural pesticides (Altman and Campbell, 1977). GE crops often have lower nutrient uptake and physiological efficiency (Gordon, 2006, 2007).

2.2.2.1 Glyphosate Effects on Nontarget Organisms

Nontarget effects of glyphosate are especially prominent to honeybees, earthworms, pollinators, mycorrhizae, nitrogen-fixing organisms, organisms involved in nutrient mineralization, and natural biological disease control organisms in the environment. These organisms may be damaged by chemical drift, root exudates, release during residue decay, or desorption of residual glyphosate from soil. As a consequence, it is not uncommon to see Ca, Cu, Co, Fe, Mg, Mn, Mo, Ni, and Zn deficiencies intensify in plants grown on soils that were once considered fully sufficient for these nutrients when they are treated with glyphosate. Toxicity of glyphosate to *Lactobacillus* and *Bifidobacterium* bacteria in the bee honey crop was shown to cause colony collapse disease and predispose bees to other diseases and mites (Balbuena, Tison, Hahn, et al., 2015; Blot, Veillat, Rouze, et al., 2019; Erick, Motta, Raymann, et al., 2018; Foulk, 2009; Herbert, Vazquez, Arenas, et al., 2014; Motta, Mak, De Jong, et al., 2020; Nicola, 2019).

2.2.2.2 Glyphosate Effects on Animals and Humans

Swanson, et al. (2014) found an extremely high correlation coefficient for over 22 human diseases with the glyphosate applied to corn and soybeans. Hoy, Swanson, and Seneff (2015) found a similar relationship with wild animals. Toxicity of glyphosate to beneficial gut (GI track) organisms and predisposition of animals and humans to such diseases as chronic botulism, *Salmonella*, *Escherichia coli*, and inflammatory bowel diseases have also been reported (Krueger et al., 2014; Swanson et al., 2014) (Table 2.3). Replacement of feed, food, and water contaminated with glyphosate plus mineral supplementation to compensate for induced deficiencies, and restoration of the healthy gut microbiome often relieve the health damages caused by glyphosate. Pigs fed with GMO corn or soybeans developed severe stomach inflamation (Carman et al., 2013), and feed containing glyphosate resulted in a high rate of miscarriage and produced offspring with various deformities depending on the concentration of glyphosate in the feed (Pedersen, 2014). High rates of miscarriage,

TABLE 2.3

Some Diseases of Humans Associated with Glyphosate Exposure (after Antoniou et al., 2012; Avila-Vasqueze, et al., 2020; Fox, 2012; Guyton, et al., 2015; Samsel and Seneff, 2013a,b, 2015a,b, 2016; Seralini, et al., 2014; Swanson, Leu, Abrahamson, et al., 2014)

Allergies	Bowel Disease	Inflammatory Bowel
Alzheimer's	Cancer (various)	Irritable bowel
Anencephaly	Celiac disease	Leakey gut
Arthritis	Chronic fatigue syndrome	Nonalcoholic fatty liver disease
Asthma	Colitis	Miscarriage
Atopic dermatitis	Crohn's disease	Morgellons disease
Attention deficit	Dementia	Multiple sclerosis
Autoimmune diseases	Diabetes	Obesity
Autism	Difficile diarrhea	Pancreas abnormality
Bipolar	Gluten intolerance	Parkinson's disease
Birth defects	Indigestion	Sudden Infant Death
Bloat (fatal)	Infertility	Neurologic diseases

anencephaly, and birth defects were reported for babies in Washington, USA, after glyphosate was applied to the three rivers supplying irrigation and drinking water to Yakima, Benton, and Hamilton counties for aquatic weed control (Peterson, 2016). Similar effects have been reported in Argentina and other countries (Warren and Pisarenko, 2013; Robinson, 2010).

2.3 UNDERSTANDING THE EFFECTS OF GENETIC ENGINEERING FOR HERBICIDE TOLERANCE

Plants genetically engineered to tolerate glyphosate-based herbicides contain an alternate EPSPS enzyme (EPSPS-II) that is not blocked by glyphosate, so that the herbicide can be applied directly to the engineered plant. This alternate pathway, however, does nothing to the glyphosate that is applied; thus, it is still an active mineral-chelating agent and antimicrobial compound. Residual glyphosate reduces nutrient density of food and feed products by chelation, and it is toxic to important soil microorganisms and microorganisms in animals and humans consuming it, or those exposed to it in the environment. The reduced physiological efficiency of GE plants is often reflected in reduced yield, reduced water use efficiency, and increased disease susceptibility when compared with nonengineered parent plants (Bott, Tesfamariam, Candan, et al., 2008; Cakmak, Yazici, Tutus, et al., 2009) (Johal and Huber, 2009; Zobiole, Oliveira, Kremer, et al., 2010a,b,c; Zobiole, Oliveira, Huber, et al., 2010; Zobiole, Bonini, Oliveira Jr., et al., 2010). Chelation of essential mineral nutrients is frequently observed as a yellowing or "flashing" of leaves that persists until the plant can resupply the immobilized nutrients from soil. Plant uptake and translocation of iron (Fe), manganese (Mn), and zinc (Zn) are commonly reduced as much as 80% by drift rates (1/40th of an herbicidal rate) of glyphosate (Eker, S., Yazici, A., Erenoglu, B., et al., 2006). The persistence and accumulation of glyphosate in perennial plants, soil, and root meristems can significantly reduce root growth and the development of root absorptive tissues of GE as well as nonengineered (wild-type) plants.

2.4 INCREASED PLANT, ANIMAL, AND HUMAN DISEASES

In contrast to microbial toxicity to beneficial organisms, glyphosate in soil and root exudates stimulates soil organisms that immobilize minerals and, thereby, denies access to nutrients through pathogenic activity. Plant pathogens increased by glyphosate (Table 2.4) include ubiquitous bacterial and

TABLE 2.4

Some Plant Pathogens Stimulated by Glyphosate
(after Datnoff, Elmer, Huber, 2007; Evans, Solberg, and
Huber, 2007; Huber, 2010; Johal and Huber, 2009)

Botryosphaeria dothidea	Gaeumannomyces graminis
Clavibacter michiganensis nebraskensis	Magnaporthe grisea
Corynespora cassiicola	Marasmius sp.
Fusarium avenaceum	Monosporascus cannonballus
Fusarium graminearum	Myrothecium verrucaria
Fusarium oxysporum f.sp. cubense	Phaeomoniella chlamydospora
Fusarium oxysporum f.sp. (canola)	Phytophthora sp.
Fusarium oxysporum f.sp. glycines	Pythium sp.
Fusarium oxysporum f.sp. vasinfectum	Rhizoctonia solani
Fusarium solani f.sp. glycines	Septoria nodorum
Fusarium solani f.sp. phaseoli	Thielaviopsis basicola
Fusarium solani f.sp. pisi	Xylella fastidiosa

fungal root, crown, and stalk rotting organisms; vascular colonizing organisms that disrupt vascular transport to cause wilt and dieback; and root "nibblers" that impair access or uptake of soil nutrients.

Chelation of micronutrients by glyphosate (that are the regulators, activators, and inhibitors of plant defense mechanisms) increases pathogenesis to increase the severity of many abiotic, as well as infectious, diseases. Abiotic diseases increased by glyphosate include various nutrient deficiencies, bark cracking, winterkill, and drought stress. Many of the diseases caused by pathogens listed in Table 2.4 are emerging or reemerging throughout Africa and other areas of the world. In addition to direct crop losses from pathogenesis, several diseases are also associated with mycotoxin contamination (deoxynivalenol, fumonisin, nivalenol, zearalenone, etc.) of food and feed products that are increased by glyphosate (Fernandez, M., Zentner, Basnyat, et al., 2009; Fernandez, Zentner, DePauw, et al., 2007; Fernandez, Kremer, et al., 2008). High concentrations of mycotoxins in grain may be from root infection and translocation to the edible grain.

The indiscriminant application of glyphosate throughout the environment for weed control, and as a crop desiccant, has resulted in extensive food, water, and feed contamination by this persistent, water-soluble, systemic toxicant (Battaglin, Meyer, Kuivila, et al., 2014; Eberbach, 1998). Over 32 human diseases and numerous animal diseases, approaching epidemic proportions, are associated with glyphosate intake or exposure (Table 2.3). The epidemiological curve for all of these diseases has a similar pattern that is correlated with the increased use of glyphosate herbicides (Swanson, Leu, Abrahamson, et al., 2015). Over half of these diseases involve the gut microbiome which is severely damaged by the antibiotic activity of glyphosate from dietary intake or environmental exposure. Krüger, Shehata, Schrödl, et al. (2013) showed that 0.1 ppb glyphosate in feed caused fatal chronic botulism in dairy cows and other animals. Severe kidney, liver, and other damages have been reported from glyphosate levels of below 0.1 ppb by other researchers (Seralini, Clair, Mesnage, et al., 2014; Mesnage et al., 2015).

2.5   IMPACT OF GLYPHOSATE ON THE ENVIRONMENT

Degradation of the environment is common from the indiscriminant application of glyphosate-based herbicides. Scientists of the United States Geological Survey (USGS, 2018) have reported extensive residual glyphosate and AMPA (the first degradation product of glyphosate) contamination of soil and surface, rain, and groundwater. Extensive areas in the Southeastern and Midwestern

United States have accumulated over 2 tons of residual glyphosate and AMPA per square mile (United States Geological Survey, 2018). Chronic wasting and other diseases of deer and other wildlife are commonly associated with exposure to glyphosate (Hoy, Swanson and Seneff, 2015). The source of this contamination can be direct application, chemical drift, and runoff from soil or other surfaces. Many nontarget effects have been reported. The Colony Collapse Disease (CCD) of bees is one example of damages from both the chemical-chelating effect of glyphosate in reducing the availability of physiologically active essential minerals in pollen and its direct antibiotic damage to microorganisms in the bee's honey crop (stomach) (Faulk, 2009; Herbert et al., 2014; Motta, Mak, DeJong, et al., 2020). Providing several *Lactobacillus* and *Bifidobacterium* species, along with minerals in drinking water, has offset many of the damaging effects of glyphosate and eliminated CCD to preserve the viability of honeybees as an essential pollinator for food production. Unintended effects of glyphosate are also reported for ladybird beetles and other wildlife (Freydier, and Lundgren, 2016).

Eutrophication of lakes and streams is always of environmental concern. Wisconsin scientists studying causes of eutrophication reported that one-third pound of phosphorus enters Lake Erie each year since glyphosate from surrounding farm and forest lands is often aerially applied liberally for weed and plant species control (Barrera, 2016; Chow, 2016). The resulting effect is severe eutrophication from a burgeoning growth of algae resistant to glyphosate in the absence of competition from other weeds that had been killed by the weed killer.

2.6   REMEDIATION OF PESTICIDE DAMAGE

A first step in remediation is to recognize the magnitude of the problem and the damage done. Glyphosate is a difficult chemical to analyze, and each technique needs to be carefully certified for accuracy (Brewster, Warren, and Hopkins, 1991; Chamkasem and Harmon, 2016; Granby, Johnson, and Vahl, 2003). Switching US adults and children from a conventional diet contaminated with glyphosate to a glyphosate-free similar organic diet is reported to reduce levels of urinary glyphosate and AMPA (main metabolite of glyphosate) by an average of 70% within three days (Fagan, Bohlen, Patton, et al., 2020). Non-excreted glyphosate bioaccumulates in all tissues and organs of the body (Krueger, Schrodl, Pederson, et al., 2014; Samsell and Seneff, 2013a,b, 2015a,b, 2016).

As persistent mineral chelators immobilize essential nutrients, these minerals become less available in soil, and thus crops produced on these soils have lower nutrient density. This lower nutrient content is then magnified up the food chain. Nutritional bioenrichment of crops can be accomplished through amendment of the soil or foliar applications of necessary nutrients based on soil or tissue analysis. Emphasis has traditionally been placed on the application of N, P, K, Ca, Mg, and S (Datnoff, Elmer, Huber, 2007); however, many agricultural pesticides chelate micronutrients (B, Co, Cu, Fe, Mn, Ni, Zn, and other nutrients) that need supplementation to provide adequate nutrient density of crops. Nutrient supplements for minerals such as Fe, Mg, and Mn may be taken directly, and various forms of the different nutrients can often be readily purchased at stores or online.

Modification of the soil or intestinal microbial environment (microbiome) can often be used to increase the availability of nutrients for plant uptake by favoring certain organisms and may be more effective than amending the soil with an exogenous source of specific microorganisms (Huber and Haneklaus, 2007). This assumes that the organisms needed are also indigenous to the specific soil and have not been eradicated by the antibiotic effects of residual glyphosate. Inoculation of soil or plants with mycorrhizae, or legume seeds with nitrogen-fixing organisms, has been an effective practice for many years if the soil environment is favorable for the organism to develop. Most of the plant-growth-promoting soil organisms such as *Pseudomonas*, *Bacillus*, and *Trichoderma* function by increasing the availability of various essential nutrients. Tissue nutrient levels considered sufficient for non-GMO plants may need to be increased 25%–50% higher for similar GE plants in order to compensate for the genetically reduced nutrient efficiency or immobilization by glyphosate, or other herbicides, that are applied directly to the crop. A similar approach, referred to as a fecal transplant, has been used successfully in life-and-death treatment of the *C. difficile* disease in humans (Yoon and Brandt, 2010).

Changing to a non-glyphosate-contaminated diet of organic food has facilitated the removal of 70% of urinary glyphosate after one week (Oates, Cohen, Braun, et al., 2014). Humic acids and clinoptilolite have been shown to complex with glyphosate in the GI track and remove glyphosate from the body (Gerlach, Gerlach, Schrodl, et al., 2014a, b; Prasai, Walsh, Bhattarai, et al., 2016, 2017). Residual glyphosate in soil may be "detoxified" by chelating with soil nutrients, applied gypsum, or other source of water-soluble cation for chelation (Caetano, Ramalho, Botrel, et al., 2012); however, high mineral intake has been attributed as a cause of end-stage kidney disease in agricultural workers exposed to glyphosate in Sri Lanka and central America (Jayasumana, Gunatilake, and Senanayake, 2015). Modification of the soil environment through tillage, crop sequence, or other cultural practices, or in some limited trials, with biological amendments, may also increase the full degradation of glyphosate in soil.

2.7 CONCLUSIONS

Synthetic pesticides have profound and far-reaching impacts on soil, crop, animal, human, and environmental health. Their safety and impact on soil, crop, animal, and human health have often been disregarded or inadequately evaluated through a betrayal of public trust by regulatory agencies or the manufacturers themselves (Druker, 2014, 2015; ENSSER, 2012; Gillam, 2017; Rowell, 2003). The herbicide glyphosate is a strong, broad-spectrum mineral chelator and toxic antibiotic to beneficial organisms. It reduces nutrient availability in soil and physiological function in plants and animals. This synthetic amino acid bioaccumulates and is persistent in many soils (Torstensson, Lundgren, and Stenstrom, 1989). Remediation involves the use of alternative practices to manage weeds, and nutrient amendment to compensate for the decreased availability and function of nutrients. Serious deterioration of soil, crop, animal, and environmental health has occurred in Africa and many other regions from the indiscriminate use and distribution of this synthetic analogue of an essential amino acid. Future historians may well look back upon our time and write not about how many pounds of pesticides we did or didn't apply, but how willing we were to sacrifice our children and jeopardize future generations with the extensive application of synthetic chelating, antibiotic, persistent, endocrine hormone disrupting chemicals and to proceed with this massive experiment, which we call *genetic engineering*, that is based on *false promises* and *flawed science*, just to benefit the bottom line of a commercial enterprise.

REFERENCES

Abraham, W. 2010. Glyphosate formulations and their use for the inhibition of 5-enolpyruvylshikimate-3-phosphate synthase. US 7771736. United States Patent Office.

Ackermann, W., Coenen, M., Schrodl, W., Shehata, A.A., and Krueger, M. 2015. The influence of glyphosate on the microbiota and production of botulinum neurotoxin during ruminal fermentation. *Curr. Microbiol.* 70: 374–382. doi:10.1007/s00284-014-0732-3.

Altman, J. and Campbell, C.L. 1977. Effects of herbicides on plant diseases. *Annu. Rev. Phytopathol.* 15: 361–385.

Antoniou, M., Habib, M.E.M., Howard, C.V., Jennings, R.C., Leifert, C., Nodari, R.O., Robinson, C.J., and Fagan, J. 2012. Teratogenic effects of glyphosate-based herbicides: divergence of regulatory decisions from scientific evidence. *J. Environ. Anal. Toxicol.* S4:006. doi:10.4172/2161-0525.S4-006.

Avila-Vasqueze, M., Difilippo, F., Lean, B. M., and Maturano, E. 2020. Risk of asthma and environmental exposure to glyphosate in an ecological study. *Authorea*, August 13, 2020. doi:1022541/au.159734524.47178780.

Balbuena, M.S., Tison, L., Hahn, M.L., Greggers, U., Menzel, R., and Farina, W.M. 2015. Effects of sub-lethal doses of glyphosate on honeybee navigation. *J. Exp. Biol.*, dev-117291. http://jeb.biologists.org/content/early/2015/07/09/dev.117291.short.

Barrera, L. 2016. Scientists: Glyphosate contributes to phosphorus runoff in Lake Erie. *Crop Prot. Nutrient Manag.* June 11, 2016.

Battaglin, W.A., Meyer, M.T., Kuivila, K.M., and Dietze, J.E. 2014. Glyphosate and its degradation product AMPA occur frequently and widely in U. S. soils, surface water, groundwater, and precipitation. *J. American Water Res. Assoc.* 50: 275–290. doi:10.1111/jawr.12159.

Bernards, M.L., Thelen, K.D., Penner, D., Muthukumaran, R.J., and McCracker, J.L. 2005. Glyphosate inter-action with manganese in tank mixtures and its effect on glyphosate absorption and translocation. *Weed Sci.* 53: 787–794.

Blot, N., Veillat, L., Rouze, R., and Delatte, H. 2019. Glyphosate, but not its metabolite AMPA, alters the hon-eybee gut microbiota. Plos One 14(4): e0215466. https://doi.org/10.1371/journal.pone.0215466.

Bott, S., Tesfamariam, T., Candan, H., Cakmak, I., Roemheld, V., and Neumann, G. 2008. Glyphosate-induced impairment of plant growth and micronutrient status in glyphosate-resistant soybean (Glycine max L.). *Plant Soil* 312: 185–194.

Bott, S., Tesfamariam, T., Kania, A., Eman, B., Aslan, N., Roemheld, V., and Neumann, G. 2011. Phytoxicity of glyphosate soil residues re-mobilised by phosphate fertilization. *Plant Soil* 315: 2–11.

Brewster, D.W., Warren, J., and Hopkins II, W.E. 1991. Metabolism of glyphosate in Sprague-Dawley rats: Tissue distribution, identification, and quantitation of glyphosate-derived materials following a single oral dose. *Fund. Appl. Toxicpl.* 17: 43–51.

Caetano, M.S., Ramalho, T.C., Botrel, D.F., da Cunha, E.F.F., and de Mellow, W.C. 2012. Understanding the inactivation process of organophosphorus herbicides: A DT study of glyphosate metallic complexes with Zn^{2+}, Ca^{2+}, Mg^{2+}, Cu^{2+}, Co^{3+}, Fe^{3+}, Cr^{3+}, and Al^{3+}. *Int. J. Quantum Chem.* 112: 2752–2762.

Cakmak, I., Yazici, A., Tutus, Y., and Ozturk, L. 2009. Glyphosate reduced seed and leaf concentrations of cal-cium, magnesium, manganese, and iron in non-glyphosate resistant soybean. *Eur. J. Agron.* 31: 114–119.

Carman, J.A., Vlieger, H.R., Ver Steeg, L.J., Sneller, V.E., Robinson, G.W., Clinch-Jones, C.A., Haynes, J.I., and Edwards, J.W. 2013. A long-term toxicology study on pigs fed a combined genetically modified (GM) soy and GM maize diet. *J. Org. Syst.* 8(1): 38–54.

Chamkasem, N. and Harmon, T. 2016. Direct determination of glyphosate, glufosinate, and AMPA in soybean and corn by liquid chromatography/tandem mass spectrometry. *Anal. Bioanal. Chem.* 408: 4995–5004. doi:10.1007/s00216-9597-66.

Cheng, M.-W., 2005. Manganese Transition States During Infection and Early Pathogenesis in Rice Blast., M.S. Thesis, Purdue University, West Lafayette, IN.

Chow, L. 2016. Glyphosate sprayed on GMO crops linked to Lake Erie's toxic algae bloom. Ecowatch Jul. 05, 2016.

Clair, E., Linn, L. Travert, C., Amiel, C., Séralini, G.-E., Panoff, J.-M. 2012. Effects of Roundup(®) and Glyphosate on Three Food Microorganisms: Geotrichum candidum, Lactococcus lactis subsp. cremoris and *Lactobacillus delbrueckii* subsp. bulgaricus. *Curr Microbiol.* 2012 Feb 24.

Cobine, P.A., Cruz, L.F., Navarrete, F., Duncan, D., Tygart, M., De La Fuente, L. 2013. Xylella fastidiosa differentially accumulates miner elements in biofilm and planktonic cells. *PLoS ONE* 8(1): e54936. doi:10.1371/journal.pone.oo54936.

Datnoff, L.E., Elmer, W.H., and Huber, D.M. (eds.). 2007. *Mineral Nutrition and Plant Disease.* APS Press, St. Paul, MN. 278 pages.

Dominguez, A., Brown, G.G., Sautter, K.D., de Oliveira, C.M.R., de Vasconcelos, E.C., Niva, C.C., Bartz, M.L.C., and Bedano, J.C. 2016. Toxicity of AMPA to the earthworm *Eisenia andrei* bouche, 1972 in tropical artificial soil. *Scientific Rep.* 6: 19731. doi:10.1038/srep 19731.

Druker, S.M. 2014. Why the FDA's policy on genetically engineered foods is irresponsible and illegal. *Alliance Bio-integ.* 2014.

Druker, S.M. 2015. Altered Genes, Twisted Truth; How the Venture to Genetically Engineer Our Food Has Subverted Science, Corrupted Government, and Systematically Deceived the Public. Clear Water Press, Salt Lake City, UT.

Eberbach, P. 1998. Applying non-steady-state compartmental analysis to investigate the simultaneous degrada-tion of soluble and sorbed glyphosate (N-(phosphonomethyl) glycine) in four soils. *Pestic. Sci.* 52: 229–240.

Eker, S., Ozturk, L., Yazici, A., Erenoglu, B., Roemheld, V., and Cakmak, I. 2006. Foliar-applied glyphosate substantially reduced uptake and transport of iron and manganese in sunflower (*Helianthus annuus* L.) plants. *J. Agric. Food Chem.* 54: 10019–10025.

ENSSER. 2012. Questionable biosafety of GMOs, double standards and, once again, a "shooting-the-messenger" style debate. ENSESER Statement on Seralini et al. (2012 publication and reactions evoked.)

Erick, V., Motta, S., Raymann, K., and Moran, N.A. 2018. Glyphosate perturbs the gut microbiota of honey bees. *PNAS* 115: 1035–10310.

Evans, I.R., Solberg, E., and Huber, D.M. 2007. Copper and plant disease. pp. 177–188 (Chapter 12) In: L.E. Datnoff, W.H., Elmer, and D.M. Huber (eds.) *Mineral Nutrition and Plant Disease*. APS Press, St. Paul, MN.

Fagan, J., Bohlen, L., Patton, S., and Klein, K. 2020. Organic diet intervention significantly reduces urinary glyphosate levels in U.S. children and adults. *Environ. Res.* doi:10.1016/j.envres.2020.109898.

Farenhorst, A., McQueen, D.A.R., Saiyed, I., Hilderbrand, C., Li, S., Lobb, D.A., Messing, P., Schumacher, T.E., Papiernik, S.K., and Lindstrom, M.J. 2009. Variations in soil properties and herbicide sorption coefficients with depth in relation to PRZM (Pesticide Root Zone Model) calculations. *Geoderma* 150: 267–277.

Faulk, K.E. 2009. Identifying the role of glyphosate-containing herbicides on honeybee mortality rates and colony collapse disorder. Envir. Sci. Paper. Jr. Science, Engineering and Humanities Symp., Univ. Missouri, St. Louis.

Fernandez, M.R., Kremer, R.J., Zentner, R.P., Johnson, E.N., Kutcher, H.R. and McConkey, B.J. 2008. Effect of glyphosate on Fusarium root infection of pea crops grown in rotation with spring wheat in the semi-arid Canadian prairies. Agri-Food Canada.

Fernandez, M.R., Zentner, R.P., Basnyat, P., Gehl, D., Selles, F. and Huber, D. 2009. Glyphosate associations with cereal diseases caused by Fusarium spp. in the Canadian Prairies. *Eur. J. Agron.* 31: 133–143.

Fernandez, M.R., Zentner, R.P., DePauw R.M., Gehl, D., and Stevenson, F.C. 2007. Impacts of crop production factors on Fusarium head blight in barley in Eastern Saskatchewan. *Crop. Sci.* 47: 1585–1595.

Fox, M.W. 2012. Pet Foods with Genetically Modified Ingredients (GMOs) Put Dogs & Cats at Risk. Manuscript ipan@erols.com.

Freydier, L. and Lundgren, J.G. 2016. Unintended effects of the herbicides 2,4-D and dicamba on lady beetles. *Ecotoxicology* 25: 1270–1277. doi:10.1007/s10646-016-1680-4.

Ganson, R.J. and Jensen, R.A. 1988. The essential role of cobalt in the inhibition of the cytosolic isozyme of 3-deoxy-D-arabino-heptulosonate-7-phosphate synthase from *Nicotiana sylvestris* by glyphosate. *Arch. Biochem. Biophys.* 260: 85–93.

Gaupp-Berghausen, M., Hofer, M., Rewald, B. and Zaller, J.G. 2015. Glyphosate-based herbicides reduce the activity and reproduction of earthworms and lead to increased soil nutrient concentrations. *Sci. Repts.* 5: 12886. doi:10.1038/srep12886.

Gerlach, H., Gerlach, A., Schrodl, W., Schottdorf, B., Haufe, S., Helm, H., Shehata, A.A. and Krueger, M. 2014a. Oral application of charcoal and humic acids to dairy cows influences Clostridium botulinum blood serum antibody level and glyphosate excretion in urine. *J. Clin. Toxicol.* 186: 2161–0495. doi:10.4172/2161-0495.186.

Gerlach, H., Gerlach, A., Schrodl, W., Schottdorf, B., Haufe, S., Schottdorf, B., Shehata, A.A. and Krueger, M. 2014b. Oral application of charcoal and humic acids to dairy cows influence selected gastrointestinal microbiota, enzymes, electrolytes, and substrates in the blood of dairy cows challenged with glyphosate in GMO feeds. *J. Environ. Anal. Toxicol.* 4: 256. doi:10.4172/2161-0525.1000256.

Gillam, C. 2017. *White Wash: The Story of a Weed Killer, Cancer, and the Corruption of Science*. Island Press, Washington, DC.

Glass, R.L. 1984. Metal complex formation by glyphosate. *J. Agric. Food Chem.* 32:1249–1253.

Gordon, W.B. 2006. Manganese nutrition of glyphosate-resistant and conventional soybeans. *Better Crops* 91: 12–13. Great Plains Soil Fertility Conf. Proc. Denver, CO, March 7–8, 2006: 224–226.

Gordon, W.B. 2007. Does (the) glyphosate gene affect manganese uptake in soybeans? *Fluid J. Early Spring* 15: 12–13.

Granby, K., Johannesen, S. and Vahl, M. 2003. Analysis of glyphosate residues in cereals using liquid chromatography-mass spectrometry (LC-MS/MS). In: Taylor & Francis (eds). *Food Additives and Contaminants. Chemistry, Analysis, Control, Exposure, and Risk Assessment*. 20(8): 692–698.

Guyton, K.C. et al., 2015. Carcinogenicity of tetrachlorvinphos, parathion, malathion, diazinon, and glyphosate. World Health Organization-IARC, Lyon, France. doi:10.1016/S1470-2045(15)70134-8.

Herbert, L.H., Vazquez, D.E., Arenas, A., and Farina, W.M. 2014. Effects of field-realistic doses of glyphosate on honeybee appetitive behavior. *J. Exp. Biol.* 217: 3457–3464.

Hoy, J., Swanson, N., and Seneff, S. 2015. The high cost of pesticides: Human and animal diseases. *Poult. Fish Wildl. Sci.* 3:1–19. doi:10.4172/2375-446X.1000132.

Huber, D.M. 2010. Ag chemical and crop nutrient interactions – current update. Proc. Fluid Fert. Forum, Scottsdale, AZ February 14–16, 2010. Vol. 27. *Fluid J.* 18(3): 14–16.

Huber, D.M. and Haneklaus, S. 2007. Managing nutrition to control plant disease. *Landbauforschung Volkenrode* 57(4): 313–322.

Huber, D.M. and McKay-Buis, T.S. 1993. A multicomponent analysis of the take-all disease of cereals. *Plant Dis.* 77: 437–447.

Jayasumana, C., Gunatilake, S., and Senanayake, P. 2015. Drinking well water and occupational exposure to herbicides is associated with chronic kidney disease in Padavi[Sripura, Sri Lanka. *Environ. Health* 14: 6.

Johal, G.S. and Huber, D.M. 2009. Glyphosate effects on diseases of plants. *Eur. J. Agron.* 31: 144–152.

Johal, G.S., and Rahe, J.E. 1984. Effect of soilborne plant-pathogenic fungi on the herbicidal action of glyphosate on bean seedlings. *Phytopathology* 74: 950–955.

Kanissery, R.G., Welsh, A., and Sims, G.K. 2014. Effect of soil aeration and phosphate addition on the microbial bioavailability of 14C-glyphosate. *J. Environ. Qual.* doi:10.2134/jeg2014.08.0331.

Kobylecka, J., Ptaszyniski, B., and Zwolinnska, A. 2000. Synthesis and properties of complexes of lead(II), cadmium(II), and zinc(II) with N-phosphonomethylglycine. *Monatshefte für Chemie* 131(1): 1–11.

Kremer, R.J. and Means, N.E. 2009. Glyphosate and glyphosate-resistant crop interactions with rhizosphere microorganisms. *Eur. J. Agron.* 31: 153–161.

Krueger, M., Schrodl, W. Pedersen, I., and Shehata, A.A. 2014. Detection of glyphosate in malformed piglets. *J. Environ. Anal. Toxicol.* 4: 5. doi:10.4172/2161-0525.1000230. ISSN: 2161-0525.

Krueger, M., Schledorn, P., Schrodl, W., Hoppe, H.-W., Lutz, W., and Sheta, A.A. 2014. Detction of glyphosate residues in animals and humans. *J. Environ. Anal. Toxicol.* 4: 2.

Krüger M., Shehata A.A., Schrödl, W., Rodloff, A. 2013. Glyphosate suppresses the antagonistic effect of Enterococcus spp. on *Clostridium botulinum. Anaerobe* 20: 74–78.

Kurenbach, B., Marjoshi, D., Amabile-Cuevas, C.F., Ferguson, G.C., Godsoe, W., Gibson, P., and Heinemann, J.A. 2015. Sublethal exposure to commercial formulations of the herbicides dicamba, 2,4-dichlorophenoxyacetic acid, and glyphosate cause changes in antibiotic susceptibility in Escherichia coli and Salmonella enterica serovar Typhimurium. *mBio* 6: 2. doi:10.1128/mBio.00009.15.

Levesque, C.A. and Rahe, J.E. 1992. Herbicide interactions with fungal root pathogens, with special reference to glyphosate. *Ann. Rev. Phytopathol.* 30: 579–602.

Londager and Madsen, H.E.L., et al. 1978. Stability constants of copper, zinc, manganese, calcium and magnesium complexes of glyphosate. *Acta Chem. Scand. A* 32: 79–83.

Mallory-Smith, C.A. and Retzinger Jr., J. 2003. Revised classification of herbicides by site of action for weed resistance management strategies. *Weed Tech.* 17: 605–619.

Marasmius, H. 2011. *Marchner's Mineral Nutrition of Higher Plants.* Academic Press, Cambridge, MA.

Martell, A.E. and Smith, R.M. 1974. *Critical Stability Constants.* Plenum Press, New York, Vol. 1, 5 (first supplement) 6 (second supplement).

Mesnage, R., Arno, M., Costanzo, M., Malatesta, M., and Séralini, G.-E. 2015. Transcriptome analysis reflects rat liver and kidney damage following chronic ultra-low dose Roundup exposure. *Environ. Health.* 14(1): 70. doi:10.1186/s12940-015-0056-1.

Motekaitis, R.J. and Martell, A.E. 1985. Metal chelate formation by N-phosphonomethyl glycine and related ligands. *J. Coord. Chem.* 14: 139–149.

Motta, E.V., Mak, M., De Jong, T.K., Powell, J.E., O'Donnell, A., Suhr, K.J., and Moran, N.A. 2020. Oral and topical exposure to glyphosate in herbicide formulation impact the gut microbiota and survival rates of honey bees. *Appl. Environ. Microbiol.* https://aem.asm.org/content/early/2020/07/07/AEM.01150-20. abstract.

Mucha A, Drag M, Dalton JP, and Kafarski P. 2010. Metallo-aminopeptidase inhibitors. *Biochimie* 92: 1509–529.

Nicola, P. 2019. The impacts of glyphosate-based herbicides on bumble bee productivity and parasite load. A thesis submitted for the degree of Master of Science, School of Biological Sciences, Queen's University, Belfast.

Nilsson, G. 1985. Interactions between glyphosate and metals essential for plant growth. In: Grossbard E. and Atkinson, D. (eds.). *The Herbicide Glyphosate.* Butterworth, London, pp. 35–47.

Oates, L., Cohen, M., Braun, L., Schhembri, A., and Taskova, R. 2014. Reduction in urinary organophosphate pesticide metabolites in adults after a week-long organic diet. *Environ. Res.* 132: 105–111.

Ozturk, L., Yazici, A., Eker, S., Gokmen, O., Roemheld, V., and Cakmak, I. 2008. Glyphosate inhibition of ferric reductase activity in iron deficient sunflower roots. *New Phytol.* 177: 899–906.

Pandy, A.K., Pandy, S.D., and Misra, V. 2000. Stability constants of metal-humic acid complexes and its role in environmental detoxification. *Ecotoxic. Environ. Saf.* 47: 195–200. doi:10.1006/essa.2000.1947.

Pedersen, I.B. 2014. An on farm study showing deformities, abortions, and fertility problems in pigs linked to GM soya and Roundup. *J. Environ. Anal. Toxicol.* 4: 5. doi:10.4172/2161-0525.1000230. ISSN: 2161-0525.

Peterson, B.H. 2016. Glyphosate, brain-damaged babies, and Yakima Valley – a river runs through it. Farm Wars, March 6, 2014.

Prasai, T.P. Walsh, K.B., Bhattarai, S.P., Midmore, D.J., Van, T.T.H., Moore, R.J., and Stanley, D. 2016. Biochar, bentonite, and zeolite supplemented feeding of layer chickens alters intestinal microbiota and reduces Campylobacter load. *PLOS One.* April 26, 2016. doi:10.1371/journal.pone.0154061.

Prasai, T.P. Walsh, K.B., Bhattarai, S.P., Midmore, D.J., Van, T.T.H., Moore, R.J., and Stanley, D. 2017. Zeolite food supplementation reduces abundance of enterobacteria. *Microbiol. Res.* 195: 24–30.

Robinson C. 2010. Argentina's Roundup human tragedy. *Sci. Soc.* 48, 30–31.

Rowell, A. 2003. *Don't Worry It's Safe to Eat: The True Story of GM Food, BSE & Foot and Mouth.* Earthscan Publications, Sterling, VA.

Samsell, A. and Seneff, S. 2013a. Glyphosate's suppression of cytochrome P450 enzymes and amino acid biosynthesis by the gut microbiome: Pathways to modern diseases. *Entropy* 15:1-x manuscripts. doi:10.3390/el40x000x.

Samsel, A. and Seneff, S. 2013b. Glyphosate, Pathways to modern diseases II: Celiac sprue and gluten intolerance. *J. Interdis. Toxicol.* 6: 159–184.

Samsel, A. and Seneff, S. 2015a. Glyphosate, pathways to modern diseases III: Manganese, Neurological diseases, and associated pathologies. *Surg. Neurol. Int.* 6:45. http://www.ncbi.nlm.nnih.gov/pmc/articles/PMC4392553/.

Samsel, A. and Seneff, S. 2015b. Glyphosate, pathways to modern diseases IV: Cancer and related pathologies. *J. Biol. Phys. Chem.* 15: 121–159. doi:10.4024/11SA15R.jbpc.15.03.

Samsel, A. and Seneff, S. 2016. Glyphosate, pathways to modern diseases V: Amino acid analogue of glycine in diverse proteins. *J. Biol. Phys. Chem.* 16: 9–46.

Schafer, J.R, Hallett, S.G., and Johnson, W.G. 2010. Role of soil-borne fungi in the response of giant ragweed (Ambrosia trifida) biotypes to glyphosate. *Proc. Northcentral Weed Sci. Soc.* 65.

Schafer, J.R., Hallett, S.G., and Johnson, W.G. 2012. Glyphosate-resistant 'superweeds' may be less susceptible to diseases. ScienceDaily July 17, 2012.

Schafer, J.R., Hallett, S.G., and Johnson, W.G. 2013. Soil microbial root colonization of glyphosate-treated giant ragweed (Ambrosia trifida) horseweed (Conyza canadensis), and common lambs quarters (Chenopodium album) biotypes. *Seed Sci.* 61:289–295.

Schrödl, W., Krüger, S., Konstantinova-Müller, T., Shehata, A.A., Rulff, R., and Krüger, M. 2014. Possible effects of glyphosate on mucorales abundance in the rumen of dairy cows in Germany. *Curr. Microbiol.* doi:10:1007/s00284-014-0656-y.

ScienceDaily. May 19, 2016. Antibiotics that kill gut bacteria also stop growth of new brain cells. https://www.sciencedaily.com/releases/2016/05/160519130105.htm.

Seralini, G-E., Clair, E., Mesnage, R., Gress, S., et al. 2014. Republished study: Long term toxicity of a Roundup herbicide and a Roundup-tolerant genetically modified maize. *Environ. Sci. Eur.* 26:14–31.

Shehata, A., Schrodl, W., Aldin, Q.Q., Hafez, H.M., and Krueger, M. 2012. The effect of glyphosate on potential pathogens and beneficial members of poultry microbiota in vitro. *Curr. Microbiol.* Doi:10.1007/s00284-012-0277-2.

Shehata, A.A., Kühnert, M., Haufe, S., and Krüger, M. 2013. Neutralization of the antimicrobial effect of glyphosate by humic acid in vitro. *J. Chemosphere* 10: 258–261.

Subramaniam, V. and Hoggard, P.E. 1988. Metal complexes of glyphosate. *J. Agric. Food Chem.* 36:1326–1329.

Swanson, N. L., Leu, A., Abrahamson, J., and Wallet, B. 2014. Genetically engineered crops, glyphosate and the deterioration of health in the United States of America. *J. Organ. Syst.* 9:6–37.

Thompson, I.A. and Huber, D.M. 2007. Manganese and plant disease. pp. 139–153 (Chapter 10). In: L.E. Datnoff, W.H. Elmer, D.M. Huber (eds.) *Mineral Nutrition and Plant Disease.* APS Press, St. Paul, MN.

Torstensson, N.T., Lundgren, L.N. and Stenstrom, J. 1989. Influence of climatic and edaphic factors on persistence of glyphosate and 2,4-D in forest soils. *Ecotoxicol. Environ. Saf.* 18, 230–239.

U.S. Geological Survey, 2018; https://waterdata.*usgs*.gov/)

United States Patent Office. 1964. Aminomethylenephosphinic acids, salts thereof, and process for their production. Patent No. 3,160,632, Dec. 8. 1964.

U.S. Patent 5902804 (United States Patent Office). Pharmaceutical composition for inhibiting the growth of viruses and cancers. (glyphosate derivatives).

U.S. Patent 7771736 B2 (United States Patent Office). (Monsanto) Glyphosate formulations and their use for the inhibition of 5-enolpyruvyl-shikimate-3-phosphate synthase. (Glyphosate as an antibiotic for enteric organisms.)

United States Department of Agriculture. 2015. National Agricultural Statistical Service (USDA-NASS).

Warren, M. and Pisarenko, N. 2013. Birth defects, cancer in Argentina linked to agrochemicals: AP investigation. The Associated Press, October 20, 2013. http://www.ctvnews.ca/health/birth-defects-cancer-in-argentina-linked-to-agrochemicals-ap-investigation-1.1505096.

Yamada, T., Kremer, R.J., Camargo e Castro, P.R., and Wood, B.W. 2009. Glyphosate interactions with physiology, nutrition, and diseases of plants: Threat to agricultural sustainability? *Eur. J. Agron.* 31: 111–113.

Yoon, S.S., and Brandt, L.J. 2010. Treatment of refractory/recurrent C. difficile-associated disease by donated stool transplanted via colonoscopy: a case series of 12 patients. *J. Clinic. Gastroenterol.* 44:562–566.

Zobiole, L.H.S., Bonini, E.A., Oliveira Jr., R.S., Kremer, R.J., and Ferrarese-Filho, O. 2010. Glyphosate affects lignin content and amino acid production in glyphosate-resistant soybean. *Acta Physiol. Plant.* 32:831–837.

Zobiole, L.H.S., Oliveira Jr., R.S., Huber, D.M., Constantin, J., Castro, C., Oliveira, F.A., Oliveira Jr., A. 2010. Glyphosate reduces shoot concentrations of mineral nutrients in glyphosate-resistant soybeans. *Plant Soil* 328:57–69.

Zobiole, L.H.S., Oliveira Jr., R.S., Kremer, R.J., Constantin, J., Yamada, T., Castro, C., Oliveira, F.A., and Oliveira Jr., A. 2010a. Effect of glyphosate on symbiotic N_2 fixation and nickel concentration in glyphosate-resistant soybeans. *Appl. Soil Ecol.* 44:176–180.

Zobiole, L.H.S., Oliveira Jr., R.S., Kremer, R.J., Constantin, J., Bonato, C.M., and Muniz, A.S. 2010b. Water use efficiency and photosynthesis as affected by glyphosate application to glyphosate-resistant soybean. *Pestic. Biochem. Physiol.* 97:182–193.

Zobiole, L.H.S., Oliveira Jr., R.S., Kremer, R.J., Muniz, A.S., and Oliveira Jr., A. 2010a. Nutrient accumulation and photosynthesis in glyphosate resistant soybeans is reduced under glyphosate use. *J. Plant Nutr.* 33:1860–1873.

3 Health, Nutrition, and Sustainability: Precious Commodities in Jeopardy from Agricultural Pesticides

Arden Andersen
Independent Family Practice and Occupational
Medicine Physician and Consultant

CONTENTS

3.1 INTRODUCTION

There are many catch phrases and politically charged statements in the media today regarding our state of health, our environment, our healthcare system, emerging diseases, rogue virus, political correctness, sustainability, and public health. There seem to be two distinct and conflicting positions for each one of these: the industry position and the scientific position. Certainly, industry would like everyone to believe that "they" are "scientifically" based and vetted at every turn and only profit from legitimate technologies after thorough scientific rigor and scrutiny. Science or more correctly "scientists" know better. Although many "scientists" do work in, for, and with the industry, they must be ever rhetorical in skewing their "scientific discoveries" so as to conform to the economic desires their industry patrons dictate. This fact is well illustrated in the book, *Science for Sale* by David Lewis, Ph.D., former EPA research scientist and University of Georgia Professor. It lays out clearly that governmental agency policy conforms to, supports, and enforces industry business directives [1].

3.2 CONFLICTS OF PRACTICE AND HEALTH

Specifically at issue in this book regarding health, nutrition, and sustainability, Dr. Lewis blew the whistle on the idea that the application of municipal sewage waste applied to agriculture was/is directly causing ill-health, not only to the farmers and communities exposed to the dried waste dust and groundwater contamination, but also through contamination of the food crops grown on said "fertilized" fields. He proved that the drugs, industrial chemicals, and most of the pathogens found in city sewage passed relatively unchanged through waste treatment plants and into the applied waste products.

These products are given numerous names such as biosolids, organic fertilizer, "Nutri-green," NitroHumus, Milorganite, Granulite, and Nu-Earth, to name a few. A concerted misinformation campaign was waged against Dr. Lewis and his research at EPA to discredit his work and him personally. Governmental agency research that conflicts with or disproves such agency "policy" is altered, discredited, and shredded while the scientists are reprimanded, discredited, fired, or sent to early retirement.

Industry hijacking of "science" is not a new phenomenon nor is "peer" condemnation of scientific discovery. Galileo was condemned by "The Church" for suggesting that the sun, not the earth, was the center of our solar system. Alexander Gordon, MD, Oliver Wendell-Holmes, MD, and Ignis Semmelweis, MD, were all vilified by the medical profession for suggesting that physicians wash their hands between working in the morgue and delivering babies in the delivery ward as physicians were causing the extremely high death rate of postpartum mothers due to "childbed fever," also known as puerperal fever. Pasteur, in 1879, over 100 years after Gordon's warnings, showed that strep bacteria could be cultured in most cases of puerperal fever, and the medical profession finally acknowledged their grave error and literally began to "clean up" their act [2].

Hannaman, Still, and Palmer met with similar distain, public slandering, and professional discrediting. Modern medicine, with its great technologies, misses the fundamentals of physics that Hannaman grasped and those of structure and function that Still and Palmer mastered. Modern doctors choose to be ignorant of medical physics called acupuncture that has been practiced for 2000 years by the Chinese.

Albert Abrams, MD, who was a prodigal pathologist at what is today Stanford University and was mentored by Rudolph Virchow, the father of pathology, went from renowned to rejected when he discovered that every organ, tissue, and cell of the human body has a bio-resonant frequency, which was detectable and alterable with the correct instrumentation. It turns out that it's just physics, but it threatened the medical establishment's status quo [3,4]. Robert Becker, MD, an orthopedic surgeon, later discovered that nonhealing fractures could be healed with specific frequency stimulation. This research was further delineated by Abraham Liboff, Ph.D., Stephen Smith, Ph.D. and others. Becker's work got too close to regenerating human limbs with tuned electromagnetic force, EMF, was funded by the U.S. Navy and classified, never again to see the light of day. In 1987, Stephen Smith was interviewed stating that the regeneration of mammalian limbs was already possible, but all funding was withdrawn because the bigger and more powerful prosthetics industry could not make money on limb regeneration.

Industry includes men and women choosing to make unethical decisions, and it is not just limited to astronomy and medicine. Nicola Tesla was labeled a charlatan and dangerous inventor by Thomas Edison for his invention of the alternating current (AC) electric system. Edison had invented the DC system and wanted to continue his profitable monopoly empire on this system. Had it not been for Westinghouse financially backing Nicola Tesla and building the first AC hydroelectric power station at Niagara Falls, Edison's DC system would have prevailed for years to come as well as his monopolistic profits [5]. Henry Ford was chastised for building the horseless carriage and assembly line. The Wright Brothers were scoffed at, laughed at, and chastised for suggesting that man could fly before they proved it at Kitty Hawk, NC.

Philip S. Callahan, Ph.D., my Ph.D. advisor and mentor, demonstrated that insects communicate and home in on their food via infrared, ultraviolet, and radiowave radiations. His work was classified and funded by the Department of Defense throughout his tenure at the University of Florida (UF). What really brought the scorn of the entomology department at UF was his demonstration (as did a French biologist, Francis Chaboussou) that insects only attack sick plants because such a revelation and demonstration could lead to significant defunding by pesticide companies supporting UF agricultural research [6,7]. Such attacks on science and truth in science continue today with character assassination, literal assassination attempts, journal snubbing, and defunding campaigns including many of the writers in this book: Huber, Hayes, Seneff, Carmen, Seralini, Kremer, and others. Unfortunately, it seems that human nature has not changed much in 10,000 years, where profit exploitation, power, influence, and greed are possible, and truth, science, and truth in science become the casualties.

3.3   AGRICULTURE DETERMINES THE NUTRITIONAL QUALITY OF SOCIETY

What about agriculture today and particularly the quality of the food farmers produce and we eat? In 1922, Rudolph Steiner stated that the nutrient value of the food had so declined that it was adversely affecting people's ability to learn [8]. Charles Northern, MD, read into the Congressional record in 1936 that GI diseases were correlated with the declining nutrient values of our food [9].

Why does this matter? We reportedly already produce more food than the world consumes, attributing most starvation to a distribution problem, not a production problem. As true as this may be, health is much more than quantity of food or filler of the belly.

Nutrition is the reason for the consumption of food. It is the primary source for all nutrition on this planet for every living organism. Nutrition, however, has a varied, sometimes marred, and comprehensive definition. The word nutrition is used in varying contexts related to soil fertilization and plant feeding, cattle feedlot and dairy rations, hospital, nursing homes and school lunch menus, IV nutrient bags, supplement formulations, and fad diets.

Nutrition, in the context of a registered dietician (RD), is significantly different than in the context of a Certified Clinical Nutritionist (CCN). The former's focus is primarily about fats, proteins, and carbohydrates that is more in line with the commodity industry lobby/Federal Government policy parameters than factual nutritional biochemistry. The latter is all about factual nutritional biochemistry regardless of government policy or industry business model. The CCN scrutinizes all fats, proteins, and carbohydrates for their value in the nutritional biochemistry of the individual patient/client. See IAACN.org.

Speaking of actually growing food crops, nutrition as defined and practiced by Don Huber and Michael McNeill, both writing later in this book, is significantly different than the nutrition one will encounter in the practices of most any typical Cooperative Extension Service agronomist, and certainly different than any agronomist working for the local cooperative fertilizer company, chemical company, or farm association such as the American Corn Growers Association and American Soybean Growers Association. The difference is quite simple. Nutrition is really about the state of health of any living organism. Regarding that state of health, when the nutrition is truly balanced, the organism is not susceptible to disease. In fact, it is nutritional balancing that will reverse states of disease and pre-disease to return the organism to health. At this point, pesticides, termed "plant protection agents," which are really just chemical weapons, are no longer needed.

3.4   THE DYNAMIC INTERACTION OF NUTRITION AND HEALTH

There is a nutritional continuum between health, pre-disease, disease, and death that helps in understanding the dynamic functioning of nature. Death is quite easily defined and understood by everyone as simply, the organism is no longer viable in any sense of the word. Disease is fairly well defined by medicine and agriculture to mean the organism has some sort of malady that puts it into a state of pending or progressing death/doom over time. That may be a physiologic disease such as cancer, multiple sclerosis, Parkinson's, diabetes, or coronary artery disease or may be infective where the organism is "attacked" from a bacterial, viral, or fungal infection such as by coronavirus, HIV, cholera, influenza, Johne's disease, *Salmonella*, botulism, powdery mildew, Panama disease, and citrus greening. This category on the continuum is the foundation of the most profitable and corrupt industry on earth – the drug and chemical weapon (pesticide) industry.

Pre-disease is the state of existence where the organism is either in the early stages of disease, yet undiscovered or susceptible to invading infective agents that simply haven't occurred yet. From a biochemical physiological perspective, pre-disease is not much different than disease as it may simply be a matter of technology lag or pathogen exposure that differentiates the two. Pre-disease is the initial state below health when the organism's biochemistry begins to become unbalanced, sluggish, disturbed, or fouled. It is about nutritional balance at the molecular and compound level, long before the cellular or tissue level.

It is at the fundamental fourth-dimensional molecular structures and most importantly at the initiation and generation of electromagnetic signatures, actually at the bio-photon level according to Fritz-Albert Popp, Ph.D. Popp demonstrated that bio-photonic communication within and between each living cell preempted biochemistry; bio-photonic communication was the determinant of subsequent biochemical physiology [10,11].

Callahan showed that it is the bio-photonic signature riding on UV and RF signals that is scanned by insects to determine if the molecules and compounds corresponding to these signatures are suitable for their diet or not. This is truly where health and pre-disease are differentiated at their most fundamental or rudimentary level. If these signatures represent excess or free-spacing inorganic nitrogen, excess simple amino acids, excess simple sugars, cellular waste products, or "leaked" simple cations, insects will attack to feast on this food. Callahan's work has been furthered and brought into practical practice by Thomas Dykstra, Ph.D., who founded Dykstra Laboratories, Inc. in Gainesville, Florida, to understand the specific mechanisms of insect olfaction and to develop insect traps by integrating bioelectromagnetics (www.dykstralabs.com).

Killing the "garbage crew" does not correct the bio-photonic signature nor the biochemistry associated with it. That is only corrected with appropriate nutritional management and not by dietetics or genetic engineering of the crop variety, although both of these approaches are very profitable for the chemical and drug industries.

In the context or mindset of the typical dietician, extension service agronomist, industry agronomist, veterinarian, and physician, nutrition is merely a moderator of functional status of the person, animal, or plant. It is used to maximize the immediate output of athletic performance, animal, or crop commodity under the narrow spectrum of nutrition that these "professionals" acknowledge. If disease is encountered, a search and destroy mission must be undertaken to identify the invading disease agent and subsequently kill it with some chemical agent.

Alternatively, in a new approach, a genetically engineered stem cell implant must be administered or a vaccine developed to thwart the invading evil pathogen. Cropping models are entirely built around this model whether it be with genetic engineering for the crop to produce its own synthetic chemical weapon or, more commonly, for the crop to tolerate the indiscriminate application of the latest herbicide weed chemical weapon.

Nutritional research in this context is entirely centered on reductionist isolation of individual biochemical processes that then can be synthesized in the lab to develop a patentable synthetic product. This is the model that has gotten us, our planet, our environment to the point of filth, pollution, erosion, and environmental collapse in which we find ourselves. It is purely an extraction model, entirely unsustainable in the real world of nature.

3.5   THE INTRICATE LINK BETWEEN AGRICULTURE AND HEALTH

Sustainability has become a catch phrase for both the environmental lobby and the chemical industry and is rarely used scientifically by either. It is used by farmers and retailers as "sustainably grown" to market their products. It is used by Cooperative Extension Service agents to convince chemical farmers and the public that their "government" research and white papers are scientifically based, leading-edge, environmentally safe, regenerative, and "sustainable."

Agriculture and medicine are intimately linked. The companies that manufacture the chemical weapons (crop protection agents) that cause or contribute to human and animal cancers are the same companies that manufacture the drugs to treat these cancers. Medicine is equally bound to the pharmaceutical industry as the entire medical system beyond emergency care is geared toward "disease treatment." "Prevention" is merely "early detection," e.g., colonoscopy, endoscopy, mammography, biopsy, and CT scans.

Consequently, the medical establishment has become the third leading cause of death as documented in a July 26, 2000 JAMA article (#284). Twenty years later, that ranking stands. An article in the British Medical Journal 2019 documented that "… one in twenty patients are exposed to preventable harm in medical care" [12].

Real prevention, e.g., diet overhaul, is never mentioned in medical circles, official CDC, or other government documents. A look closer reveals that the USDA drives these recommendations as they are tasked to regarding its recommendations via government-funded nutritional programs such as SNAP, food stamps, school lunch programs, and their "Food Guide Pyramid." The USDA has a direct conflict of interest as it is tasked with promoting agribusiness and commodities while at the same time promoting health. USDA officials compromise. A letter from the National Pork Board to the 2010 Dietary Guidelines Advisory Committee stated, "Pork Board cautions the Committee against making any decisions which would limit the ability of Americans to choose the most nutrient-rich foods in the meat group." That is a business statement, not a nutritional statement. Article after article in the medical literature shows that the higher the meat consumption, the higher the cancer risk, cardiovascular risk, and diabetes risk [13,14].

According to the World Health Organization, by 2020, chronic diseases would account for almost three-quarters of all death worldwide. The majority will be in developing countries, a significant change from infectious disease and starvation. Yet, like infectious disease and starvation, chronic disease is preventable. WHO acknowledges that modern dietary patterns and lifestyle behaviors are the cause of chronic disease.

Significantly, traditionally plant-based diets have been replaced by high-fat, energy-dense diets that are largely animal-based. This change is driven by the financial interests of the commodities industry, the Chicago Board of Trade, and the supporting pesticide/GMO industry. Further, antibiotics are not nearly as profitable as chronic disease drug regimes. These global changes are reflective of a business plan, not science [15].

3.6 FOOD IS ABOUT PROVIDING SUSTENANCE FOR THE CONSUMER

We have the knowledge and the experience to solve all these problems around the world. What we have lacked is the will to make it happen. That will must counter the prevailing economic inertia of Americanized agriculture, medicine, and industry. The only reason we don't go back to a hemp-based fiber society is because consumers lack the will to counter the chemical-/oil-based synthetic fiber industry. My home state Kansas used to be the number two hemp fiber producer in the United States.

Farmers and the farm industry spend a lot of money advertising to the public about how they are sustainable, feed the world, and care about the environment. That's for another discussion, but that the food primarily comes from the farm is where we will start. Food is about providing sustenance for its consumers, whether it is us as people or our animals. That sustenance is truly about nutrition, the 60-plus minerals we need, and the thousands of vitamins, antioxidants, phytonutrients, enzymes, amino acids, and yet unidentified items that nutritious food contains.

Certainly, there is the fat, protein, and carbohydrate content that is focused upon by dieticians and industry marketeers because the simplification of nutrition to these three generalized components allows the industry to synthesize and market a wide variety of "food" items with very little nutritional value. These would include items such as white bread and white sugar, French fries, most dried breakfast cereals, many packaged sweets, and many canned vegetables as they need added sugar and salt to get people to eat them.

What really needs to be addressed is what Steiner and Northern alluded to nearly a century ago that vitamins, minerals, amino acids, phytonutrients, essential oils, and antioxidants should be the focus. Repeated studies over the last several decades indicate a steadily declining nutrient density of the food crops coming off our farms. In other words, the actual nutrient value of a carrot or apple or cucumber per carrot, apple or cucumber of the same size and weight, has declined steadily. Several studies have concluded this fact, yet none of them have pointed to the cause nor the solution. These studies include Donald Davis et al. at the University of Texas published in December 2004 in the

Journal of the American College of Nutrition. Another is the Kushi Institute analysis of nutrient data from 1975 to 1997 on 12 vegetables that all show a decline in mineral and vitamin contents. Another study is the British version published in the British Food Journal on data from 1930 to 1980 finding that the nutrient contents of 20 vegetables have declined [16]. Yet another study published in the National Hog Farmer (Swine Information Service, No. E25, 1968) looked at over 4,000 grain samples taken from 11 midwestern states that showed a significant decline in grain mineral content.

This decline in nutrient values at the farm gate presents a bit of a quandary for consumers because they must be ever diligent to eat a variety of foods to source the full spectrum of nutrient needs. Vitamins and minerals are the mainstay for our bodies to function, grow, repair cells and tissues, fight off infections, alleviate environmental stressors, and detoxify endogenous and exogenous compounds. Certainly, diet is the most important factor in the development of chronic diseases such as diabetes, cardiovascular disease, kidney disease, arthritis, autoimmune diseases, dementia, and cancer. All of these diseases have been shown to be both induced by the Standard American Diet (SAD) and reversed by replacing the SAD with a plant-based vegetarian or Mediterranean diet. There are multiple journal articles on this subject accessed at NutritionFacts.org for those that wish to research the topic further. The documentary "The Game Changers" is an excellent overview from the world's most elite athletes.

Accompanying the generally reduced nutrient levels are the absence of nutrients vital to immune function, brain development, and liver detoxification; and the added plethora of agrichemicals that include many endocrine disruptors, mineral chelators that steal vital trace nutrients and oxidizers to further the nutrient decline in one's body and increasing already unbalanced vital functions.

3.7   CHEMICAL DISRUPTERS OF NUTRITION AND FUNCTION

Observers wonder why we have such an increase in all childhood disorders from autism to ADD, and cancer to gender dysphoria. It is pretty straightforward when one looks at what these chemicals do in an ever-declining nutrient environment. Every chemical, whether found in the food directly, in the air we breathe, the water we drink, or the clothing we wear, further taxes our oxidation–reduction system and our Phase I, II, and III detoxification systems – all of which require full nutrition to function correctly and fully. It's a matter of understanding Krebs' cycle, its biochemistry.

Various pesticides are suspected or proved to act as endocrine disruptor compounds (EDCs) [20].

Endocrine-disrupting chemicals (EDCs) represent a broad class of exogenous substances that cause adverse effects in the endocrine system, mainly by interacting with nuclear hormone receptors (NRs). Examples include binding of the mycoestrogen α-zearalanol to estrogen receptors, the covalent interaction of organotins with retinoid X- and peroxisome proliferator-activated receptors, and the cooperative binding of two chemicals to the pregnane X receptor. Several hypotheses could further explain low-concentration effects of EDCs with weaker affinity towards NRs. (19) More than 50 pesticide active ingredients have been identified as endocrine disruptors by the European Union and endocrine disruptor expert Theo Colborn, PhD. Endocrine disruption is the mechanism for several health effect endpoints." [24]

   Endocrine disruptors function by: (i) Mimicking the action of a naturally-produced hormone, such as estrogen or testosterone, thereby setting off similar chemical reactions in the body; (ii) Blocking hormone receptors in cells, thereby preventing the action of normal hormones; or (iii) Affecting the synthesis, transport, metabolism and excretion of hormones, thus altering the concentrations of natural hormones. Endocrine disruptors have been linked to attention deficit hyperactivity disorder (ADHD), Parkinson's and Alzheimer's diseases, diabetes, cardiovascular disease, obesity, early puberty, infertility and other reproductive disorders, and childhood and adult cancers. Exposure to elevated concentrations of diazinon, chlorpyrifos and chlorpyrifos-methyl may be associated with endometriosis" (25) while exposure to select pesticides may be associated with impaired beta-cell function and poorer glycemic control among youth with diabetes [26].

Dr. Hayes addresses his findings with atrazine and endocrine disruption in frogs. It has profound ramifications for the gender dysphoria we see in humans that can be directly attributed to agriculture;

specifically, farmers spraying these endocrine-disrupting herbicides and other pesticides into our environment and on our food [17].

Risk assessments and some epidemiologic studies have looked at the exposure and toxicology of a single compound; however, two other considerations must be included: the presence of pesticide by-products and the cumulative exposure to multiple pesticide residues [21]. Besides known factors, exposure to EDCs has also been associated with several health conditions to include problems during pregnancy and/or at birth, early puberty, menstrual irregularities, polycystic ovary syndrome (PCOS), endometriosis, breast cancer, or early menopause (premature ovarian insufficiency or failure). Reproductive effects in humans and wildlife originate from linking exposure to endocrine-disrupting chemicals (EDCs) in the womb to declining sperm counts and increasing prevalence of undescended testes, testicular cancer, and urinary duct malformation in males [21]. The effects of EDC exposure during early life may be activated or become worse due to additional EDC exposure throughout a woman's life. Increased understanding of EDCs has changed the ways toxic actions are viewed. Traditionally, toxicology has primarily focused on chemical concentration, i.e., "the dose makes the poison." It is now clear that the state (or life stage) of the targeted organism is also critical. Implementing "timing" into toxicological and regulatory sciences is a great challenge, but will certainly lead to more protective chemical regulations [22].

These data suggest that DES alters uterine development and consequently adult reproductive function by modifying the enhancer landscape at ERα binding sites near estrogen-regulated genes [23].

Triazophos (TAP) was a widely used organophosphorus insecticide in developing countries. TAP could produce specific metabolites, triazophos-oxon (TAPO) and 1-phenyl-3-hydroxy-1,2,4-triazole (PHT), and non-specific metabolites, diethylthiophosphate (DETP) and diethylphosphate (DEP). TAP might interfere with the endocrine function of the adrenal gland. It might also bind strongly with glucocorticoid receptors and thyroid hormone receptors. TAPO might disrupt the normal binding of androgen, estrogen, progesterone and adrenergic receptors to their natural hormone ligands. DEP might affect the biosynthesis of steroid and thyroid hormones. In addition, DEP might disrupt the binding and transport of thyroid hormones in blood and the normal binding of thyroid hormones to their receptors. These results suggested that TAP and DEP might have endocrine disrupting activities and that they were potential endocrine disrupting chemicals. These results emphasize the need for a comprehensive evaluation of the toxicity of organophosphorus chemicals and their metabolites. [17]

Studies of EDCs have challenged traditional concepts in toxicology for decades, especially the dogma of "the dose makes the poison," because EDCs can have effects at low doses that are not predicted by effects at higher doses. Whether low doses of EDCs influence certain human disorders is no longer conjecture, because epidemiological studies show that environmental exposures to EDCs are associated with human diseases and disabilities. We conclude that when nonmonotonic dose–response curves occur, the effects of low doses cannot be predicted by the effects observed at high doses. Thus, fundamental changes in chemical testing and safety determination are needed to protect human health [27].

"The incidence of kidney disease has increased rapidly in recent years. One major possible reason for this increase in nephrosis is from foodborne toxins" [18]. Another disaster in the waiting is the increasing number of young adults developing Alzheimer's dementia. A new study from BlueCross BlueShield titled, "Early-onset dementia and Alzheimer's rates grow for younger Americans," February 27, 2020, looked at adults aged 30–64. In 2013, there were 4.2 per 10,000 adults in this age range with a diagnosis of either form of dementia; by 2017, there were 12.6 or a 200% increase. The average age was 49. This equates to 78,000 adults. Why was BCBS interested? They insure these people and have to pay the medical costs for such a diagnosis.

What's the cause? It's not purely genetics, and even if it were genetics, why is this generation now developing the disease two to three decades earlier than their parents or grandparents? To answer this question, we have to look at causes and contributing secondary causes. Swedish scientists showed that dementia can stem from traumatic brain injury over 30 years from the TBI event [28]. We know that there are many precipitating and correlative mechanisms to brain damage leading to dementia. One is TBI. TBI etiology is brain inflammation. Inflammation, we are learning, is the root of most disease mechanisms, so this is not really surprising [29]. Diet appears to be the primary determinant of dementia rates in any population [30].

Diet is about the food people choose to eat, and more and more is from processed and packaged sources. A recent analysis of the global packaged food market expects continued global growth, with snack foods and beverages leading the way [31].

One significant problem with processed and packaged foods is the compromised nutrient value of these foods. Most are made with bleached flower, which, by definition, has the minerals and vitamins removed and fiber removed. Many are sweetened with high-fructose corn syrup or refined sugar, both also devoid of minerals and vitamins even though the modern farming advocates, farmers, and the chemical industry routinely state that agriculture is so much more productive today than a century ago. It is in volume, certainly. It is in farm size, machinery size, and technological innovations applied to the farm and to the sorting, processing, and packaging of farm products, but not to nutrient content.

3.8 PROLIFERATION OF CHEMICAL RESIDUE CONTAMINATION IN MODERN FOOD

What is equally as concerning as the decline in nutrient value of our food is the toxic residues accompanying our food today. Agriculture dumps a plethora of chemical weapons directly onto our food and into the environment in which food is grown, all in the name of "plant protection products." The EPA only evaluates such chemicals in a vacuum of each chemical individually and not in the collective soup of the multitude of chemicals possible.

Industry, including farmers, regularly contends that pesticides are safe and would not be approved for use if they weren't; yet, not one of them is studied for safety in combination. At the same time, industry and farmers acknowledge that combining pesticides increases their killing power for targeted diseases [32]. When we put that into human impact, we know from basic biochemistry principles and clinical experience that drugs – chemicals – work synergistically; that the combination of drugs requires lower doses to get the desired therapeutic results than if used alone. Pesticide research, as previously stated, often finds the same outcome when looking at the potential combination of pesticides relative to toxicity.

In a study by Coleman et al., fungicides showed more toxicity when combined with chemicals than using alone. Their finding means that scientific evaluation of pesticides and environmental chemicals will have to be significantly overhauled. It means that we can no longer depend upon the classic "dose makes the disease" belief for evaluating these poisons. More importantly we must understand that there is often a "U"-shaped curve of toxicity response and a synergistic negative relationship when chemicals are combined. It should be recognized that they always are combined in the environment.

Sustainability is truly about survivability, and survivability is tied directly to health, pre-disease, and disease. With young people getting sicker and sicker, our women and men experiencing higher and higher rates of infertility, miscarriages, and subsequent children with birth defects; a change must happen or we as a species will not survive. The turnaround starts in the soil, with regeneration of the soil, its minerals, and microbiology. It progresses to producing food with increased nutrient density, naturally free of pests and disease, without the "NEED" for chemical weapons. It is already being done. It just needs to become the majority rather than the minority method. By default this system of agriculture regenerates soil, environment, food, and consumers. It is sustainable production and creates sustainability for us all [33].

The following chapters outline the scientific facts of the dire state of health, nutrition, and sustainability in which we find ourselves. Some will see this pessimistically as the end of times as they read in the Book of Revelations. Perhaps, but I see it simply as nature's way of telling us what to expect when we fail to learn, heed, and follow the laws of nature. Our battle is not with nature.

Nature is truly on our side. Rather, our battle is with the greed and ego of people/industries gaining the most profits from a sick planet via the sale of chemical weapons, drugs, dirty vaccines, surgical interventions, and Super Fund cleanup projects.

When farmers follow the basics principles of sustainable, regenerative, scientific agriculture, they often significantly reduce their synthetic fertilizer needs and may entirely eliminate herbicide, fungicide, and insecticide use, while increasing crop yields with GE-free genetics. Proper animal husbandry also increases production efficiency, quality, and quantity of animal products. Properly cared for animals seldom need antibiotics, wooden hoof orthotics, assistance with calving/lambing/fouling, chemical dewormers, and organophosphate delousing rubs, nor as many dirty vaccines. These animals produce more and a cleaner, higher nutrient dense product. Clinicians also can learn how to reverse nearly every acute and chronic disease afflicting human kind.

The foundation of any program, from soil, plants, animals, and humans, is nutrition. Nutrition in this context, however, means comprehensive nutrition taking into account every known mineral nutrient, every known amino acid, every known vitamin, fatty acid, carbohydrate, fiber, and protein. It is an understanding that the biochemistry of living organisms, humans included, is fundamentally nutritional biochemistry. Drugs and chemicals merely alter, skew, interfere with, block, or accelerate nutritional biochemical reactions, thus, altering physiology.

For example, if a person has congestive heart failure, CHF, meaning that the heart muscle is not able to keep up with its pumping demand, fluid gets backed up in the lungs and extremities. Drug intervention seeks to off-load fluids with diuretics, increase heart muscle contraction with something like digoxin, reduce blood pressure and heart demand with a beta blocker and also an ACE inhibitor to further lower blood pressure. The entire protocol is to address the consequent symptoms of CHF, not the actual biochemical cause. Fundamentally, the consequent cause is that the heart muscle is running out of energy to keep up with demand. That is a biochemical problem beginning with the production of ATP.

To address this we give CoQ10, magnesium, carnitine, taurine, B complex, and hawthorn. We may need to increase nitrous oxide to dilate the blood vessels to get more blood to the muscle fibers. We may need to give some of these nutrients intravenously to get sufficient therapeutic levels over time to return normal physiologic function. Insufficient nutrition is the reason the heart goes into congestive failure, and therapeutic nutrition is the reason it returns to normal function.

Drugs only delay the inevitable terminal failure. It is paramount that one understands that drugs and pesticides are not nutrients. They are nutrient antagonists most of the time and, thus, when considered in combination with the steady and continuous decline in food nutrient density at the farm gate, pesticide contamination/residue in and on our food is simply unacceptable and, more importantly, unnecessary.

Further, to detoxify these drugs and chemicals, our liver requires a full complement and therapeutic level of nutrients to do its detoxification job. If it's short nutrients, then chemicals prevail and lead to disease developing outcomes such as autoimmune diseases, cancer, fatty liver disease, chronic fatigue, environmental hypersensitivity, allergies, and neurological decline. We must return the nutrient density to our food, starting with the farm, if we are ever to really be sustainable, and truly address crop weed, disease, and insect problems currently limiting production and heath.

Nutrition also means taking into account the source and quality of these nutrients and the bioavailability and biochemical pathway in which they are assimilated, utilized, and recycled. It is knowing and understanding that mineral content alone is not sufficient information to determine the value of a product for fertilization. Municipal sewage waste is the perfect example of a product to avoid at all costs on one's farm.

Nutrition is understanding and applying the difference between calcium carbonate and calcium ascorbate; between magnesium chloride and magnesium oxide; between amino acids derived from animal sources compared to amino acids derived from plants; and the differences in their biological specificity and immunological and inflammatory effects in the human body that lead to the development or reversal of chronic diseases.

Nutrition in agriculture means understanding the nutritional and biological consequences of anhydrous ammonia as a nitrogen source vs. the consequences of *Azorhizobium*, *Azoarcus*, and *Azospirillum* bacteria produced nitrogen. It means understanding the physiological difference between using manganese sulfate vs. manganese glycinate or glucoheptonate, or the need to overcome glyphosate herbicide's immobilization of manganese by using some phosphite in the nutrient mix.

It is understanding that limiting the farmer to N-P-K and a few trace elements at standard tissue testing reference ranges guarantees the continued "need" for insecticides, fungicides, miticides, bacteriocides, and creates a demand for more genetic engineering of "resistant" varieties. A good reference book for plant nutrition basics is "Mineral Nutrition and Plant Disease," by Datnoff, Elmer, and Huber, 2007. A good professional reference book on human nutrition is "Learning to Thrive in a Toxic World: A Reference for Healthcare Practitioners and Patients," by Lisa Everett Andersen.

REFERENCES

1. Lewis, D.L. 2014. *Science for Sale*. Skyhorse Publishing. New York.
2. Chamberlain G. 2006. British maternal mortality in the 19th and early 20th centuries. *J. R. Soc. Med.* 99(11): 559–563. doi:https://doi.org/10.1258/jrsm.99.11.559.
3. Jewish Museum of the American West. 2020. Dr. Albert Abrams: Controversial Doctor of San Francisco. Accessed on June 28, 2020. http://www.jmaw.org/abrams-jewish-san-francisco/.
4. Bioregulatory Medicine Institute. 2020. History – Albert Abrams. Accessed online June 28, 2020. https://www.brmi.online/albert-abrams.
5. Lantero, A. 2014. "The War of the Currents: AC vs. DC Power." November 18, 2014. Energy.gov. Accessed online June 28, 2020 at https://www.energy.gov/articles/war-currents-ac-vs-dc-power.
6. Callahan, P.S. 1975. *Tuning Into Nature*. Acres USA, Austin, TX.
7. Chaboussou, F. 2005. Healthy Crops: A New Agricultural Revolution. Jon Carpenter, 28 April 2005. Charlbury, U.K.
8. Pfeiffer, E. 2020. Bio-Dynamic Gardening and Farming. Mercury Press, Spring Valley, New York. Accessed online June 28, 2020 at https://www.biodynamics.com/files/eventdownload/NewDirectionsInAgriculture_EPfeiffer-2.pdf.
9. U. S. Senate Document 264, 74th Congress, 2nd Session. 2020. "MODERN MIRACLE MEN". Accessed online June 28, 2020 at http://relaxwithrhonda.com/pdfs/US_Senate_Document_264.pdf.
10. International Institute of Biophysics. 2020. Interview with Fritz-Albert Popp, Accessed online June 28, 2020 at https://www.youtube.com/watch?v=trycaZzwMeo.
11. Popp, F.-A. 2020. About the Coherence of Biophotons. Fritz-Albert Popp International Institute of Biophysics (Biophotonics) Raketenstation, 41472 Neuss, Germany. Accessed online June 28, 2020 at https://www.academia.edu/1901658/About_the_Coherence_of_Biophotons_Fritz-Albert_Popp_International_Institute_of_Biophysics_Biophotonics_Raketenstation_41472_Neuss_Germany.
12. BMJ. 2019. Prevalence, severity, and nature of preventable patient harm across medical care settings: systematic review and meta-analysis. *BMJ* 2019; 366. doi: https://doi.org/10.1136/bmj.l4185. (Published 17 July 2019) *BMJ* 2019; 366:l4185.
13. Sinha, R, Cross, A.J., Graubard, B.I., Leitzmann, M.F., Schatzkin, A. 2009. Meat intake and mortality: A prospective study of over half a million people. *Arch. Intern. Med.* 169(6):562–571. doi:https://doi.org/10.1001/archinternmed.2009.6.
14. Zheng, Y., Li, Y., Satija, A., Pan, A., Sotos-Prieto, M., Rimm, E., Willett, W.C., Hu, F.B. 2019. Association of changes in red meat consumption with total and cause-specific mortality among U.S. women and men: two prospective cohort studies. BMJ, online June 12, 2019. doi: https://doi.org/10.1136/bmj.l2110.
15. World Health Organization. 2020. The global burden of chronic disease. Accessed online July 4, 2020 at https://www.who.int/nutrition/topics/2_background/en/.
16. Scientific American. 2011. Dirt poor: have fruits and vegetables become less nutritious? Because of soil depletion, crops grown decades ago were much richer in vitamins and minerals than the varieties most of us get today. https://www.scientificamerican.com/article/soil-depletion-and-nutrition-loss/.
17. Yang, I.F.-W., Li, Y.-X., Ren, F.-Z., Luo, J., Pang, G.-F. 2019. Assessment of the endocrine-disrupting effects of organophosphorus pesticide triazophos and its metabolites on endocrine hormones biosynthesis, transport and receptor binding *in silico*. *Food Chem. Tox.* 133: 110759. Accessed online June 28, 2020 at https://doi.org/10.1016/j.fct.2019.110759.

18. Yang, X., Li, Y., Zheng, L., He, X., Luo, Y., Huang, K., Xu, W. 2019. Glucose-regulated protein 75 in foodborne disease models induces renal tubular necrosis. *Food Chem. Tox.* 133: 110720. Accessed online June 28, 2020 at https://doi.org/10.1016/j.fct.2019.110720.

19. Balaguer, P., Delfosse, V., Grimaldi, M., Bourguet, W. 2017. Structural and functional evidences for the interactions between nuclear hormone receptors and endocrine disruptors at low doses. *Comptes Rendus Biol.* 340 (9–10): 414–420. Accessed online June 28, 2020 at https://doi.org/10.1016/j.crvi.2017.08.002.

20. Combarnous, Y. 2017. Endocrine disruptor compounds (EDCs) and agriculture: The case of pesticides. *Comptes Rendus Biol.* 340 (9–10): 406–409. Accessed June 28, 2020 at https://doi.org/10.1016/j.crvi.2017.07.009.

21. Mnif, W., Hassine, A.I., Bouaziz, A., Bartegi, A., Thomas, O., Roig, B. 2011. Effect of endocrine disruptor pesticides: A review. *Int. J Environ. Res. Pub. Health.* 8(6):2265–2303. doi: https://doi.org/10.3390/ijerph8062265. Epub 2011 Jun 17. PMID: 21776230; PMCID: PMC3138025.

22. FREIA and the Health and Environment Alliance (HEAL). 2020. Endocrine Disrupting Chemicals (EDCs) and Women's Reproductive Health. Factsheet written by FREIA, and endorsed by the International Federation of Gynecology and Obstetrics (FIGO) and the International Federation of Fertility Societies (IFFS). Email: info@freiaproject.eu Web: www.freiaproject.eu Twitter: @freiaprojectEU ©FREIA February 2020.

23. Jefferson, W.N., Kinyamu, H.K., Wang, T., et al. 2018. Widespread enhancer activation via ERα mediates estrogen response in vivo during uterine development. *Nucleic Acids Res.* 46(11):5487–5503. doi:https://doi.org/10.1093/nar/gky260.

24. Saffron, J. 2020. The environment influences brain development, experts say. Environ. Factor. March 2020. Accessed online June 28, 2020 at https://factor.niehs.nih.gov/ 2020/3/science-highlights/brain-development/index.htm.

25. Li, A.J., Chen, Z., Kannan, K. 2019. Association of urinary metabolites of organophosphate and pyrethroid insecticides, and phenoxy herbicides with endometriosis. *Environ. Intern.* 136: 105456. Accessed online June 28, 2020 at https://doi.org/10.1016/j.envint.2019.105456.

26. Kaur, I.N., Starling, A.P., Calafat, A.M., Sjodin, A., Clouet-Foraison, N., Dolan, L.M., Imperatore, G., Jensen, E.T., Lawrence, J.M., Ospina, M., Pihoker, C., Taylor, K., Turley, C., Dabelea, D., Jaacks, L.M. 2020. Longitudinal association of biomarkers of pesticide exposure with cardiovascular disease risk factors in youth with diabetes. *Environ. Res.* 181: 108916. Accessed online June 28, 2020 at https://doi.org/10.1016/j.envres.2019.108916.

27. Vandenberg, L.N., Colborn, T., Hayes, T.B., Heindel, J.J., Jacobs Jr., D.R., Lee, D.-H., Shioda, T., Soto, A.M., vom Saal, F.S., Welshons, W.V., Zoeller, R.T., Myers, J.P. 2012. Hormones and endocrine-disrupting chemicals: low-dose effects and nonmonotonic dose responses. *Endocr. Rev.* 33(3): 378–455.

28. Nordstro, M.A., Nordstro, M.P. 2018. Traumatic brain injury and the risk of dementia diagnosis: A nationwide cohort study. *PLoS Med.* 15(1): e1002496. doi: https://doi.org/10.1371/journal. pmed.1002496.

29. Hunter, P. 2012. The inflammation theory of disease. The growing realization that chronic inflammation is crucial in many diseases opens new avenues for treatment. *EMBO Rep.* 13(11):968–970. doi: https://doi.org/10.1038/embor.2012.142.

30. Grant, W.B. 2014. Trends in diet and Alzheimer's disease during the nutrition transition in Japan and developing countries. *J. Alzheimers Dis.* 38(3):611–620. doi: https://doi.org/10.3233/JAD-130719.

31. Grand View Research, Inc. 2019. Packaged Food Market Size, Share & Trends Analysis Report By Application, Regional Outlook, Competitive Strategies, And Segment Forecasts, 2019 to 2025. Grand View Research, Inc. 201 Spear Street 1100. San Francisco, CA. 94105, U. S.

32. Prakash, G., Pandian, R.T.P., Rajashekara, H., Sharma, P., and Singh, U.D. 2015. Assessment of efficacy of pesticides applied singly and in combination on rice sheath blight disease development. *Indian Phytopathol.* 68: 248–253.

33. Coleman, M.D., O'Neil, J.D., Woehrling, E.K., Ndunge, O.B.A., Hill, E.J., Menache, A., et al. 2012. A preliminary investigation into the impact of a pesticide combination on human neuronal and glial cell lines *In Vitro. PLoS ONE* 7(8): e42768. doi: https://doi.org/10.1371/journal.pone.0042768.

4 Agricultural Pesticides and the Deterioration of Health

Stephanie Seneff
MIT Computer Science and Artificial Intelligence Laboratory

CONTENTS

4.1 INTRODUCTION

The continent of Africa is blessed with agricultural practices that are mostly based on small family-run farms. This is different from other continents, particularly North and South Americas, where agriculture is dominated by industrialized monocrop megafarms that use far more chemical-based fertilizers and pesticides. Nonetheless, as African farmers become more comfortable with the use of pesticides, particularly if it appears to lower costs and/or increase yields, there is a danger that what will follow is a rise in many chronic health issues that currently beset the populations of industrialized nations, such as diabetes, obesity, fatty liver disease, cancer and birth defects. Often the small family farmers are not properly trained to use protective gear to minimize the risk of exposure to the chemicals they are working with, and this places them and their family members at a greater risk to acute toxicity.

It is appropriate to start this chapter by assessing the specific pesticides that are most commonly used in Africa and then to consider the risk that these pesticides may pose to the population. The risks are both through direct exposure to the farmers working with the chemicals and through indirect exposure to the consumers who eat contaminated foods. Some countries in Africa import much of their food, and often these imports are cheap processed foods produced elsewhere with heavy use of pesticides. Their highly processed Western diet is reflected in their increased risk to diabetes and obesity.

A detailed analysis of pesticide use in sub-Saharan Africa concluded that although pesticide use generally increases yield, it comes at a cost in terms of increased health expenditures and time lost from work due to sickness.[1] However, it is not easy to link a specific pesticide to a specific disease because farmers are generally exposed to multiple pesticides in the course of their field work. Statistics on the use of pesticides in Africa are presented in Table 4.1, reporting both by volume in tons and by monetary value in US dollars.[2] Herbicides make up the bulk of the pesticides used, representing nearly 80% of the cost and 88% of the total volume. The remainder are about equally divided between fungicides (Mancozeb and Metalaxyl) and insecticides (neonicotinoids and pyrethroids). The primary herbicides used include glyphosate, atrazine, 2,4-D, and paraquat, among which the use of glyphosate is much heavier than those of the other three (representing 64% of the total volume of herbicides). Glyphosate is probably preferred over other herbicides due to its easy availability,

TABLE 4.1

Pesticides Imported Into Africa: Total Monetary Value and Volume

Formulation	Value (million $) (%)	Volume (1000 Tons) (%)
Glyphosate	128.67 (47.5%)	66.1 (56.5%)
Paraquat	62.53 (23%)	25.97 (22.2%)
2,4-D	14.78 (5.5%)	7.29 (6.2%)
Atrazine	10.19 (3.8%)	3.13 (2.7%)
Total Herbicide	216.4 (79.8%)	102.5 (87.6%)
Fungicide (Mancozeb + Metalaxyl)	28.3 (10.4%)	6.88 (5.9%)
Insecticide (Neonicotinoid and Pyrethroid)	25.1 (9.26%)	7.59 (6.5%)
Total	271.12 (100%)	116.96 (100%)

Source: Bertrand (2019).

Bertrand, P.G. "Uses and misuses of agricultural pesticides in Africa: neglected public health threats for workers and population." In M. Larramendy and S. Soloneski, Ed. *Pesticides – Use and Misuse and Their Impact in the Environment.* July 17, 2019. DOI: 10.5772/intechopen.78909.

effectiveness against all varieties of weeds, and perceived nontoxicity to humans. However, recent evidence from independent investigators has revealed that glyphosate is much more toxic to human health than the manufacturer has been claiming, a subject we will return to later in this chapter.

Genetically engineered (GE) food crops have been very slow to penetrate the African market. Until very recently, only Sudan and South Africa embraced GE crops. South Africa is a major producer of GE maize, including both Roundup resistance and the Bt gene to protect from the boll worm and other insects. Sudan is a large producer of Bt cotton; although cotton is not normally considered a food crop, cottonseed oil is widely used in the food processing industry.[3] Nigeria is the biggest economy on the continent, and they have recently approved biotech cowpeas to resist the *Maruca* pests. Cowpeas are a major food in Nigeria and its neighboring countries.[4] The approval of GE crops in Nigeria in 2019 may open the door for other GE crops in Nigeria and in the neighboring countries, which may follow their lead. Currently, no country in Africa besides South Africa has adopted any Roundup-resistant crops, which keeps exposure levels to glyphosate from food sources well below the levels that Americans are exposed to.

More than 75% of the food grown in Africa is produced by small-scale farmers, who often do not use protective gear when handling the pesticides. They may also not be aware that pesticides should not be stored in their homes and the containers should not be reused for other purposes. This can lead to high-level exposures to young family members and pregnant women.[5]

The main export countries for pesticides in Africa are Nigeria, South Africa, Ghana, Ivory Coast, Egypt, and Kenya.[6] South Africa in particular has a long history of glyphosate usage dating back to the 1970s, particularly on GE glyphosate-resistant maize crops that were first introduced in the early 2000s.

To gain perspective on the amount of glyphosate used in Africa, it is enlightening to compare the number to the use of glyphosate in the United States, estimated to be about 280 million pounds annually (127,000 metric tons). This is about twice the amount used in Africa, but the total population of Africa is 3.6 times as much as the US population. Thus, per person, Africa consumes only about 1/7 as much glyphosate as the United States.

There have been few studies where glyphosate levels in African-sourced food products have been tested, in part because of the false perception that it is safe. In 2017, Koortzen measured a total of 81 food products from stores in South Africa containing either maize or soybean or both as major constituents. In total, 66.7% of the products were found to contain glyphosate, at levels ranging from 27 to 2,257 parts per billion. Over half of the maize products were contaminated, as were all of the soybean products. The highest levels were found in soy protein products.[7]

Because glyphosate stands out as being by far the most used pesticide in Africa, I will devote most of the discussion in this chapter to glyphosate. But first I will discuss the commonly seen health issues linked to the highly restricted herbicides, insecticides, and fungicides that are commonly used in Africa.

4.2   HEALTH EFFECTS OF REGULATED PESTICIDES

Large statistical studies have shown that pesticide applicators are at increased risk to multiple cancers, including prostate cancer, non-Hodgkin lymphoma, leukemia, multiple myeloma, and breast cancer.[8] It is often difficult to establish a direct link between a single pesticide and a cancer, because applicators are often exposed to multiple herbicides and insecticides in the course of their work.

The herbicides atrazine, paraquat, and 2,4-D are highly regulated in many countries, usually requiring a permit prior to purchase, or even banned outright. All three are endocrine disruptors, mimicking the hormones estrogen and androgen and hence, disturbing human development. 2,4-D is specifically linked to thyroidal disorders.[9]

Atrazine is well established as an endocrine disruptor, primarily through the work of Prof. Tyrone Hayes and colleagues.[10] It disturbs sexual development in frogs, inducing complete feminization and chemical castration of male frogs. Paraquat's usage is highly restricted, due primarily to its acute pulmonary and cutaneous toxicities. Paraquat has also been linked to Parkinson's disease, although studies are not conclusive.[11]

Like 2,4-D, the fungicide Mancozeb has been shown to suppress thyroid hormone expression in animal experiments, and if this effect occurs in humans, it can lead to developmental problems due to thyroid insufficiency in a pregnant woman.[12]

The class of neonicotinoid insecticides was originally considered to show little toxicity to species other than their target species, although they have recently come into the limelight due to suspicions that they are the cause of bee colony collapse syndrome. In response to these concerns, Europe took action to ban the use of neonicotinoids in agriculture in 2013.[13]

Pyrethroids are the fourth most used class of insecticides worldwide and are generally considered to be much more toxic to insects than to vertebrates. Although originally they were not recognized as endocrine disruptors, more recent research has shown that they interfere with endocrine signaling in both fish and mammals by blocking, mimicking, or synergizing endogenous hormones.[14]

4.3   GLYPHOSATE AND THE DETERIORATION OF HEALTH

Glyphosate has been on the market for over 45 years, and its usage has increased dramatically over time, particularly following the introduction of GE Roundup-Ready crops in the late 1990s. A seminal study conducted in 2014 by Nancy Swanson et al. revealed stunning correlations over time between glyphosate usage on core crops in the United States and the rise in incidence, prevalence, or mortality rates in a large number of debilitating diseases.[15] Table 4.2 lists the correlations and p-values for several of the diseases identified by Swanson et al., including diabetes, obesity, Alzheimer's, autism, and many cancers. Many of these same diseases become increasingly prevalent in countries around the world after the population starts to adopt a Western diet based primarily on heavily processed foods derived from crops that are treated with glyphosate.

These authors wrote in their conclusion as follows:

> Although correlation does not necessarily mean causation, when correlation coefficients of over 0.95 (with p-value significance levels less than 0.00001) are calculated for a list of diseases that can be directly linked to glyphosate, via its known biological effects, it would be imprudent not to consider causation as a plausible explanation.

Most people are unwilling to accept the idea that a single chemical could be causal in so many diseases, but it is now beginning to be recognized that glyphosate has some unique toxic effects

TABLE 4.2

Selected Results from the Swanson et al. Paper on Correlations Over Time between Glyphosate Usage on Core Crops and Various Diseases

Condition	Correlation Coefficient	Probability
Hyperlipidemia/Hypercholesterolemia (death)	0.973	7.9E-9
Diabetes (prevalence)	0.971	9.2E-9
Obesity	0.962	1.7E-8
Hypertension (deaths)	0.923	1.6E-7
Kidney failure	0.978	6.0E-9
Autism (prevalence)	0.989	3.6E-7
Dementia (deaths)	0.994	1.8E-9
Inflammatory bowel disease	0.938	7.1E-8
Thyroid cancer (incidence)	0.988	7.6E-9
Liver cancer (incidence)	0.960	4.6E-8
Pancreatic cancer (incidence)	0.973	2.0E-8
Kidney cancer (incidence)	0.973	2.0E-8

that could explain a causal relationship. One growing concern is the possibility that glyphosate is disrupting the balance of the gut microbiome, because microbes do possess the shikimate pathway, which glyphosate blocks in plants, and they use it to produce the aromatic amino acids for the host. These amino acids are not only among the building blocks of proteins but are also precursors to many important biologically active molecules, such as the neurotransmitters serotonin, melatonin, dopamine, and epinephrine; the skin-tanning agent melanin; thyroid hormone; and certain B vitamins such as niacin and riboflavin.[16]

Glyphosate has been shown experimentally to suppress gut colonization of the symbiotic bacteria *Lactobacillus* and Firmicutes, while inducing an overgrowth of pathogenic species. At the same time, it induced an inflammatory response, increasing messenger RNA expression of the cytokines interleukin-1β, interleukin-6, tumor necrosis factor-α (TNF-α), and nuclear factor-κB (NF-κB). Furthermore, it induced an ion imbalance and oxidative stress.[17] Glyphosate has also been shown to suppress cytochrome P450 enzymes (CYPs) in the liver.[18] These enzymes are essential for activating vitamin D, metabolizing vitamin A, producing bile acids, and detoxifying many other environmental toxins, including various pharmaceutical drugs.

Glyphosate is also a very strong metal chelator, and it is particularly effective at binding to manganese and cobalt, making them unavailable. A study on cows in eight different dairy farms in Denmark exposed to high levels of glyphosate in their food consistently found levels of manganese and cobalt in the blood that were far below the minimum of the expected range.[19] Iron, zinc, and magnesium also bind tightly to glyphosate. These minerals are essential as catalysts for many different enzymes. The impaired availability of minerals to the gut microbiome, combined with the suppression of the shikimate pathway, can lead to gut dysbiosis, a driver behind many diseases.[20]

In the United States, there have been several successful cases of lawsuits where the plaintiff, claiming that glyphosate caused non-Hodgkin's lymphoma, was awarded a large sum of money in a jury trial. There are now thousands of pending cases with similar claims awaiting trial.

4.4 DOES GLYPHOSATE SUPPRESS PEPCK?

Glyphosate has been shown to kill weeds primarily by disrupting an enzyme called 5-enolpyruvylshikimate-3-phosphate (EPSP) synthase in the shikimate pathway. This pathway is critical in plants and microbes but not found in human cells, and this is a primary motivation for the claim that glyphosate is nontoxic to humans. Exactly how glyphosate disrupts this enzyme is open for debate, but the primary mechanism that has been proposed is through glyphosate occupying the site where phosphoenolpyruvate (PEP) normally binds.[21]

Phosphoenolpyruvate carboxykinase (PEPCK) is an enzyme that is expressed in multiple cell types in humans, and it plays an essential regulatory role in metabolism. It depends on manganese as a catalyst, so glyphosate's chelation of manganese could suppress its activity.[22] However, more importantly, it has remarkable parallels with EPSP synthase; in that, it also binds with PEP, and, in both cases, the PEP binding site contains a highly conserved glycine residue and two positively charged amino acids that serve to anchor PEP's negatively charged phosphate anion in place. Notably, studies have shown that a genetic mutation where the glycine residue is substituted by alanine in EPSP synthase results in a version of the enzyme that is completely insensitive to glyphosate's damaging effects.[23] In fact, the GE crops that are resistant to glyphosate are endowed with a microbial version of EPSP synthase that has alanine instead of glycine at the binding site.[24]

Together with several colleagues, I have proposed in multiple published studies that glyphosate, acting as a glycine analogue, can substitute by mistake for glycine during protein synthesis and that this easily explains the observed effects on EPSP synthase.[25] The conditions at the active site in EPSP synthase are ideal for glyphosate substitution, because it can settle its extra methylphosphonate unit comfortably into the space where the phosphate anion of PEP is supposed to fit. These ideal conditions are reproduced in PEPCK, which leads me to suspect that glyphosate also suppresses PEPCK, and this effect would have major implications for human metabolism.

PEPCK is central to the proper function of the mitochondrial tricarboxylic acid (TCA) cycle, which generates ATP as the energy source for the cell. In particular, PEPCK orchestrates the balance of intermediate metabolites in the cycle. PEPCK specifically converts the TCA cycle intermediate oxaloacetate into PEP. PEP can then serve as a source for gluconeogenesis – the synthesis of glucose from other nutritional sources. When PEPCK is deficient, gluconeogenesis becomes impaired, and this is a cause of neonatal hypoglycemia. It becomes necessary to upregulate the set point for blood glucose when this happens, in order to avoid going into a coma when blood sugar falls too low, for example, during strenuous exercise. This results in hyperglycemia, a precursor to type 2 diabetes.

4.5 FATTY LIVER DISEASE AND ITS SEQUELAE

Mice with nonfunctional cytoplasmic PEPCK are not viable, and mice with a 90% reduction in PEPCK activity survive beyond birth but suffer from hepatic steatosis (fatty liver disease).[26] The concentration of the TCA cycle intermediate malate increases by ninefold, and this interferes with the metabolism of fats through the TCA cycle, resulting in lipid storage in the liver.

Nonalcoholic fatty liver disease (NAFLD) has become a worldwide health crisis, with nearly 24% of the world's population suffering from this condition. A study of the global epidemiology of NAFLD revealed that the regional prevalence was estimated to be 23.4% in Asia, 23.7% in Europe, 31.8% in the Middle East, 24.1% in North American, 30.5% in South America, and only 13.5% in Africa.[27] It is interesting to note that the region with by far the lowest reported prevalence was Africa, likely reflecting their reduced use of glyphosate in food production.

NAFLD is a critical metabolic indicator of other chronic conditions known to be highly prevalent in industrialized countries, including type 2 diabetes, obesity, metabolic syndrome, coronary heart disease, dyslipidemia, and hypertension.[28] According to a meta-analysis of studies from 20 different countries, nearly 75% of individuals with NAFLD are obese and 58% of people with diabetes suffer from NAFLD.[29] Accumulation of fat in the liver is associated epidemiologically with liver damage (e.g., fibrosis), insulin resistance, elevated blood lipids, and hypertension, as well as inflammation and elevated levels of liver transaminases (markers of liver disease).[30]

A mitochondrial version of PEPCK is expressed in the pancreatic beta cells, and it regulates insulin production. When this enzyme is defective, insulin release becomes impaired, leading to type 1 diabetes.[31] The prevalence of type 1 diabetes has been increasing globally in recent years, despite the belief that it is a genetic disease. A recent study based in Europe revealed a 3.4% increase in type 1 diabetes each year over the past 25 years.[32]

4.6 GLYPHOSATE AND FATTY LIVER DISEASE

While to my knowledge no studies have investigated whether glyphosate reduces the activity of PEPCK, a large number of studies, spanning multiple decades, have demonstrated that glyphosate is remarkably toxic to the liver, and, in fact, the liver is one of the first organs to be obviously affected by glyphosate exposure. One study, in particular, published in 2017 showed that rats exposed orally to very low doses of glyphosate (below regulatory limits) developed fatty liver disease.[33]

Mice with a genetic mutation disabling cytoplasmic PEPCK developed fatty liver disease, with elevations in serum levels of alanine, aspartate, lactate, and triglycerides and reduced levels of hepatic glycogen and blood sugar.[34] A study published in 2019 looked specifically at liver markers in adult male rats exposed to Roundup through oral administration every day over a 14-day period at doses of 0, 5, 10, 25, 50, 100, and 250 mg kg^{-1} body weight. They found similar results as most other studies – e.g., markers of oxidative stress and liver fibrosis.[35] However, importantly, they also noticed a depletion of liver glycogen following exposure, in a dose–response relationship. This is highly significant, because it would be a key marker of PEPCK deficiency. Because gluconeogenesis is disrupted, glycogen stores have to be used up to maintain an adequate level of blood sugar.

Remarkably, a study on fish exposed to environmentally realistic levels of the glyphosate formulation, Roundup, published in 2019 found altered metrics that were consistent with the profile of the PEPCK-deficient GE mice.[36] The fish were found to have elevated serum lactate and triglycerides, as well as elevations in alanine transaminase (ALT) and aspartate transaminase (AST), enzymes that synthesize alanine and aspartate, respectively. These enzymes become upregulated when PEPCK is defective, as an alternative pathway to metabolize oxaloacetate. These fish also had reduced levels of glycogen in the liver and muscles and reduced blood glucose, consistent with impaired gluconeogenesis.

A study published in 2018 looked at the effects of a high dose of glyphosate on albino rats.[37] Animals were administered 25, 50, and 100 mg kg^{-1} body weight of Roundup daily for 15 days. The effects on the liver are described as follows: "increase of enzymes activities of ALT and AST, cellular infiltration, many signs of nucleus degeneration, focal necrosis, rarified cytoplasm, disorganization of cellular organelles, and deposition of lipid droplets."

A study conducted in 2011 involved exposing albino mice to a single dose of glyphosate at 50 mg kg^{-1} of body weight, then sacrificing them 72 hours later, and measuring a number of parameters that are well-established indicators of liver and kidney diseases.[38] Glyphosate induced a significant increase in serum AST and ALT, urea, and creatinine. These are all indicators of liver and kidney damage.

Nonalcoholic steatohepatitis (NASH) is a more advanced form of NAFLD, which includes hepatitis (liver inflammation) and liver cell damage. It progresses to liver fibrosis (scarring) and, ultimately, liver cancer in extreme cases. A study on humans published in 2020 showed evidence of an association between NASH and glyphosate exposure. The study involved 63 patients with biopsy-proven NASH, and it was shown that patients had significantly higher levels of glyphosate residues in their urine compared to controls ($p < 0.008$). Furthermore, those patients with advanced fibrosis had significantly higher levels than those patients without advanced fibrosis ($p < 0.001$).[39]

4.7 DIABETES AND OBESITY

In Africa, there is an "obesity divide," where over half of the population in the countries bordering the Mediterranean Sea and South Africa are "overweight" or "obese" (body mass index > 25), whereas fewer than one-third of the people in the bulk of the sub-Saharan countries are fat (see Table 4.3). This can be compared with other countries around the world where glyphosate is heavily

TABLE 4.3

Prevalence of Overweight or Obesity among Adults in Selected Countries in Africa in 2016 (BMI > 25), According to the Centers for Disease Control

Country	% BMI > 25	Country	% BMI > 25
Libya	66.8	Ghana	32.0
Egypt	63.5	Nigeria	28.9
Algeria	62.0	Angola	27.5
Morocco	60.4	Kenya	25.5
South Africa	53.8	Ethiopia	20.9

used in agriculture: Brazil (56.5), Argentina (62.7), United Kingdom (63.7), Australia (64.5), and United States (67.9).

Nineteen million Egyptians are obese (BMI > 30), accounting for 35% of the adult population, the highest rate in the world. Social stress due to political upheaval is part of the problem.[40] Poverty drives people toward food choices that are unhealthy, heavily processed foods with poor nutritional value, and likely to be heavily contaminated with glyphosate. As of 2018, one-sixth of Egypt's merchandise imports were food products.

Due to an arid climate and a shortage of arable land (only 1.7% of the total area), as well as civic unrest, Libya, just to the West of Egypt and also on the Mediterranean coast, depends heavily on food imports to meet its nutritional needs, and these are mostly processed foods coming from Europe. The prevalence of obesity in Libya increased from 12.6% in 1984 to 30.5% in 2009, to 42.4% in 2016, thus more than tripling in three decades.[41]

Industrialized chemical-based agriculture has had a long history in South Africa. Glyphosate is the main herbicide used in South Africa, where it was first registered for use in the 1970s. Glyphosate-tolerant varieties of cotton and soybeans were first introduced in 2001/2002, and this was followed quickly with GE maize in 2003/2004. Cultivation of these crops has increased significantly since that time. Glyphosate is also used in field crops, timber, horticulture, sugar, and grapes.[42]

A large population study published in 2018 based in Malawi revealed much lower rates of diabetes, obesity, and hypertension than those that are typical in the industrialized world, but the rates are rising at an alarming rate, particularly among urban populations. Age-adjusted diabetes prevalence in Malawi was found to be 5.4% in urban areas and only 2.6% in rural areas.[43] This compares to 9% of people over 30 with diabetes in South Africa as of 2009, a nearly twofold increase since 2000.[44]

4.8 CONCLUSION

Agricultural practices that rely on the heavy use of pesticides are increasingly being implicated in the failing health of nations that adopt such practices. Rising rates of diabetes, obesity, fatty liver disease, autism, dementia, reproductive disorders, and cancer can all be attributed in part to exposures to toxic pesticides in the food, water, and the atmosphere near the fields where such pesticides are applied. It is of paramount importance that we move toward an agricultural model that does not rely on pesticides for food production. Africa is a shining example to other continents in their reduced use of these chemicals, although the situation is rapidly changing for the worse.

Glyphosate is by far the most used pesticide in Africa, in part because it is perceived as being practically nontoxic to humans. In this chapter, we have provided an argument that this concept that glyphosate is safe may be untrue, particularly when you consider epidemiological data, both temporally and spatially. Specifically, in Africa, the countries where the food supply is less contaminated with glyphosate have the lowest rates of diabetes and obesity. While correlation does not always mean causation, glyphosate has been directly linked to fatty liver disease, which in turn is associated with many other modern chronic diseases. An argument can be made that glyphosate could cause these diseases, in part through its hypothesized substitution for glycine during protein synthesis.

The continent of Africa is poised to play a leadership role in the return to renewable organic agricultural practices that had been widespread before the emergence of the "Green Revolution." While the goal of the green revolution was to increase crop yield and reduce the cost of food production, it was based on the overuse of pesticides and synthetic fertilizers, which come at a great societal and environmental cost. Most countries in sub-Saharan Africa still have a strong foundation in agricultural practices based on the small family farm. It is important to educate farmers to appreciate the need to avoid the use of toxic pesticides that are contributing to the deterioration of human health. This will motivate them to reduce or even eliminate pesticide use on their own farms, assuring the health and well-being of themselves and their family members, as well as offering high-quality products to the astute consumers. The number of consumers who recognize the value of organic food and are willing to pay more for this added value is rapidly growing throughout the world, and this is a motivating factor as well for farmers to "do well by doing good."

REFERENCES

1 Sheahan, M., Barrett, C.B., Goldvale, C. "Human health and pesticide use in sub-Saharan Africa." *Agricultural Economics* 48, No. S1 (2017): 27–41.
2 Bertrand, P.G. "Uses and misuses of agricultural pesticides in Africa: neglected public health threats for workers and population." In M. Larramendy and S. Soloneski, Ed. *Pesticides -- Use and Misuse and Their Impact in the Environment*. July 17, 2019. DOI: 10.5772/intechopen.78909.
3 Abdallah, N.A. "The story behind Bt cotton: Where does Sudan stand?" *GM Crops Food* 5, No. 4 (2014) 241–243.
4 Gakpo, J.O. "Nigeria's approval of GMO crops boosts Africa's hopes for the technology." *Cornell Alliance for Science*. April 2, 2019. https://allianceforscience.cornell.edu/blog/2019/04/nigerias-approval-gmo-crops-boosts-africas-hopes-technology/.
5 We-Empower, Inc. "Synthetic Pesticides in Africa: the Good, the Bad, and the Ugly." February 28, 2019 https://www.agrilinks.org/users/we-empower#profile-main.
6 Bertrand, P.G. "Uses and misuses of agricultural pesticides in Africa: Neglected public health threats for workers and population." In M. Larramendy and S. Soloneski, Ed. *Pesticides – Use and Misuse and Their Impact in the Environment*. July 17, 2019. DOI: 10.5772/intechopen.78909.
7 Koortzen, B.J. "Presence of glyphosate in food products in South Africa of which maize and soybean is the primary constituents." Dissertation submitted in fulfilment of the requirements for the degree Magister Medical Scientiae (Human Molecular Biology), Faculty of Health Sciences, Department of Haematology and Cell Biology, University of the Free State, 2017.
8 Alavanja, M.C., Ross, M.K., Bonner, M.R. "Increased cancer burden among pesticide applicators and others due to pesticide exposure." *CA Cancer J Clin* 63, No. 2 (2013): 120–42.
9 Goldner, W.S., Sandler, D.P., Yu, F., et al. "Hypothyroidism and pesticide use among male private pesticide applicators in the agricultural health study." *J Occup Environ Med* 55, No. 10 (2013): 1171–1178.
10 Hayes, T.B., Khoury, V., Narayan, A. et al. "Atrazine induces complete feminization and chemical castration in male African clawed frogs (Xenopus laevis)." *Proc Natl Acad Sci USA* 107, No. 10 (2010): 4612–4617.
11 Vaccari, C., El Dib, R., de Camargo, J.L.V. "Paraquat and Parkinson's disease: A systematic review protocol according to the OHAT approach for hazard identification." *Syst Rev* 6 (2017): 98.
12 Axelstad, M. Boberg, J., Nellemann, C. et al. "Exposure to the widely used fungicide mancozeb causes thyroid hormone disruption in rat dams but no behavioral effects in the offspring." *Toxicol Sci* 120, No. 2 (2011): 439–446.
13 McDonald-Gibson, C. "'Victory for Bees' as European Union bans neonicotinoid pesticides blamed for destroying bee population." The Independent. April 29, 2013. https://www.independent.co.uk/environment/nature/victory-for-bees-as-european-union-bans-neonicotinoid-pesticides-blamed-for-destroying-bee-8595408.html.
14 Brander, S.M., Gabler, M.K., Fowler, N.L., Cannon, R.E. Schlenk, D. "Pyrethroid pesticides as endocrine disruptors: Molecular mechanisms in vertebrates with a focus on fishes." *Environ Sci Technol* 50, No. 17 (2016): 8977–8992.

15 Swanson, N.L., Leu, A., Abrahamson, J., et al. "Genetically engineered crops, glyphosate and the deterioration of health in the United States of America." *J Org Syst* 9 (2014): 6–37.

16 Samsel, A. Seneff, S. "Glyphosate's suppression of cytochrome p450 enzymes and amino acid biosynthesis by the gut microbiome: Pathways to modern diseases." *Entropy* 15 (2013): 1416–1463.

17 Tang, Q., Tang, J., Ren, X., Li, C. "Glyphosate exposure induces inflammatory responses in the small intestine and alters gut microbial composition in rats". *Environ Pollut* 261(2020): 114129.

18 Fathi, M.A., Han, G., Kang, R., et al. "Disruption of cytochrome P450 enzymes in the liver and small intestine in chicken embryos in ovo exposed to glyphosate." *Environ Sci Pollut Res Int* 27, No. 14 (2020): 16865–16875.

19 Krüger, M. Schrödl, W., Neuhaus, J. Shehata, A.A. "Field investigations of glyphosate in urine of Danish dairy cows." *J Environ Anal Toxicol* 3, No. 5 (2013): 1000186.

20 Wilkins, L.J., Monga, M., Miller, A.W. "Defining dysbiosis for a cluster of chronic diseases." *Sci Rep* 9 (2019): 12918.

21 Schönbrunn, E., Eschenburg, S., Shuttleworth, W.A., et al. "Interaction of the herbicide glyphosate with its target enzyme 5-enolpyruvylshikimate 3-phosphate synthase in atomic detail." *Proc Natl Acad Sci USA*. 98, No. 4 (2001): 1376–1380.

22 Baly, D.L., Keen, C.L., Hurley, L.S. "Pyruvate carboxylase and phosphoenolpyruvate carboxykinase activity in developing rats: Effect of manganese deficiency." *J Nutr* 115, No. 7 (1985): 872–879.

23 Eschenburg, S., Healy, M.L., Priestman, M.A., Lushington, G.H., Schönbrunn, E. "How the mutation glycine96 to alanine confers glyphosate insensitivity to 5-enolpyruvyl shikimate-3-phosphate synthase from Escherichia coli." *Planta* 216 (2002): 129–135.

24 Funke, T., Han, H., Healy-Fried, M.L., Fischer, M., Schönbrunn, M. "Molecular basis for the herbicide resistance of roundup ready crops." *PNAS* 103, No. 35 (2006): 13010–13015.

25 Samsel, A., Seneff, S. "Glyphosate, pathways to modern diseases V: Amino acid analogue of glycine in diverse proteins." *J Biol Phys Chem* 16 (2016): 9–46. Gunatilake, S., Seneff, S., Orlando, L. "Glyphosate's synergistic toxicity in combination with other factors as a cause of chronic kidney disease of unknown origin." *Int J Environ Res Public Health* 16, No. 15 (2019): 2734.

26 She, P., Shiota, M., Shelton, K.D., Chalkley, R., Postic, C., Magnuson, M.A. "Phosphoenolpyruvate carboxykinase is necessary for the integration of hepatic energy metabolism." *Mol Cell Biol* 20, No. 17 (2000): 6508–6517.

27 Younossi, Z.M., Koenig, A.B., Abdelatif, D., Fazel, Y., Henry, L., Wymer, M. "Global epidemiology of nonalcoholic fatty liver disease – meta-analytic assessment of prevalence, incidence, and outcomes." *Hepatology* 64 (2016): 73–84.

28 Fabbrini, E., Sullivan, S., Klein, S. "Obesity and nonalcoholic fatty liver disease: Biochemical, metabolic and clinical implications." *Hepatology* 51, No. 2 (2010): 679–689.

29 Younossi, Z.M., Koenig, A.B., Abdelatif, D., Fazel, Y., Henry, L., Wymer, M. "Global epidemiology of nonalcoholic fatty liver disease – meta-analytic assessment of prevalence, incidence, and outcomes." *Hepatology* 64 (2016): 73–84.

30 Dongiovanni, P., Stender, S., Pietrelli, A., et al. "Causal relationship of hepatic fat with liver damage and insulin resistance in nonalcoholic fatty liver." *J Intern Med* 283, No. 4 (2018): 356–370.

31 Stark, R., Kibbey R.G. "The mitochondrial isoform of phosphoenolpyruvate carboxykinase (PEPCK-M) and glucose homeostasis: Has it been overlooked?" *Biochim Biophys Acta* 1840, No. 4 (2014): 1313–1330.

32 Patterson, C.C., Harjutsalo, V., Rosenbauer, J., et al. "Trends and cyclical variation in the incidence of childhood type 1 diabetes in 26 European centres in the 25 year period 1989–2013: A multicentre prospective registration study". *Diabetologia* 62 (2019): 408–417.

33 Mesnage, R., Renney, G., Séralini, G.E. "Multiomics reveal non-alcoholic fatty liver disease in rats following chronic exposure to an ultra-low dose of roundup herbicide." *Sci Rep* 7 (2017): 39328.

34 Hakimi, P., Johnson, M.T., Yang, J., et al. "Phosphoenolpyruvate carboxykinase and the critical role of cataplerosis in the control of hepatic metabolism." *Nutr Metab (Lond)* 2 (2005): 33.

35 Pandey, A., Dhabade, P., Kumarasamy, A. "Inflammatory effects of subacute exposure of roundup in rat liver and adipose tissue." *Dose Response* 17, No. 2 (2019): 1559325819843380.

36 de Moura, F.R., da Silva Lima, R.R., da Cunha A.P.S., da Costa Marisco, P., Aguiar, D.H. "Effects of glyphosate-based herbicide on pintado da amazônia: Hematology, histological aspects, metabolic parameters and genotoxic potential." *Environ Toxicol Pharmacol* 56 (2017): 241–248.

37 Saleh, S.M., Elghareeb, T.A., Ahmed, M.A.I., Mohamed, I.A., Ezz El-Din, H.A. "Hepato-morpholoy and biochemical studies on the liver of albino rats after exposure to glyphosate-roundup." *J Basic Appl Zool* 79 (2018): 48.

38 Cavuşoğlu, K., Yapar, K., Oruç, E., Yalçın, E. "Protective effect of ginkgo biloba L. leaf extract against glyphosate toxicity in Swiss albino mice." *J Med Food* 14, No. 10 (2011): 1263–1272.

39 Mills, P.J., Caussy, C., Loomba, R. "Glyphosate excretion is associated with steatohepatitis and advanced liver fibrosis in patients with fatty liver disease." *Clin Gastroenterol Hepatol* 18, No. 3 (2020): 741–743.

40 Al-Daydamouni, S. "The Socio-Economic Problem of Obesity in Egypt." The Arab Weekly. Jan. 1, 2019. https://thearabweekly.com/socio-economic-problem-obesity-egypt.

41 Lemamsha, H., Randhawa, G., Papadopoulos, C. "Prevalence of overweight and obesity among Libyan men and women." *Biomed Res Int* 2019 (2019): 8531360.

42 Gouse, M. "Assessing the Value of Glyphosate in the South African Agricultural Sector. Department of Agricultural Economics, Extension and Rural Development." Department of Agricultural Economics, Extension and Rural Development. University of Pretoria. November 2014.

43 Price, A.J., Crampin, A.C., Amberbir, A., et al. "Prevalence of obesity, hypertension, and diabetes, and cascade of care in sub-Saharan Africa: A cross-sectional, population-based study in rural and urban Malawi." *Lancet Diabetes Endocrinol* 6 (2018) 208–222.

44 Pheiffer, C., Pillay-van Wyk, V., Joubert, J.D., et al. "The prevalence of type 2 diabetes in South Africa: A systematic review protocol." *BMJ Open* 8 (2018): e021029.

5 Synthetic Pesticides and the Brain

Prahlad K. Seth
Biotech Park

Vinay K. Khanna
CSIR-Indian Institute of Toxicology Research

CONTENTS

5.1 INTRODUCTION

Pesticide usage has a long history and goes back to 1000 BC when sulfur was used as a fumigant for the first time in China and later as fungicide in the 18th century in Europe [1]. Use of arsenic and other inorganic compounds including lead, calcium arsenate, copper sulfate, thiram, lime, and biological substances such as nicotine and pyrethrum came in practice with the time to control pest outbreaks [1]. The discovery of dichlorodiphenyltrichloroethane (DDT), an organochlorine compound, by Paul Hermann Müller, a Swiss chemist in 1939, was a major breakthrough leading to the development of synthetic pesticides [2]. With advancements in chemistry and passage of time, new generations of pesticides were introduced with wide applications to control food-and vector-borne diseases in humans [3,4]. Use of pesticides has been effective in horticulture, forestry, and home gardens for control of pests and insects and in veterinary practices to prevent the spread of diseases in animals [5–7].

Pesticides have been classified in several ways. Based on the action on the target organism, the pesticides have been grouped as acaricides, algicides, bactericides, fungicides, herbicides, insecticides, molluscicides, rodenticides, weedicides, etc. [8,9]. Likewise, based on the chemical type and structure, synthetic pesticides have been classified as organochlorines, organophosphates (OPs), carbamates, pyrethroids, and neonicotinoids [8].

Global use of pesticides is expected to increase from 2 million ton/year to 3.5 million ton/year by 2020 [10,11]. China is the largest producer, consumer, and exporter of pesticides followed by the United States, Argentina, Thailand, Brazil, Italy, France, Canada, Japan, and India. Pesticide usage in African continent is the lowest and accounts for only 2%–4% of the global market [10,11]. However, use of pesticides in South Africa has been reported to be the highest in sub-Saharan Africa [12]. There is no consistent pattern of consumption and use of pesticides; rather, it varies significantly in various countries and different regions of the world [3,10]. Due to their adverse effects on ecosystem and humans, some pesticides have been banned in the developed countries. Restricted use of DDT and some other pesticides has been permitted in developing nations in public health programs to control malaria and other vector-borne diseases as alternates to these pesticides are expensive [13].

According to World Health Organization (WHO), around 3 million cases of pesticide poisoning with over 2,00,000 deaths have been reported every year [14]. Exposure to high levels of pesticides has also been reported in individuals attempting suicide[14–16]. Out of 800,000 cases of suicide, a substantial number of individuals (1,10,000–1,68,000) die every year due to pesticide poisoning[14,17,18]. Higher incidences of pesticide poisoning and suicide have been reported from the developing countries, largely attributable to lack of awareness about their adverse effects, and inadequate training for their usage and handling [19,20].

While pesticides have been reported to affect several body functions, their effects on the brain have received much attention and concern. Pesticide neurotoxicity can be manifested as severe signs on acute exposure at high concentrations or subtle effects on chronic exposure at low concentrations. Neuronal degeneration associated with behavioral deficits on exposure to pesticides has been reported frequently with increasing risk to develop neurological and psychological abnormalities in exposed individuals [21,22]. A number of epidemiological studies have been carried out to understand the risk of pesticide exposure and development of Parkinson's disease (PD), Alzheimer's disease (AD), amyotrophic lateral sclerosis (ALS), developmental neurological abnormalities, and psychiatric disturbances among exposed individuals [22,23]. Table 5.1 summarizes some of the neurological disorders associated with pesticide exposure. Pesticides impact both adult and developing brains, but the outcomes of exposure during the development are more severe and are still

TABLE 5.1
Neuropsychiatric Diseases Associated with Pesticide Exposure

Disease	Pesticides	References
Alzheimer's Disease	Organochlorines (DDT, DDE, Dieldrin, HCH, etc.) Organophosphates (Chlorpyrifos, Parathion, Malathion, etc.)	[28,29,30,31]
Cognitive Deficits	Organochlorines (DDT, DDE Dieldrin, HCH, etc.) Synthetic Pyrethroids (Deltamethrin, Cypermethrin, Permethrin, etc.) Organophosphates (Chlorpyrifos, Parathion, Malathion, etc.) Carbamates (Aldicarb, Maneb, etc.)	[29,32–41]
Parkinson's Disease	Organochlorines (DDT, DDE, Dieldrin, Lindane, Endosulfan, etc.) Organophosphates (Malathion, Chlorpyrifos, Dimethoate, Monocrotophos, etc.)	[42–51]
Amyotrophic Lateral Sclerosis (ALS)	Organochlorine Pesticides (DDT, DDE, Dieldrin, HCH, Lindane, Endosulfan, etc.) Organophosphates (Parathion, Malathion, Glyphosate) Synthetic Pyrethroids (Imiprothrin, Phenotorin, Resmethrin, Phthalthrin, etc.)	[52,53]
Autism	Organophosphates (Chlorpyrifos, Diazinon, Parathion, Malathion, etc.) Synthetic Pyrethroids (Deltamethrin, Cypermethrin, Permethrin, etc.)	[26,54–56]

(Continued)

TABLE 5.1 (*Continued*)

Neuropsychiatric Diseases Associated with Pesticide Exposure

Disease	Pesticides	References
ADHD	Organophosphates (Chlorpyrifos, Parathion, Malathion, etc.) Synthetic Pyrethroids (Imiprothrin, Phenotorin, Resmethrin, Phthalthrin, etc.)	[32,57–61]
Neuropsychosis (Anxiety, Depression, suicidal Tendency, etc.)	Organochlorines (Hexachlorophene, etc.) Organophosphates (Chlorpyrifos, Parathion, Malathion, etc.)	[62,63]

not well understood. Low-level exposure to pesticides during pregnancy has been reported to lead to various forms of abnormalities including mental retardation, cognitive deficits, attention deficit, autism, dyslexia, and neurodevelopmental disorders that are often visible in exposed individuals later in life [24–27]. Children and infants are more sensitive and vulnerable to pesticide toxicity as compared to adults [24,26]. This points out the need for more well-planned studies to assess the impact of pesticides during early brain development and outcomes later in life.

As the need to protect the food grains and economic crops and control vector-borne diseases is essential, the use of pest control agents will continue. Newer and safer pesticides are therefore being introduced with a greater thrust on natural and biological pesticides. Thus, there will be a need to assess their safety and neurotoxicity. Presently only few biomarker-based neuroepidemiological studies are possible due to lack of availability of well-categorized biomarkers. Therefore, significant research to identify biomarkers suitable to assess exposure and effect as well as progression of pesticide-induced neurotoxicity or disease, which could predict them, is required.

The adverse effects of pesticides on environment, animals, and health have been dealt in this book and elsewhere in detail. Therefore, the present chapter will focus on neurotoxicity of the selected group of synthetic pesticides and their association with neuropsychiatric diseases.

5.2 DEVELOPMENT AND VULNERABILITIES OF THE BRAIN

The brain is divided into distinct regions that control important physiological functions of the body. It undergoes rapid structural and functional changes during pregnancy and in neonatal period, and is sensitive to prenatal programming [64,65]. The ontogeny of the brain and its vulnerabilities have been reviewed by several investigators [66–68]. Brain development in humans starts prenatally and continues after birth through adolescence (Figure 5.1). The process of brain development is quite complex and involves multiple steps in an orchestrated manner [69,70]. During this time, brain cells proliferate and migrate to appropriate locations in the brain architecture, differentiate into different cell types, and make connections to form neuronal circuits. The process of myelination on neurons starts *in utero*, soon after cell proliferation and migration, and continues at a fast speed up to 5 years of age [69,71,72]. Neurotransmitters and their receptors and other regulators develop on a specific timeline. Apoptosis also takes place at appropriate time during the process of development and helps in refining and consolidation of neuronal circuits [69,70,73,74].

Developing fetuses and infants are sensitive to environmental chemicals, which may disrupt the specific developmental processes and cause a variety of neurodevelopmental disabilities [26,27,75]. The greater sensitivity of developing brain to chemicals is due to pharmacokinetic differences. Small body size, different ratio of fat, muscle, and water, higher breathing and metabolic rate per body weight, and immature blood–brain barrier in comparison to adults add to their further vulnerability to insults [67,76–78].

The rate of chemicals to cross the placental barrier varies and depends on their size and lipophilicity [79]. Several pesticides have been reported to cross the placental barrier and reach

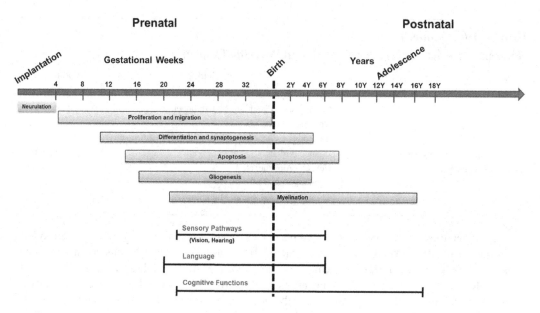

FIGURE 5.1 Key processes and functions associated with pre-and postnatal human brain development.

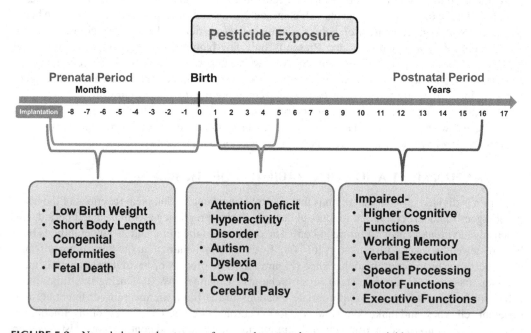

FIGURE 5.2 Neurobehavioral outcome of pre- and postnatal exposures to pesticides.

the fetus [79–81]. Exposure to pesticides at any stage of brain development is thus crucial and may affect its functioning leading to behavioral and functional abnormalities (Figure 5.2). Partitioning of pesticides and other chemicals from maternal blood into breast milk further enhances vulnerability of infants to develop neurotoxicity [82–84]. While genetic inheritance has an important role in fetal brain development, environmental factors including exposure to pesticides, industrial chemicals, pharmaceuticals, nutrition, infectious agents, and other psychosocial indices are reported to exert profound influence on the brain development by altering the signaling in specific brain regions [85,86]. Genetic predisposition to environmental chemicals leading to these neurological disorders has led researchers to study the interplay between genes and environment [66,86].

5.3 NEUROBEHAVIORAL EFFECTS OF ORGANOCHLORINE PESTICIDES

Use of organochlorine pesticides has been enormous in agriculture and public health programs to control pests and insects. With reports that the organochlorine pesticides have long persistence in the environment and cause adverse effects on aquatic organisms and wildlife, use of DDT was banned in the United States in 1972 followed by other countries. Consumption of other organochlorine pesticides especially those of cyclodiene group also came down with the reports exhibiting the presence of their residues in different environmental matrices and adverse effects on human health. Although many of the organochlorine pesticides (aldrin, DDT, dieldrin, endosulfan, etc.) were banned in South Africa long back [87], residues of organochlorine pesticides have been found in environmental matrices and dietary products in South Africa [88–90]. Therefore, besides exposure during manufacturing and handling, environmental exposure to organochlorine pesticides is also possible due to their persistence in the environment.

Neurological and neuropsychological deficits in individuals following occupational and environmental exposure to organochlorine pesticides have often been observed and reviewed from time to time [91–93]. Cognitive deficits on chronic exposure to DDT have been reported in spray workers [94,95]. Association of higher DDT levels in the brain with AD has been reported [28]. Also, exposure to pesticides including organochlorines has been reported to enhance the risk of PD among occupational workers [42–44]. High levels of dieldrin in the blood of individuals associated with its manufacturing have been reported in a study from India [45]. Further, the presence of dieldrin in autopsied brain samples of PD cases in separate case control studies indicates that organochlorine pesticides may affect the dopaminergic neurons in the substantia nigra [28,46,47]. This was further strengthened with the reports of increased levels of other organochlorine pesticides (γ-HCH, lindane) in the brain of PD patients [47]. In a study from India, the risk of PD has been associated with increased levels of organochlorine pesticides in the environment [96]. Involvement of heptachlor epoxide, a metabolite of organochlorine pesticide in causing Lewy body pathology, has been suggested recently (Honolulu-Asia-Aging Study) [48]. The analysis of data from population study (NHANES) to assess the impact of organochlorine pesticides on cognitive functions showed that increase in serum dichlorodiphenyldichloroethylene (DDE, a metabolite of DDT) levels enhances the risk of cognitive impairment on nonoccupational exposure [97]. Consistent with this, an increased risk of AD was found to be associated with high serum DDE levels [29]. High blood levels of β-HCH, dieldrin, and pp'-DDE were associated with an increased risk of AD [30]. It has been further suggested that vulnerability to DDE to cause cognitive deficits may be more in carriers of an apolipoprotein (APOE) ε4 allele [29]. Recently, the presence of blood DDE, pp'-DDE, and hexachlorobenzene in PD cases has been reported in a case control study from Greece [49].

In the 1970s, an epidemic due to chlordecone (Kepone) exposure at a manufacturing plant in Virginia (United States) caused neurological deficits (tremor, opsoclonus, and nervousness) in workers and some of their spouses. The neurological symptoms were found to subside with reduced levels of chlordecone in the blood on follow-up after 6 years [98]. Poisoning from endosulfan, another cyclodiene class of organochlorine pesticide, has been reported to affect the functions of basal ganglia and occipital cortex leading to seizures, psychosis, cortical blindness, and rigidity of limbs [99]. Also, recurrent epileptic seizures have been reported in few individuals (one male and two females) who consumed endosulfan to commit suicide [100]. High levels of endosulfan residues in environmental matrices in Kasaragod district of Kerala in India have been reported to cause adverse neurological effects in exposed individuals and congenital abnormalities in progeny among the residents [101–103]. The buildup of endosulfan residues was due to its widespread aerial spray to protect cashew plantations. These reports led to ban of this pesticide in Kerala and later all over India. Cases of mental retardation (age group 0–30 years; 74.5% males and 74.1% females) have been reported 7 years after the ban on the use of endosulfan in that area [104]. Hexachlorophene, used as a skin disinfectant in soap, has been associated with adverse neurological defects [62]. Bilateral optical atrophy on chronic exposure to hexachlorophene has also been reported [62,105]. Risk to develop amyotrophic lateral sclerosis on exposure to organochlorine pesticides (aldrin, dieldrin, DDT, and toxaphene) has been reported on analyzing the data from the Agricultural Health Study from the United States [52].

The presence of residues of DDT, HCH, and other chlorinated pesticides in human breast milk further raises health risk concerns for infants and children [106,107]. These reports show that despite the ban, organochlorine pesticides continue to pose severe health risks particularly in the developing world due to their persistence in the environment.

5.4 NEUROBEHAVIORAL EFFECTS OF ORGANOPHOSPHATE PESTICIDES

Historically, OPs were used as nerve poisons and warfare agents during World War II. Neuropsychiatric problems including impairment in cognitive functions with other symptoms were associated with the use of OPs in US veterans involved in the Gulf War [108]. With the reports of adverse effects of organochlorine pesticides on environment and human health and their long persistence due to low biodegradability, OP pesticides were introduced to control insects and pests in agriculture, homes, and public health programs. Although OPs degrade fast as compared to organochlorines, acute and chronic exposures to OPs in occupational and nonoccupational settings have been frequently reported. Their indiscriminate use often poses a major public health issue associated with high morbidity and mortality [109–111]. Aerotoxic syndrome that includes central nervous system (CNS) symptoms including headache, dizziness, sensory changes, nerve pain, tremors, and cognitive impairment have been reported in air crew members and certain passengers due to OP exposure [112,113]. Because of their free availability, OP pesticides have been abused for suicide, and a number of suicide poisoning cases have been reported from developing nations [114–116].

OPs primarily target cholinesterase at the neuromuscular junction and enhance acetylcholine levels that may affect the muscarinic and nicotinic receptors in the periphery and CNS [117,118]. Muscle weakness, CNS depression, and coma are the common cholinergic signs of acute OP poisoning. Besides, hypertension, gastrointestinal upset, bronchospasm, sweating, lacrimation, bradycardia, fasciculations have also been reported on acute exposure.

OPs, unlike other pesticides, are reported to cause delayed neurotoxicity [119,120]. Paresis of the proximal limb and facial muscles, and cranial nerve palsies are the symptoms of intermediate syndrome, observed between 24 and 96 hours after the exposure. Delayed neuropathy leading to muscle weakness, ataxia in lower limbs, flaccid paralysis, and spasticity, is frequently reported 2–3 weeks after acute exposure to OPs due to inhibition of neurotoxic esterase [119]. Higher acute or sub-chronic exposure to OP has been associated with organophosphate-induced delayed neuropathy (OPIDN) in adults [121]. The OPIDN generally was seen 1–6 weeks after exposure to OP and persists for months, years, or indefinitely [122].

Psychological disturbances including depression, anxiety, and suicidal thoughts have been reported in a study on workers on acute exposure to OPs [36]. Neuropsychological deficits on acute poisoning with OPs remained persistent even 2 years after exposure in a study group of 36 men [123]. A report of neuropsychological and brain-evoked potential deficits following 6–8 months after the terrorist attack with Sarin gas in Tokyo Subway [124] further illustrates the persistence of long-term neurobehavioral impairments in individuals on exposure to OPs. In view of detrimental effects of OPs on the brain, data of different studies undertaken on human subjects have been analyzed from time to time by different investigators [50,63,125]. Distal weakness in muscles associated with paralysis was observed on acute exposure to OPs [51,125]. Restlessness, delirium, convulsions, and coma were common symptoms on acute exposure to OPs [125]. Both acute and chronic exposures to OPs were found to cause neuropsychiatric abnormalities in affected individuals [119]. Severe poisoning with OPs may cause extrapyramidal symptoms [126,127]. Short-term memory and decreased visomotor speed have also been reported in OPs-exposed individuals [113,128,129]. While reviewing the data of 24 studies to understand the impact of OPs on neurological functions, adverse CNS effects in exposed individuals were observed in most of the studies [130].

Occupational exposure to OP pesticides has been found to cause neurobehavioral abnormalities as revealed by impairment in response movement and coordination, sustained attention, visual perception, and memory [31,131,132]. In a cross-sectional study, long-term exposure to OP pesticides was found to affect the neuropsychological performance in exposed individuals [133]. Interestingly, impaired neuropsychological performance in OP pesticide–exposed individuals was associated with the duration of exposure but not with age [133]. Slower reaction time on neuropsychological assessment was observed in farmers using OPs for a long time [129]. However, no significant effect on visomotor skills, memory, and mood was observed [129]. Epidemiological studies have found cognitive deficits including impairment in working memory, attention, motor speed, and coordination in OP-exposed individuals [134–136].

Decreased IQ and verbal skills in OP-exposed workers have been frequently reported and associated with low acetylcholinesterase activity [37]. Further, impaired functioning in the left frontal lobe in OP-exposed individuals was also linked with decreased acetylcholinesterase activity [37]. Neurobehavioral deficits in Egyptian adolescent applicators were associated with repeated exposure to OPs and linked with increased urinary levels of 3,5,6-trichloro-2-pyridinol (TCPy, a chlorpyrifos metabolite and biomarker of exposure) and decreased blood butyrylcholinesterase (BChE, biomarker of effect). Urinary levels of TCPy were found to be increased following pesticide application and returned to baseline at the end of the application season. Decreased activity of blood BChE activity also exhibited a trend of recovery after several months of pesticide exposure [38,57].

Vulnerability of children to develop neurobehavioral and neuropsychological deficits following exposure to organophosphates is more due to several factors, which have been reviewed from time to time [121,137]. Several reports indicate that *in utero* exposure to OP pesticides may affect the neurobehavioral development in children [39,138–141]. Anomalies of primitive reflexes in infants have been found to be associated with prenatal levels of urinary OP metabolites of mothers [138]. Maternal levels of chlorpyrifos above the detection limit were associated with reduced head circumference of infants [142,143]. Reduced activity of maternal paraoxonase, an enzyme involved in the hydrolysis of paraoxon and certain other OPs including chlorpyrifos, was also linked in this study [143]. In another study, prenatal chlorpyrifos exposure significantly affected the birth weight and length of the baby [144]. Cognitive performance and motor development were also found to be impaired in children prenatally exposed to chlorpyrifos on follow-up [58,141]. Acute exposure to OP in children has been found to cause miosis, excessive salivation, muscle weakness, hyperreflexia, and hypertonia involving sympathetic, parasympathetic, and CNS [39,125,145].

Effects on memory, executive functions, motor skills, and other behaviors have also been reported in children exposed to OPs postnatally [159,146–148]. To assess the impact of prenatal and postnatal exposures to OPs on cognitive abilities, urinary di-alkyl-phosphate (DAP) levels were measured in school-age children in a study from the United States [40]. Urinary DAP levels in children (7-year-old) prenatally exposed to OPs were associated with impaired intellectual development. Effects on mental development and perceptual reasoning were observed in children prenatally exposed to OPs. These changes were found to start to begin at the age of 12 months and continued through early childhood and correlated well with urinary DAP levels of mothers. Changes in children were more intense in cases their mothers who had PON1 Q192R QQ genotype associated with slow catalytic activity for chlorpyrifos [149]. The role of PON1 gene in the toxicity of OPs is well demonstrated [149–151]. Adverse effects on mental development in 2-year-old Mexican-American children were also associated with prenatal exposure to OPs, largely chlorpyrifos and malathion [39]. In a recent study, while no association of maternal urinary DAP concentrations with lower child IQ scores (at 6 years of age) was observed, an inverse relation of urinary DAPs in late pregnancy was associated with nonverbal IQ of children [152]. Higher levels of urinary total DAP as compared to general population has been reported in women residing in agricultural regions [153–155]. Interestingly, total DAP levels in urinary samples collected after delivery were higher than in samples collected during pregnancy, indicating a higher risk of infants to OP exposure [153].

5.5 NEUROBEHAVIORAL EFFECTS OF SYNTHETIC PYRETHROIDS

Although pyrethroids are considered safe, their widespread use and the presence of their residues in the environment and dietary products point out a possible risk of exposure in humans and nontarget organisms [156–158]. Exposure to pyrethroids could occur through dietary and non-dietary sources at homes as well [54]. High levels of pyrethroid metabolites detected in the urine in population-based biomonitoring studies undertaken in the United States, Germany, and China further exhibit the risk of human exposure to pyrethroids [32,159].

Pyrethroids primarily target voltage-dependent sodium channels and the prolonged repetitive activity leads to tremors, hypersensitivity, choreoathetosis, and salivation [160–163]. While modeling the risk of pyrethroids in humans, symptoms of pyrethroid toxicity were found to be like those observed in mammals except the reports of death on exposure [164]. The cyano-pyrethroids (fenvalerate, deltamethrin, cypermethrin) unlike non-cyano-pyrethroids (allethrin, tetramethrin, and phenothrin) are reported to cause transient skin paresthesia among the exposed workers [164].

Dizziness, headache, nausea, anorexia, fatigue, and muscular fasciculation have been observed in cases of acute pyrethroid (deltamethrin, fenvalerate, cypermethrin) poisoning [165]. Convulsive attacks, disturbances of consciousness (drowsiness, cloudiness, and coma), and signs of pulmonary edema have been reported on extreme exposure to pyrethroids [160,165]. Besides, gastrointestinal disturbances and pulmonary edema, CNS abnormalities (confusion, coma, and seizures) were reported on acute poisoning with permethrin formulation in a study from Taiwan [166]. Also, signs of nervous system disorders have been reported following acute exposure even after 2 years of exposure [167]. Long-term exposure to pyrethroids has been found to impair cognitive performance with no significant changes in motor indices assessed by computerized Behavioral Assessment and Research System in a cross-sectional study undertaken on workers involved in public health program from Bolivia [33]. Chronic inhalation of pyrethroids has been associated with high risk to develop amyotrophic lateral sclerosis [52,53].

As human serum is devoid of carboxylesterase, metabolism of pyrethroids cannot take place in the blood [117]. Consistent with the notion, a greater peak (approximately twofold) of deltamethrin in the human brain was predicted based on the physiologically based pharmacokinetic modeling while developing a rat–human diffusion-limited model [168].

Pyrethroid metabolites have been detected in the urine of pregnant women [144,169]. Studies have also found that 67% of a cohort of preschool children had detectable levels of 3-phenoxybenzoic acid (a pyrethroid metabolite) in their urine [55,170]. The presence of pyrethroid metabolites in the urine of elementary-age children, primarily due to residential exposure, has also been reported [171,172] and exhibits the vulnerability of children to adverse effects of pyrethroids.

Developmental disorders that may range from severely impaired analytical abilities and/or motor skills, to decrease in concentration and memory, have been reported in children constantly exposed to pesticides including pyrethroids [146]. Besides the CNS, pyrethroid may also affect the peripheral nervous system. The epidemiological finding that increased urinary levels of 3-PBA were associated with plausibility to develop ADHD in children was found to be consistent with the experimental studies on mice to assess the impact of developmental exposure to deltamethrin [60,173]. Further, boys were found to be more vulnerable to pyrethroid exposure to develop ADHD [61]. In a mother–child cohort study, low-level exposure to deltamethrin, a type II synthetic pyrethroid, was found to cause cognitive deficits in children [34,35]. The cognitive deficits in exposed children were associated with high levels of cis-DBCA, a selective metabolite of deltamethrin, and 3-PBA, a common metabolite of pyrethroids in the urine of these children [34]. The finding was associated with increased urinary levels of 3-PBA. Symptoms of learning disability and ADHD reported by patients were not associated with postnatal pyrethroid exposure in NHANES study on US children [32]. Risk to develop autism in children on prenatal exposure to pyrethroids has also been reported [56].

5.6 NEUROBEHAVIORAL EFFECTS OF CARBAMATES

As compared to organochlorine and organophosphate pesticides, carbamates appear to be rather safe as no serious adverse effects have been reported among industrial or agriculture workers. Like organophosphates, carbamates also act by inhibiting the acetylcholinesterase [174–176]. CNS disturbances have been observed in individuals exposed to carbamates accidentally or intentionally. Headache, confusion, disorientation, and nausea associated with cholinesterase inhibition were reported in a person on acute exposure to Baygon (carbamate-containing spray) while spraying it in a cellar [177]. These symptoms lasted for around 20 minutes. Clonus and intense weakness on the right side of the body were observed on neurological examination of a case presented with aldicarb (carbamate) poisoning. Although these changes were associated with cholinesterase inhibition, there was no effect on sensation and deep tendon reflexes. Further, the oxime therapy, which is effective to manage OP poisoning, helped to improve the clinical state [178]. Loss of sensorium and vomiting were reported in an interesting case of transdermal carbamate poisoning while using it on the body to control mites [179]. These effects were managed on treatment with atropine.

Only few studies are available on effects of carbamates on brain during development. No effects of carbamates have been reported in infants and children. Both nicotinic (stupor, coma, hypertoxicity) and muscarinic (diarrhea, vomiting) effects have been reported in 26 children (17 boys, 9 girls) with a history of carbamate (methomyl or aldicarb) poisoning. These affected children were managed effectively with oxime therapy and repeated dose of atropine sulfate [180].

5.7 NEUROBEHAVIORAL EFFECTS OF NEONICOTINOIDS

Neonicotinoids have emerged as a new class of insecticides preferred over carbamates, organochlorines, OPs, and synthetic pyrethroids. These are systemic insecticides absorbed by the plants and get distributed in its different parts. Being agonists of nicotinic acetylcholine receptors, they have been found to target insects selectively. Residues of neonicotinoids identified in water, fruits, and other dietary produce suggest the risk of human exposure. A case–control study from Japan assessed the association of urinary N-desmethylacetamiprid (DMAP), a common metabolite of six chlorinated neonicotinoids (imidacloprid, nitenpyram, acetamiprid, thiacloprid, clothianidin, and thiamethoxam) with neurological functions [41]. Besides, general fatigue, pain in abdomen, and chest symptoms were also associated with urinary DMAP levels. These effects persisted for a long period from several days to months even after the cessation of consumption of produce. Although limited reports on occupational exposure to neonicotinoids are available [181], CNS effects in cases of acute neonicotinoid poisoning have been reported from Taiwan [182], Iran [183], and India [184]. Severe neuropsychiatric symptoms have also been reported in cases of acute neonicotinoid poisoning from India [185,186].

5.8 CONCLUSIONS

The synthetic pesticides pose a wide range of adverse effects on human brain functions following acute and long-term exposures. Due to their severe toxicity to the ecosystem, use of several organochlorine pesticides (such as aldrin, chlordane, and heptachlor), some OP pesticides (such as ethyl parathion, tetradifon, chlorfenvinphos), and carbamate (aldicarb) has been banned in several countries. In certain developing countries, restricted use of DDT and few other pesticides for public health purposes has been permitted. Accidental exposure, unregulated use, or use without proper safety gears and abuse of pesticides for suicide have been the major cause of large-scale human poisoning. As the need to protect the food grains and crops and control vector-borne diseases is essential, use of pest control agents will continue. Newer and safer pesticides are therefore being introduced with a bigger thrust on natural and biological agents as pesticides.

Several studies exhibit that due to the widespread environmental contamination and presence of pesticide residues in food, the brain is exposed to them right from the conception to childhood and even beyond. The outcomes of such exposure range from a variety of cognitive deficits to major neurological abnormalities. Pesticide exposure has been shown to cause ADHD and autism, and to affect several higher cognitive functions.

In view of such severe consequences of pesticide exposure on the developing brain, well-planned regular monitoring of pesticide exposure is required. Most of the monitoring studies have been done by measuring urinary levels of pesticide metabolites. Significant research will be required to develop and validate suitable biomarkers to assess the exposure and effect of pesticides for biomonitoring studies. Also, studies are required to identify the interaction of pesticides with disease genes through which it makes individuals vulnerable to neurological disorders such as PD, AD, and ADHD.

ACKNOWLEDGMENTS

PKS is thankful to the National Academy of Sciences India for the award of a Senior Scientist Platinum Jubilee Fellowship and Biotech Park for providing laboratory facilities. VKK acknowledges the support of CSIR-Indian Institute of Toxicology Research, Lucknow. Help of Ms Anima Kumari and Ms Anugya Srivastava is gratefully acknowledged in preparation of this chapter.

REFERENCES

1. Costa, L.G., *Toxicology of pesticides: A brief history*, in *Toxicology of Pesticides*. 1987, Springer. pp. 1–10.
2. Harada, T., et al., Toxicity and carcinogenicity of dichlorodiphenyltrichloroethane (DDT). *Toxicological Research*, 2016. **32**(1): pp. 21–33.
3. Alavanja, M.C., Pesticides use and exposure extensive worldwide. *Reviews on Environmental Health*, 2009. **24**(4): p. 303.
4. Aktar, W., D. Sengupta, and A. Chowdhury, Impact of pesticides use in agriculture: Their benefits and hazards. *Interdisciplinary Toxicology*, 2009. **2**(1): pp. 1–12.
5. Kamel, F. and J.A. Hoppin, Association of pesticide exposure with neurologic dysfunction and disease. *Environmental Health Perspectives*, 2004. **112**(9): pp. 950–958.
6. Abhilash, P. and N. Singh, Pesticide use and application: An Indian scenario. *Journal of Hazardous Materials*, 2009. **165**(1–3): pp. 1–12.
7. Yadav, I.C., et al., Current status of persistent organic pesticides residues in air, water, and soil, and their possible effect on neighboring countries: A comprehensive review of India. *Science of the Total Environment*, 2015. **511**: pp. 123–137.
8. Yadav, I.C. and N.L. Devi, Pesticides classification and its impact on human and environment. *Environmental Science and Engineering*, 2017. **6**: pp. 140–158.
9. World Health Organization, *The WHO recommended classification of pesticides by hazard and guidelines to classification 2019*. 2020: World Health Organization.
10. Sharma, A., et al., Worldwide pesticide usage and its impacts on ecosystem. *SN Applied Sciences*, 2019. **1**(11): p. 1446.
11. Zhang, W., Global pesticide use: Profile, trend, cost/benefit and more. *Proceedings of the International Academy of Ecology and Environmental Sciences*, 2018. **8**(1): p. 1.
12. Dalvie, M.A. and L. London, Risk assessment of pesticide residues in South African raw wheat. *Crop protection*, 2009. **28**(10): pp. 864–869.
13. Eddleston, M., et al., Pesticide poisoning in the developing world—a minimum pesticides list. *The Lancet*, 2002. **360**(9340): pp. 1163–1167.
14. World Health Organization, *The public health impact of chemicals: Knowns and unknowns*. 2016, World Health Organization.
15. Gunnell, D., et al., The global distribution of fatal pesticide self-poisoning: Systematic review. *BMC Public Health*, 2007. **7**(1): p. 357.
16. Prüss-Ustün, A., et al., Knowns and unknowns on burden of disease due to chemicals: A systematic review. *Environmental Health*, 2011. **10**(1): p. 9.

17. Bonvoisin, T., et al., Suicide by pesticide poisoning in India: A review of pesticide regulations and their impact on suicide trends. *BMC Public Health*, 2020. **20**(1): pp. 1–16.

18. Mew, E.J., et al., The global burden of fatal self-poisoning with pesticides 2006–15: systematic review. *Journal of Affective Disorders*, 2017. **219**: pp. 93–104.

19. Prüss-Üstün, A., et al., *Preventing disease through healthy environments: A global assessment of the burden of disease from environmental risks.* 2016: World Health Organization.

20. Kamaruzaman, N.A., et al., Epidemiology and risk factors of pesticide poisoning in Malaysia: a retrospective analysis by the National Poison Centre (NPC) from 2006 to 2015. *BMJ Open*, 2020. **10**(6): p. e036048.

21. Parrón, T., et al., Association between environmental exposure to pesticides and neurodegenerative diseases. *Toxicology and Applied Pharmacology*, 2011. **256**(3): p. 379–385.

22. Yan, D., et al., *Pesticide exposure and risk of Alzheimer's disease: A systematic review and meta-analysis. Scientific Reports*, 2016. **6**(1): p. 1–9.

23. Baltazar, M.T., et al., Pesticides exposure as etiological factors of Parkinson's disease and other neurodegenerative diseases—a mechanistic approach. *Toxicology Letters*, 2014. **230**(2): pp. 85–103.

24. Fage-Larsen, B., H.R. Andersen, and N. Bilenberg, Prenatal Exposure to Pesticides and Neurodevelopmental Disorders in Children. In *64th Annual Meeting*. 2017. AACAP.

25. Sapbamrer, R. and S. Hongsibsong, Effects of prenatal and postnatal exposure to organophosphate pesticides on child neurodevelopment in different age groups: A systematic review. *Environmental Science and Pollution Research*, 2019. **26**(18): pp. 18267–18290.

26. von Ehrenstein, O.S., et al., Prenatal and infant exposure to ambient pesticides and autism spectrum disorder in children: Population based case-control study. *BMJ*, 2019. **364**: p. 1962.

27. Philippat, C., et al., Prenatal exposure to organophosphate pesticides and risk of autism spectrum disorders and other non-typical development at 3 years in a high-risk cohort. *International Journal of Hygiene and Environmental Health*, 2018. **221**(3): pp. 548–555.

28. Fleming, L., et al., Parkinson's disease and brain levels of organochlorine pesticides. *Annals of Neurology: Official Journal of the American Neurological Association and the Child Neurology Society*, 1994. **36**(1): pp. 100–103.

29. Richardson, J.R., et al., Elevated serum pesticide levels and risk for Alzheimer disease. *JAMA Neurology*, 2014. **71**(3): pp. 284–290.

30. Singh, N., et al., Organochlorine pesticide levels and risk of Alzheimer's disease in north Indian population. *Human & Experimental Toxicology*, 2013. **32**(1): pp. 24–30.

31. Hayden, K.M., et al., Occupational exposure to pesticides increases the risk of incident AD: The cache county study. *Neurology*, 2010. **74**(19): pp. 1524–1530.

32. Quirós-Alcalá, L., S. Mehta, and B. Eskenazi, Pyrethroid pesticide exposure and parental report of learning disability and attention deficit/hyperactivity disorder in US children: NHANES 1999–2002. *Environmental Health Perspectives*, 2014. **122**(12): pp. 1336–1342.

33. Hansen, M.R.H., et al., Neurological deficits after long-term pyrethroid exposure. *Environmental Health Insights*, 2017. **11**: pp. 1–11.

34. Viel, J.-F., et al., Pyrethroid insecticide exposure and cognitive developmental disabilities in children: the PELAGIE mother–child cohort. *Environment International*, 2015. **82**: pp. 69–75.

35. Chen, S., et al., Exposure to pyrethroid pesticides and the risk of childhood brain tumors in East China. *Environmental Pollution*, 2016. **218**: pp. 1128–1134.

36. Wesseling, C., et al., Symptoms of psychological distress and suicidal ideation among banana workers with a history of poisoning by organophosphate or *n*-methyl carbamate pesticides. *Occupational and Environmental Medicine*, 2010. **67**(11): pp. 778–784.

37. Korsak, R.J. and M.M. Sato, Effects of chronic organophosphate pesticide exposure on the central nervous system. *Clinical Toxicology*, 1977. **11**(1): pp. 83–95.

38. Ismail, A.A., et al., The impact of repeated organophosphorus pesticide exposure on biomarkers and neurobehavioral outcomes among adolescent pesticide applicators. *Journal of Toxicology and Environmental Health, Part A*, 2017. **80**(10–12): pp. 542–555.

39. Eskenazi, B., et al., Organophosphate pesticide exposure and neurodevelopment in young Mexican-American children. *Environmental Health Perspectives*, 2007. **115**(5): pp. 792–798.

40. Bouchard, M.F., et al., Prenatal exposure to organophosphate pesticides and IQ in 7-year-old children. *Environmental Health Perspectives*, 2011. **119**(8): pp. 1189–1195.

41. Marfo, J.T., et al., Relationship between urinary N-desmethyl-acetamiprid and typical symptoms including neurological findings: A prevalence case-control study. *PLoS One*, 2015. **10**(11): p. e0142172.

42. Firestone, J.A., et al., Pesticides and risk of Parkinson disease: a population-based case-control study. *Archives of Neurology*, 2005. **62**(1): pp. 91–95.

43. Wang, A., et al., Parkinson's disease risk from ambient exposure to pesticides. *European Journal of Epidemiology*, 2011. **26**(7): p. 547–555.

44. Das, K., et al., Role of familial, environmental and occupational factors in the development of Parkinson's disease. *Neurodegenerative Diseases*, 2011. **8**(5): pp. 345–351.

45. Nair, A. and M. Pillai, Trends in ambient levels of DDT and HCH residues in humans and the environment of Delhi, India. *Science of the Total Environment*, 1992. **121**: pp. 145–157.

46. Kanthasamy, A.G., et al., Dieldrin-induced neurotoxicity: Relevance to Parkinson's disease pathogenesis. *Neurotoxicology*, 2005. **26**(4): pp. 701–719.

47. Corrigan, F., et al., Organochlorine insecticides in substantia nigra in Parkinson's disease. *Journal of Toxicology and Environmental Health Part A*, 2000. **59**(4): pp. 229–234.

48. Ross, G.W., et al., Association of brain heptachlor epoxide and other organochlorine compounds with Lewy pathology. *Movement Disorders*, 2019. **34**(2): pp. 228–235.

49. Dardiotis, E., et al., Organochlorine pesticide levels in Greek patients with Parkinson's disease. Toxicology Reports, 2020.

50. Bhatt, M.H., M.A. Elias, and A.K. Mankodi, Acute and reversible parkinsonism due to organophosphate pesticide intoxication: Five cases. *Neurology*, 1999. **52**(7): pp. 1467–1467.

51. Wang, A., et al., The association between ambient exposure to organophosphates and Parkinson's disease risk. *Occupational and Environmental Medicine*, 2014. **71**(4): pp. 275–281.

52. Kamel, F., et al., Pesticide exposure and amyotrophic lateral sclerosis. *Neurotoxicology*, 2012. **33**(3): pp. 457–462.

53. Doi, H., et al., Motor neuron disorder simulating ALS induced by chronic inhalation of pyrethroid insecticides. *Neurology*, 2006. **67**(10): pp. 1894–1895.

54. Morgan, M.K., Children's exposures to pyrethroid insecticides at home: A review of data collected in published exposure measurement studies conducted in the United States. *International Journal of Environmental Research and Public Health*, 2012. **9**(8): pp. 2964–2985.

55. Barr, D.B., et al., Urinary concentrations of metabolites of pyrethroid insecticides in the general US population: National Health and Nutrition Examination Survey 1999–2002. *Environmental Health Perspectives*, 2010. **118**(6): pp. 742–748.

56. Barkoski, J., et al., Utero Pyrethroid Pesticide Exposure and Risk of Autism Spectrum Disorders or Other Non-Typical Development at 3 Years and Child Gene Expression in the MARBLES Longitudinal Cohort. In *ISEE Conference Abstracts*. 2018.

57. Fortenberry, G.Z., et al., Urinary 3, 5, 6-trichloro-2-pyridinol (TCPY) in pregnant women from Mexico City: Distribution, temporal variability, and relationship with child attention and hyperactivity. *International Journal of Hygiene and Environmental Health*, 2014. **217**(2–3): pp. 405–412.

58. Rauh, V.A., et al., Impact of prenatal chlorpyrifos exposure on neurodevelopment in the first 3 years of life among inner-city children. *Pediatrics*, 2006. **118**(6): pp. e1845–e1859.

59. Bouchard, M.F., et al., Attention-deficit/hyperactivity disorder and urinary metabolites of organophosphate pesticides. *Pediatrics*, 2010. **125**(6): pp. e1270–e1277.

60. Lee, W.-S., et al., Residential pyrethroid insecticide use, urinary 3-phenoxybenzoic acid levels, and attention-deficit/hyperactivity disorder-like symptoms in preschool-age children: The environment and development of children study. *Environmental Research*, 2020. **188**: p. 109739.

61. Wagner-Schuman, M., et al., Association of pyrethroid pesticide exposure with attention-deficit/hyperactivity disorder in a nationally representative sample of US children. *Environmental Health*, 2015. **14**(1): p. 44.

62. Anderson, J., et al., Neonatal spongiform myelinopathy after restricted application of hexachlorophene skin disinfectant. *Journal of Clinical Pathology*, 1981. **34**(1): pp. 25–29.

63. Muñoz-Quezada, M.T., et al., Chronic exposure to organophosphate (OP) pesticides and neuropsychological functioning in farm workers: A review. *International Journal of Occupational and Environmental Health*, 2016. **22**(1): pp. 68–79.

64. Buss, C., S. Entringer, and P.D. Wadhwa, Fetal programming of brain development: Intrauterine stress and susceptibility to psychopathology. *Science Signaling*, 2012. **5**(245): p. pt7.

65. Sandman, C.A., et al., Prenatal programming of human neurological function. *International Journal of Peptides*, 2011.doi:10.1155/2011/837596.

66. Lenroot, R.K. and J.N. Giedd, The changing impact of genes and environment on brain development during childhood and adolescence: Initial findings from a neuroimaging study of pediatric twins. *Development and Psychopathology*, 2008. **20**(4): p. 1161.

67. Eskenazi, B., et al., Pesticide toxicity and the developing brain. *Basic &Clinical Pharmacology &Toxicology*, 2008. **102**(2): pp. 228–236.

68. Ingber, S.Z. and H.R. Pohl, Windows of sensitivity to toxic chemicals in the motor effects development. *Regulatory Toxicology and Pharmacology*, 2016. **74**: pp. 93–104.

69. Stiles, J. and T.L. Jernigan, The basics of brain development. *Neuropsychology Review*, 2010. **20**(4): pp. 327–348.

70. Budday, S., P. Steinmann III, and E. Kuhl, Physical biology of human brain development. *Frontiers in Cellular Neuroscience*, 2015. **9**: p. 257.

71. Holland, D., et al., Structural growth trajectories and rates of change in the first 3 months of infant brain development. *JAMA Neurology*, 2014. **71**(10): pp. 1266–1274.

72. Healy, J., *Your child's growing mind: Brain development and learning from birth to adolescence.* 2011: Harmony.

73. Yamaguchi, Y. and M. Miura, *Programmed cell death in neurodevelopment. Developmental Cell*, 2015. **32**(4): pp. 478–490.

74. Roth, K.A. and C. D'Sa, *Apoptosis and brain development. Mental Retardation and Developmental Disabilities Research Reviews*, 2001. **7**(4): pp. 261–266.

75. Pediatrics, A.A.O., *Prenatal pesticide exposure and neurodevelopmental disorders. AAP Grand Rounds*, 2015. **33**(2): p. 16.

76. Aldridge, J.E., et al., *Heterogeneity of toxicant response: Sources of human variability. Toxicological Sciences*, 2003. **76**(1): pp. 3–20.

77. Liu, J. and E. Schelar, *Pesticide exposure and child neurodevelopment: Summary and implications. Workplace Health & Safety*, 2012. **60**(5): pp. 235–242.

78. Landrigan, P.J. and L.R. Goldman, *Children's vulnerability to toxic chemicals: A challenge and opportunity to strengthen health and environmental policy. Health Affairs*, 2011. **30**(5): pp. 842–850.

79. Yin, S., et al., *Transplacental transfer mechanism of organochlorine pesticides: An in vitro transcellular transport study. Environment International*, 2020. **135**: p. 105402.

80. Magnarelli, G. and N. Guiñazú, Placental toxicology of pesticides. *Recent Advances in Research on the Human Placenta* (J. Zheng Ed.). InTech, Rijeka, Croatia, 2012: p. 95–119.

81. Saxena, M., et al., Placental transfer of pesticides in humans. *Archives of Toxicology*, 1981. **48**(2–3): pp. 127–134.

82. Mead, M.N., Contaminants in human milk: Weighing the risks against the benefits of breastfeeding. 2008, National Institute of Environmental Health Sciences.

83. Acosta-Maldonado, B., et al., Effects of exposure to pesticides during pregnancy on placental maturity and weight of newborns: A cross-sectional pilot study in women from the Chihuahua State, Mexico. *Human & Experimental Toxicology*, 2009. **28**(8): pp. 451–459.

84. Pajewska-Szmyt, M., E. Sinkiewicz-Darol, and R. Gadzała-Kopciuch, The impact of environmental pollution on the quality of mother's milk. *Environmental Science and Pollution Research*, 2019. **26**(8): pp. 7405–7427.

85. Hegarty, J.P., et al., Genetic and environmental influences on structural brain measures in twins with autism spectrum disorder. *Molecular Psychiatry*, 2020. **25**(10): pp. 2556–2566.

86. van der Meulen, M., et al., Genetic and environmental influences on structure of the social brain in childhood. *Developmental Cognitive Neuroscience*, 2020. **44**: p. 100782.

87. Daff, D., *Agricultural Remedies that Are Banned or Restricted for Use in the Republic of South Africa.* 2017.

88. Olisah, C., O.O. Okoh, and A.I. Okoh, Occurrence of organochlorine pesticide residues in biological and environmental matrices in Africa: A two-decade review. *Heliyon*, 2020. **6**(3): p. e03518.

89. Thompson, L.A., et al., *Organochlorine pesticide contamination of foods in Africa: Incidence and public health significance. Journal of Veterinary Medical Science*, 2017. **79**: p. 751–764.

90. Thompson, L., et al., Concentrations and human health risk assessment of DDT and its metabolites in free-range and commercial chicken products from KwaZulu-Natal, South Africa. *Food Additives & Contaminants: Part A*, 2017. **34**(11): pp. 1959–1969.

91. Costa, L.G., The neurotoxicity of organochlorine and pyrethroid pesticides. *Handbook of Clinical Neurology* (Marcello, L., Margit, L.B., Eds.) Elsevier, Amsterdam, 2015: pp. 135–148.

92. Saravi, S.S.S. and A.R. Dehpour, Potential role of organochlorine pesticides in the pathogenesis of neurodevelopmental, neurodegenerative, and neurobehavioral disorders: A review. *Life Sciences*, 2016. **145**: pp. 255–264.

93. Medehouenou, T.C.M., et al., Exposure to polychlorinated biphenyls and organochlorine pesticides and risk of dementia, Alzheimer's disease and cognitive decline in an older population: A prospective analysis from the Canadian Study of Health and Aging. *Environmental Health*, 2019. **18**(1): pp. 1–11.

94. deJoode, B.V.W., et al., Chronic nervous-system effects of long-term occupational exposure to DDT. *The Lancet*, 2001. **357**(9261): pp. 1014–1016.

95. Misra, U., D. Nag, and C.K. Murti, A study of cognitive functions in DDT sprayers. *Industrial Health*, 1984. **22**(3): pp. 199–206.

96. Chhillar, N., et al., Organochlorine pesticide levels and risk of Parkinson's disease in north Indian population. ISRN neurology, 2013. **2013**: p. 371034.

97. Kim, K.-H., E. Kabir, and S.A. Jahan, Exposure to pesticides and the associated human health effects. *Science of the Total Environment*, 2017. **575**: pp. 525–535.

98. Taylor, J., Neurological manifestations in humans exposed to chlordecone and follow-up results. *Neurotoxicology*, 1982. **3**(2): pp. 9–16.

99. Pradhan, S., et al., Selective involvement of basal ganglia and occipital cortex in a patient with acute endosulfan poisoning. *Journal of the Neurological Sciences*, 1997. **147**(2): pp. 209–213.

100. Kutluhan, S., et al., Three cases of recurrent epileptic seizures caused by Endosulfan. *Neurology India*, 2003. **51**(1): p. 102.

101. Akhil, P. and C. Sujatha, Prevalence of organochlorine pesticide residues in groundwaters of Kasaragod District, India. *Toxicological & Environmental Chemistry*, 2012. **94**(9): pp. 1718–1725.

102. Beevi, S.N., et al., Pesticide residues in soils under cardamom cultivation in Kerala, India. *Pesticide Research Journal*, 2014. **26**(1): pp. 35–41.

103. Harikumar, P., et al., Persistence of endosulfan in selected areas of Kasaragod district, Kerala. *Current Science*, 2014. **106**: pp. 1421–1429.

104. Embrandiri, A., et al., An epidemiological study on the health effects of endosulfan spraying on cashew plantations in Kasaragod District, Kerala, India. *Asian Journal of Epidemiology*, 2012. **5**(1): p. 22.

105. Slamovits, T.L., R.M. Burde, and T.G. Klingele, Bilateral optic atrophy caused by chronic oral ingestion and topical application of hexachlorophene. *American Journal of Ophthalmology*, 1980. **89**(5): pp. 676–679.

106. Pirsaheb, M., et al., Organochlorine pesticides residue in breast milk: A systematic review. *Medical Journal of the Islamic Republic of Iran*, 2015. **29**: p. 228.

107. Jayaraj, R., P. Megha, and P. Sreedev, Organochlorine pesticides, their toxic effects on living organisms and their fate in the environment. *Interdisciplinary Toxicology*, 2016. **9**(3–4): pp. 90–100.

108. Sullivan, K., et al., Neuropsychological functioning in military pesticide applicators from the Gulf War: Effects on information processing speed, attention and visual memory. *Neurotoxicology and Teratology*, 2018. **65**: pp. 1–13.

109. Faiz, M.S., S. Mughal, and A.Q. Memon, Acute and late complications of organophosphate poisoning. *Journal of College of Physicians and Surgeons Pakistan*, 2011. **21**(5): pp. 288–90.

110. Jaga, K. and C. Dharmani, Sources of exposure to and public health implications of organophosphate pesticides. *Revistapanamericana de saludpública*, 2003. **14**: pp. 171–185.

111. Julien, R., et al., Pesticide loadings of select organophosphate and pyrethroid pesticides in urban public housing. *Journal of Exposure Science &Environmental Epidemiology*, 2008. **18**(2): pp. 167–174.

112. Michaelis, S., et al., *Aerotoxic syndrome: A new occupational disease? Public Health Panorama*, 2017. **3**(02): pp. 198–211.

113. Naughton, S.X. and A.V. Terry Jr., Neurotoxicity in acute and repeated organophosphate exposure. *Toxicology*, 2018. **408**: pp. 101–112.

114. London, L., et al., Suicide and exposure to organophosphate insecticides: Cause or effect? *American Journal of Industrial Medicine*, 2005. **47**(4): pp. 308–321.

115. Pandit, V., et al., A case of organophosphate poisoning presenting with seizure and unavailable history of parenteral suicide attempt. *Journal of Emergencies, Trauma and Shock*, 2011. **4**(1): p. 132.

116. Gunnell, D. and M. Eddleston, *Suicide by Intentional Ingestion of Pesticides: A Continuing Tragedy in Developing Countries*. 2003. Oxford University Press, Oxford.

117. Richardson, J.R., et al., Neurotoxicity of pesticides. *ActaNeuropathologica*, 2019. **138**: p. 1–20.

118. Costa, L.G., et al., Neurotoxicity of pesticides: A brief review. *Frontiers in Bioscience*, 2008. **13**(4): pp. 1240–1249.

119. Jokanović, M., Neurotoxic effects of organophosphorus pesticides and possible association with neuro-degenerative diseases in man: A review. *Toxicology*, 2018. **410**: pp. 125–131.

120. Badr, A.M., Organophosphate toxicity: Updates of malathion potential toxic effects in mammals and potential treatments. *Environmental Science and Pollution Research International*, 2020. **27**: pp. 26036–26057.

121. Eskenazi, B., A. Bradman, and R. Castorina, *Exposures of children to organophosphate pesticides and their potential adverse health effects. Environmental Health Perspectives*, 1999.**107**(suppl 3): pp. 409–419.

122. Ecobichon, D.J., *Organophosphorus ester insecticides. Pesticides and Neurological Diseases*, 1994. **1994**: pp. 171–250.

123. Rosenstock, L., et al., Chronic neuropsychological sequelae of occupational exposure to organophosphate insecticides. *American Journal of Industrial Medicine*, 1990. **18**(3): pp. 321–325.

124. Yokoyama, K., et al., Chronic neurobehavioral effects of Tokyo subway sarin poisoning in relation to posttraumatic stress disorder. *Archives of Environmental Health: An International Journal*, 1998. **53**(4): pp. 249–256.

125. Peter, J.V., T.I. Sudarsan, and J.L. Moran, Clinical features of organophosphate poisoning: A review of different classification systems and approaches. *Indian Journal of Critical Care Medicine: Peer-Reviewed, Official Publication of Indian Society of Critical Care Medicine*, 2014. **18**(11): p. 735.

126. Brahmi, N., et al., Prognostic value of human erythrocyte acetylcholinesterase in acute organophosphate poisoning. *The American Journal of Emergency Medicine*, 2006. **24**(7): pp. 822–827.

127. Reji, K.K., et al., Extrapyramidal effects of acute organophosphate poisoning. *Clinical Toxicology*, 2016. **54**(3): pp. 259–265.

128. Singh, S. and N. Sharma, Neurological syndromes following organophosphate poisoning. *Neurology India*, 2000. **48**(4): pp. 308.

129. Fiedler, N., et al., Long-term use of organophosphates and neuropsychological performance. *American Journal of Industrial Medicine*, 1997. **32**(5): pp. 487–496.

130. Takahashi, N. and M. Hashizume, A systematic review of the influence of occupational organophosphate pesticides exposure on neurological impairment. *BMJ Open*, 2014. **4**(6): p. e004798.

131. Kori, R.K., et al., Neurochemical and behavioral dysfunctions in pesticide exposed farm workers: A clinical outcome. *Indian Journal of Clinical Biochemistry*, 2018. **33**(4): pp. 372–381.

132. Rodnitzky, R.L., H.S. Levin, and D.L. Mick, Occupational exposure to organophosphate pesticides: A neurobehavioral study. *Archives of Environmental Health: An International Journal*, 1975. **30**(2): pp. 98–103.

133. Roldán-Tapia, L., T. Parrón, and F. Sánchez-Santed, Neuropsychological effects of long-term exposure to organophosphate pesticides. *Neurotoxicology and Teratology*, 2005. **27**(2): pp. 259–266.

134. Farahat, T., et al., Neurobehavioural effects among workers occupationally exposed to organophosphorous pesticides. *Occupational and Environmental Medicine*, 2003. **60**(4): pp. 279–286.

135. Roldan-Tapia, L., et al., Neuropsychological sequelae from acute poisoning and long-term exposure to carbamate and organophosphate pesticides. *Neurotoxicology and Teratology*, 2006. **28**(6): pp. 694–703.

136. Blanc-Lapierre, A., et al., Cognitive disorders and occupational exposure to organophosphates: Results from the PHYTONER study. *American Journal of Epidemiology*, 2013. **177**(10): pp. 1086–1096.

137. Council, N.R., *Pesticides in the Diets of Infants and Children*. 1993: National Academies Press, Washington, D.C.

138. Engel, S.M., et al., Prenatal organophosphate metabolite and organochlorine levels and performance on the Brazelton Neonatal Behavioral Assessment Scale in a multiethnic pregnancy cohort. *American Journal of Epidemiology*, 2007. **165**(12): pp. 1397–1404.

139. Young, J.G., et al., Association between *in utero* organophosphate pesticide exposure and abnormal reflexes in neonates. *Neurotoxicology*, 2005. **26**(2): pp. 199–209.

140. Handal, A.J., et al., Occupational exposure to pesticides during pregnancy and neurobehavioral development of infants and toddlers. *Epidemiology*, 2008. **19**: pp. 851–859.

141. Rauh, V.A., et al., Prenatal exposure to the organophosphate pesticide chlorpyrifos and childhood tremor. *Neurotoxicology*, 2015. **51**: pp. 80–86.

142. Eskenazi, B., et al., Association of *in utero* organophosphate pesticide exposure and fetal growth and length of gestation in an agricultural population. *Environmental Health Perspectives*, 2004. **112**(10): pp. 1116–1124.

143. Berkowitz, G.S., et al., *In utero* pesticide exposure, maternal paraoxonase activity, and head circumference. *Environmental Health Perspectives*, 2004. **112**(3): pp. 388–391.

144. Whyatt, R.M., et al., Prenatal insecticide exposures and birth weight and length among an urban minority cohort. *Environmental Health Perspectives*, 2004. **112**(10): pp. 1125–1132.

145. Sherman, J.D., Organophosphate pesticides—neurological and respiratory toxicity. *Toxicology and Industrial Health*, 1995. **11**(1): pp. 33–39.

146. Grandjean, P., et al., Pesticide exposure and stunting as independent predictors of neurobehavioral deficits in Ecuadorian school children. *Pediatrics*, 2006. **117**(3): p. e546–e556.

147. Rohlman, D.S., et al., Neurobehavioral performance in preschool children from agricultural and non-agricultural communities in Oregon and North Carolina. *Neurotoxicology*, 2005. **26**(4): pp. 589–598.

148. Ruckart, P.Z., et al., Long-term neurobehavioral health effects of methyl parathion exposure in children in Mississippi and Ohio. *Environmental Health Perspectives*, 2004. **112**(1): pp. 46–51.

149. Engel, S.M., et al., Prenatal exposure to organophosphates, paraoxonase 1, and cognitive development in childhood. *Environmental Health Perspectives*, 2011. **119**(8): pp. 1182–1188.

150. Huen, K., et al., PON1 DNA methylation and neurobehavior in Mexican-American children with prenatal organophosphate exposure. *Environment International*, 2018. **121**: pp. 31–40.

151. Eskenazi, B., et al., PON1 and neurodevelopment in children from the CHAMACOS study exposed to organophosphate pesticides *in utero*. *Environmental Health Perspectives*, 2010. **118**(12): pp. 1775–1781.

152. Jusko, T.A., et al., Organophosphate pesticide metabolite concentrations in urine during pregnancy and offspring nonverbal IQ at age 6 years. *Environmental Health Perspectives*, 2019. **127**(1): p. 017007.

153. Bradman, A., et al., Organophosphate urinary metabolite levels during pregnancy and after delivery in women living in an agricultural community. *Environmental Health Perspectives*, 2005. **113**(12): pp. 1802–1807.

154. Kongtip, P., et al., Organophosphate urinary metabolite levels during pregnancy, delivery and postpartum in women living in agricultural areas in Thailand. *Journal of Occupational Health*, 2013. **55**: pp. 367–375.

155. Bai, X.-Y., et al., A pilot study of metabolites of organophosphorus flame retardants in paired maternal urine and amniotic fluid samples: potential exposure risks of tributyl phosphate to pregnant women. *Environmental Science: Processes & Impacts*, 2019. **21**(1): pp. 124–132.

156. Bouwman, H., B. Sereda, and H. Meinhardt, Simultaneous presence of DDT and pyrethroid residues in human breast milk from a malaria endemic area in South Africa. *Environmental Pollution*, 2006. **144**(3): pp. 902–917.

157. El-Demerdash, F.M., Lambda-cyhalothrin-induced changes in oxidative stress biomarkers in rabbit erythrocytes and alleviation effect of some antioxidants. *Toxicology in Vitro*, 2007. **21**(3): pp. 392–397.

158. Fetoui, H., et al., Exposure to lambda-cyhalothrin, a synthetic pyrethroid, increases reactive oxygen species production and induces genotoxicity in rat peripheral blood. *Toxicology and Industrial Health*, 2015. **31**(5): pp. 433–441.

159. Wu, C., et al., Urinary metabolite levels of pyrethroid insecticides in infants living in an agricultural area of the Province of Jiangsu in China. *Chemosphere*, 2013. **90**(11): pp. 2705–2713.

160. Bradberry, S.M., et al., Poisoning due to pyrethroids. *Toxicological Reviews*, 2005. **24**(2): pp. 93–106.

161. Soderlund, D.M., Molecular mechanisms of pyrethroid insecticide neurotoxicity: Recent advances. *Archives of Toxicology*, 2012. **86**(2): pp. 165–181.

162. Soderlund, D.M., Toxicology and mode of action of pyrethroid insecticides. *Hayes' Handbook of Pesticide Toxicology* (R. I. Krieger Ed.). Elsevier, Amsterdam, 2010: pp. 1665–1686.

163. Casida, J.E., Mechanisms of selective action of pyrethroids. *Annual Review of Pharmacology and Toxicology*, 1983. **23**: pp. 413–438.

164. Miyamoto, J., et al., Pyrethroids, nerve poisons: How their risks to human health should be assessed. *Toxicology Letters*, 1995. **82**: pp. 933–940.

165. He, F., Synthetic pyrethroids. *Toxicology*, 1994. **91**(1): pp. 43–49.

166. Yang, P.-Y., et al., Acute ingestion poisoning with insecticide formulations containing the pyrethroidpermethrin, xylene, and surfactant: A review of 48 cases. *Journal of Toxicology: Clinical Toxicology*, 2002. **40**(2): pp. 107–113.

167. Müller-Mohnssen, H., Chronic sequelae and irreversible injuries following acute pyrethroid intoxication. *Toxicology Letters*, 1999. **107**(1–3): pp. 161–176.

168. Godin, S.J., et al., Physiologically based pharmacokinetic modeling of deltamethrin: development of a rat and human diffusion-limited model. *Toxicological Sciences*, 2010. **115**(2): pp. 330–343.

169. Berkowitz, G.S., et al., Exposure to indoor pesticides during pregnancy in a multiethnic, urban cohort. *Environmental Health Perspectives*, 2003. **111**(1): pp. 79–84.

170. Morgan, M.K., et al., An observational study of 127 preschool children at their homes and daycare centers in Ohio: environmental pathways to cis-and trans-permethrin exposure. *Environmental Research*, 2007. **104**(2): pp. 266–274.

171. Lu, C., et al., A longitudinal approach to assessing urban and suburban children's exposure to pyrethroid pesticides. *Environmental Health Perspectives*, 2006. **114**(9): pp. 1419–1423.

172. Lu, C., et al., The attribution of urban and suburban children's exposure to synthetic pyrethroid insecticides: A longitudinal assessment. *Journal of Exposure Science & Environmental Epidemiology*, 2009. **19**(1): pp. 69–78.

173. Richardson, J.R., et al., Developmental pesticide exposure reproduces features of attention deficit hyperactivity disorder. *The FASEB Journal*, 2015. **29**(5): pp. 1960–1972.

174. Morais, S., E. Dias, and M. Pereira, Carbamates: Human exposure and health effects. *The Impact of Pesticides* (M. Jokanovi Ed.).Academic Press, Cheyenne, WY, 2012: pp. 21–38.
175. Darvesh, S., et al., Carbamates with differential mechanism of inhibition toward acetylcholinesterase and butyrylcholinesterase. *Journal of Medicinal Chemistry*, 2008. **51**(14): pp. 4200–4212.
176. Fukuto, T.R., Mechanism of action of organophosphorus and carbamate insecticides. *Environmental Health Perspectives*, 1990. **87**: pp. 245–254.
177. Muscat, P. and M. Attard, *A case report and overview of Carbamate insecticide (Baygon) poisoning.*2017.
178. Burgess, J.L., J.N. Bernstein, and K. Hurlbut, Aldicarb poisoning: A case report with prolonged cholinesterase inhibition and improvement after pralidoxime therapy. *Archives of Internal Medicine*, 1994. **154**(2): pp. 221–224.
179. Rajbanshi, L.K., Transdermal carbamate poisoning–a case of misuse. *Journal of College of Medical Sciences-Nepal*, 2016. **12**(4): pp. 187–188.
180. Lifshitz, M., et al., Carbamate poisoning and oxime treatment in children: A clinical and laboratory study. *Pediatrics*, 1994. **93**(4): pp. 652–655.
181. Calumpang, S., Applicator exposure to the insecticides deltamethrin, cypermethrin, imidacloprid, and profenofos sprayed on crops of different canopy heights. *Philippine Agricultural Scientist (Philippines)*, 2003. **86**: pp. 266–281.
182. Phua, D.H., et al., Neonicotinoid insecticides: An emerging cause of acute pesticide poisoning. *Clinical Toxicology*, 2009. **47**(4): pp. 336–341.
183. Shadnia, S. and H.H. Moghaddam, Fatal intoxication with imidacloprid insecticide. *The American Journal of Emergency Medicine*, 2008. **26**(5): pp. 634.e1–634.e4.
184. David, D., I.A. George, and J.V. Peter, Toxicology of the newer neonicotinoid insecticides: Imidacloprid poisoning in a human. *Clinical Toxicology*, 2007. **45**(5): pp. 485–486.
185. Panigrahi, A.K., D. Subrahmanyam, and K.K. Mukku, Imidacloprid poisoning: A case report. *The American Journal of Emergency Medicine*, 2009. **27**(2): pp. 256.e5–256.e6.
186. Iyyadurai, R., I.A. George, and J.V. Peter, *Imidacloprid poisoning—newer insecticide and fatal toxicity.* Journal of Medical Toxicology, 2010. **6**(1): pp. 77–78.

6 Insufficient Evidence for Pesticide Safety

Andre Leu
Regeneration International, Daintree

CONTENTS

6.1 INTRODUCTION

There is no scientific proof of safety for the current pesticides, additives, and chemicals in our food, body care products, and household products. Most are not tested, and where there is testing, it is designed to miss the vast majority of diseases at the normal rates at which they occur. The testing methodologies can only indicate if a compound causes health problems; however, they are not sensitive enough to prove safety for significant percentages of the human population.

According to the World Health Organization (WHO), there is a global epidemic of noncommunicable chronic diseases (NCDs) such as heart disease, stroke, cancer, chronic respiratory diseases, and diabetes. WHO states that NCDs diseases are the leading cause of mortality in the world. "This invisible epidemic is an under-appreciated cause of poverty and hinders the economic development of many countries. The burden is growing — the number of people, families and communities afflicted is increasing" [1].

You cannot be infected by these diseases from other people. You will not get cancer, heart disease, or diabetes from sitting next to people with these diseases. The major causes are environmental factors and lifestyle.

This means that we can prevent them by changing our habits, food, farming practices; reducing industrial pollution; and avoiding the environmental exposures and lifestyle factors that cause them.

Pesticides and chemicals are strongly implicated in this global epidemic; however, the full extent of their role is being ignored by researchers and health professionals. This is because the current best practice testing guidelines for pesticides, food additives, and chemicals are designed to miss the majority of diseases. It is important to look at these guidelines to understand why.

6.2   THE BEST PRACTICES TESTING GUIDELINES

The Organisation for Economic Co-operation and Development (OECD) Guidelines for the Testing of Chemicals are regarded as best practice for testing animals for diseases caused by chemicals such as pesticides and are similar to most good practice testing guidelines.

Guideline 451 of the OECD is used for testing chemicals, such as pesticides, causing cancers. It requires that "Each dose group and concurrent control group should therefore contain at least 50 animals of each sex." This is a group of 100 animals, with an equal amount of males and females. The guidelines also state that "At least three dose levels and a concurrent control should be used" [2].

This means that there must be a group of 100 animals, usually rats, that are the control and are not dosed with the chemical. There will be three other groups of 100 rats in each group that are given a dosage of the chemical in the highest, middle, and lowest levels. The numbers of cancers in each of the dosed groups are compared with the numbers of cancers in the control group of rats. If the numbers of cancers are the same between the treated and the control groups, then it is considered that the cancers were not caused by the chemical. It is assumed that the cancers were caused by another factor, as the control group has not been exposed to the chemical and therefore it cannot be the cause of the cancers in the control group.

This result shows that a chemical or pesticide does not cause cancer.

However, there are serious flaws in this method. One of the dosed groups of animals with one extra cancer than the control results one animal in 100 with cancer. This is the lowest theoretical rate of detection, and it means that cancer would only be detected if the pesticide caused more than 1,000 people per 100,000 people to get cancer. It would miss lower rates of cancer, which are the actual rates of cancers.

The rates of diseases are categorized by the number of people with the disease per 100,000 people. According to the Centers for Disease Control and Prevention (CDC), in the United States, the rates of common cancers such as lung cancer are 57.5 people per 100,000, colon and rectum cancers 38 per 100,000, non-Hodgkin lymphoma 18.4 per 100,000, leukemias 13.2 per 100,000, pancreatic cancer 12.8 per 100,000, and liver and intrahepatic bile duct cancers 8.3 per 100,000 [3]. CDC data for cancers of the sexual tissues in 2015 show that the rate of breast cancer was 124.8 per 100,000 woman, prostate cancer was 99.1 per 100,000 men, ovarian cancer was 11 per 100,000 woman, cervical cancer 7.6 per 100.000 woman, and testicular cancer 5.6 per 100.000 men [3].

Consequently, despite no evidence of cancer being found in the dosed groups, the study would miss a chemical that can cause the most common cancers. There is no statistically valid way to determine if the chemical in question can cause cancer in a dosed group of 100 animals, which showed no sign of cancer priorly, at rates below 1,000 people per 100,000. All of the cancers currently found in our communities will be missed.

6.3   DISEASES OTHER THAN CANCER

Guideline 408 of the OECD is used for toxicology testing for diseases. It requires that "at least 20 animals (ten female and ten male) should be used at each dose level." Like the cancer guideline 451, guideline 408 states: "At least three dose levels and a concurrent control should be used." Under guideline 408, 1 in 20 animals with a disease means that the disease could only be detected to a minimum of 5,000 cases per 100,000 people. For sex-specific diseases such as endometriosis and declines in fertility, the level of disease detection will be increased to 10,000 cases per 100,000 people. This means that if the highest dose group finds no evidence of disease, it can easily miss a chemical that can cause a disease epidemic [2].

Most importantly, the OECD guidelines cannot test for most of the diseases that afflict our communities. For example, the CDC gives the following numbers for some of the major diseases in the

United States: 1,600 people per 100,000 have a liver disease, a kidney disease is in 2,000 people per 100,000, and stroke affects 3,000 people per 100,000. These diseases will be missed by the current best practice guidelines that can only detect diseases to a minimum of 5,000 cases per 100,000 people. The only way this can be done statistically is to have a greater number of test animals [4].

6.4 SERIOUS DEFICIENCIES IN THE REGULATION OF TOXIC CHEMICALS

There are many serious deficiencies in the regulation of toxic chemicals used in our food supply, and much of the criteria used to underpin the current use patterns are based on out-of-date assumptions rather than on the latest published science [5].

The scientific credibility of pesticide regulatory authorities has to be seriously questioned when they are approving the use of pesticides on the basis of data-free assumptions.

A good example of this is the approval of formulated pesticide products as safe on the basis of just testing one of the ingredients without testing the whole formulation. Given that the other chemical ingredients are chemically active as they are added to the formulations to make the active ingredient work more effectively, the assumption that they are inert and will not increase the toxicity of the whole formulation lacks scientific credibility. Apart from limited testing criteria in the European Union, there are no requirements to test the toxicity of the whole formulation to generate credible evidence-based scientific data [5].

Regulatory authorities approve several different pesticides for a crop—such as herbicides, fungicides, and insecticides—on the basis that all of them can be used in the normal production of the crop. Consequently, multiple residues will be found in the crop; residue testing found that 47.4% of food in the United States had two or more pesticide residues. The current approval process of testing each pesticide separately is based on the assumption that if each chemical is safe individually, then the combinations of these chemicals are also safe. There are a number of published scientific studies showing that combinations of pesticide residues can cause serious adverse health outcomes due to additive or synergistic effects. The failure to test the combinations of approved pesticides for potential health risks means that regulatory authorities do not have any evidence-based data indicating that these residue combinations are safe [5].

The lack of testing for the metabolites formed by pesticides as they degrade, given that limited testing shows that many of them are more toxic and residual than the pesticide itself, is another massive data gap.

The setting of the acceptable daily intake (ADI) is another example. Given that there are hundreds of studies showing that many chemicals can be endocrine disruptors and therefore more toxic at lower doses, setting the ADI on the basis of extrapolating it from testing done at higher doses is another data-free assumption. The only way to ensure that the ADI is safe and does not act as an endocrine disruptor is to do the testing at the actual residue levels that are set for the ADI.

The special requirements of the fetus, the newborn, and the growing child in relation to developmental neurotoxicity are also subject to data-free assumptions. Currently, the pesticide testing used in the regulatory approval processes does not specifically test for any of the risks particular to these age groups, and the ADIs are set based on the testing of adolescent animals. Until testing is specifically designed to assess the dangers to the developing fetus and the very young, there is no evidence-based data specific to this age group.

It is the same with intergenerational effects. Unless testing is done over several generations, especially on organs and physiological processes, these is no data to show that the current ADIs will not cause health problems for the future generations. There are many scientific studies showing that exposure to pesticide residues cause adverse health problems in future generations, so ignoring this issue could prove dangerous.

The regulation of pesticides should be based on the data generated through credible scientific studies and testing, not on data-free assumptions as it is currently. Additional testing needs to be done for

- Mixtures and cocktails of chemicals
- The actual formulated products, not just the active ingredients
- The toxicity of pesticide metabolites
- The special requirements of fetuses, newborns, and growing children
- Endocrine disruption
- Metabolic disruption
- Intergenerational effects on all organs and physiological systems
- Developmental neurotoxicity.

Until this is done, regulatory bodies have no credible scientific evidence backing a statement that any level of pesticide residue is safe for humans or the environment.

6.5   NO PUBLISHED EVIDENCE OF PESTICIDE SAFETY IN CHILDREN

The official requirements by regulatory authorities for specific testing for pesticide-induced diseases in children are almost non-existent. The OECD guidelines state that "Young healthy adult animals of commonly used laboratory strains should be employed." The fetus, baby, and pubertal animals (i.e., children) are not tested [2].

This means that there will be no data on the safety of pesticides and other chemicals for children in the experiments that use these OECD and similar guidelines, which are the majority of tests.

The developing fetus, young children, and children going through puberty are three very critical periods in the development of humans and are completely ignored under the guidelines for diseases, endocrine disruption and cancer. These is no published scientific evidence-based testing to show that any of the current chemicals and pesticides are safe for our children, because there is no requirement to specifically test for them.

The current best practice testing guidelines will fail to detect if chemical/chemicals are causing this massive epidemic in our children. On the other hand, there is a large body of published, peer-reviewed scientific research that shows that pesticide exposure in unborn and growing children is linked to

- Autism spectrum disorders
- Lower IQs
- Attention deficit hyperactive disorder (ADHD)
- Cancers
- Thyroid disorders
- Endocrine disruption
- Immune system problems
- Lack of physical coordination
- Loss of temper—anger management issues
- Bipolar/schizophrenia spectrum of illnesses
- Depression
- Digestive system problems
- Cardiovascular disease
- Reproductive problems (as adults)
- Deformities of the genital-urinary systems
- Changes to metabolic systems, including childhood obesity and diabetes [5].

A good example is the autism epidemic in the developed world. According to the U.S. CDC, in 2014, the rates of autism were 1,680 children per 100,000 or 1 child in 59. In 2000, it was 670 children per 100,000 or 1 child in 150. This is a startling 250% increase in 14 years [6].

The OECD guideline 408 and others that use 20 animals per test group will miss the epidemic because their limit of detection of a disease is 5,000 cases per 100,000.

6.6 THE SPECIAL NEEDS OF THE DEVELOPING FETUS AND NEWBORN

Many scientific researchers have expressed concern that the current pesticide testing methodologies are grossly inadequate for children. The U.S. President's Cancer Panel (USPCP) report, written by eminent scientists and medical specialists from the U.S. Department of Health and Human Services, the National Institutes of Health, and the National Cancer Institute, stated, "They (children) are at special risk due to their smaller body mass and rapid physical development, both of which magnify their vulnerability to known or suspected carcinogens, including radiation."

This is a critical issue as there is a large body of published science showing that the fetus and the newborn are continuously being exposed to numerous chemicals. The USPCP stated,

> Some of these chemicals are found in maternal blood, placental tissue, and breast milk samples from pregnant women and mothers who recently gave birth. These findings indicate that chemical contaminants are being passed on to the next generation, both prenatally and during breastfeeding
>
> Numerous environmental contaminants can cross the placental barrier; to a disturbing extent, babies are born 'pre-polluted.' Children also can be harmed by genetic or other damage resulting from environmental exposures sustained by the mother (and in some cases, the father). There is a critical lack of knowledge and appreciation of environmental threats to children's health and a severe shortage of researchers and clinicians trained in children's environmental health. [7]

A number of studies show the link between chemical exposure, particularly exposure to pesticides, and the increase of cancer in children. The USPCP report states, "Cancer incidence in U.S. children under 20 years of age has increased" [7].

The information from USCP shows that current regulatory systems have failed to protect unborn and growing children from exposure to a massive cocktail of toxic pesticides. This has many serious implications especially the increase in a range of serious health issues in children and as adults later in life.

6.7 DEVELOPMENTAL NEUROTOXICITY

Scientific research shows that many pesticides affect the normal development of the nervous system in fetuses and children. The brain is the largest collection of nerve cells, and there are several scientific studies showing that when the fetus and the newborn are exposed to minute amounts of these pesticides, below the current limits set by regulatory authorities, they can significantly alter brain function [8].

Researchers at the Duke University Medical Center found that the developing fetus and the newborn are particularly vulnerable to amounts of pesticide that are lower than the levels currently permitted by regulatory authorities around the world. Their studies showed that the fetus and the newborn possess lower concentrations of the protective serum proteins than adults. A major consequence is developmental neurotoxicity, where the poison damages the developing nervous system. This damage interferes with the normal development of the brain and other parts of the nervous systems such as auditory nerves, optic nerves, and the autonomous nervous system resulting in lower IQs, ADHD, autism spectrum disorders, lack of physical coordination, loss of temper—anger management issues, bipolar/schizophrenia spectrum of illnesses, and depression as well as problems with eyesight and hearing.

This means that contact with chemicals at levels well below the current permitted residues in food can harm the fetus and breastfeeding children, even if the mother shows no side effects from the contact. Eating food with pesticide residues can harm young children as they are still developing their nervous systems.

6.8   BRAIN ABNORMALITIES AND IQ REDUCTIONS IN CHILDREN

Studies conducted independently by researchers at the Columbia University Center for Children's Environmental Health, the University of California, Berkeley, and the Mount Sinai School of Medicine found that fetal exposure to small amounts of organophosphate pesticides caused a range of brain abnormalities that resulted in children with reduced IQs, lessened attention spans, and increased vulnerability to ADHD [9–11].

Parents should have considerable concern that the studies found no evidence of a lower-limit threshold of exposure to organophosphates in the observed adverse impact on intelligence. This means that even very low levels of exposure could lead to reductions in a child's intelligence.

The study by Rauh et al. published in the journal Proceedings of the National Academy of Sciences of the United States of America has confirmed the findings of the previous studies. The researchers used MRI scans that revealed a large range of visible brain abnormalities present in children who had been exposed to chlorpyrifos (CPF) in utero through normal, nonoccupational uses [12].

Some of the most concerning studies show that pesticide damage can be passed on to the next generation. Not only are the offspring born with damage to the nervous system, the reproductive system, and other organs, the great grandchildren can be as well [13–15].

Researchers in a 2012 study found that pregnant rats and mice exposed to the fungicide vinclozolin during the period of reproductive organs development in the fetus found that spermatogenic cell defects, testicular abnormalities, prostate abnormalities, kidney abnormalities, and polycystic ovarian disease were significantly increased in future generations [13].

Another study showed that when pregnant rats were exposed to a combination of permethrin, a common insecticide, and DEET (N,N-diethyl-meta-toluamide), the most common insect repellent, pubertal abnormalities, testis disease, and ovarian disease (primordial follicle loss and polycystic ovarian disease) were increased in future generations [14].

The critical issue with these two studies is that small exposures to pesticides at critical times in the development of the fetus can cause multiple diseases that are passed onto future generations. It means that pregnant women eating food with minute levels of pesticides could be inadvertently exposing their children, grandchildren, and great grandchildren to permanent damage to their reproductive systems and other organs.

6.9   ENDOCRINE DISRUPTION

Children are more vulnerable than adults to the effects of endocrine disrupters because their tissues and organs are still developing and are reliant on balanced hormone signals to ensure that they develop in orderly sequences. Small disruptions in these hormone signals by endocrine-disrupting chemicals (EDCs) can significantly alter the way these body parts and metabolic systems will develop. These altered effects will not only last a lifetime; they can also be passed on to future generations [16–19].

A meta-study by United Nations World Health Organization (WHO) and the United Nations Environmental Program (UNEP) written by over 60 recognized international experts who worked throughout 2012 to contribute to the meta-analysis to ensure that it was an up-to-date compilation of the current scientific knowledge on EDCs, including pesticides, found that

...we now know that there are particularly vulnerable periods during fetal and postnatal life when EDCs alone, or in mixtures, have strong and often irreversible effects on developing organs, whereas exposure of adults causes lesser or no effects. Consequently, there is now a growing probability that maternal, fetal and childhood exposure to chemical pollutants play a larger role in the etiology of many endocrine diseases and disorders of the thyroid, immune, digestive, cardiovascular, reproductive and metabolic systems (including childhood obesity and diabetes) than previously thought possible [16].

The fetus is most vulnerable during the times when genes are turned on to develop specific organs. Small amounts of hormones give the signals to genes to start developing various body parts and systems such as the reproductive tract, the nervous system, the brain, immune system, hormone systems, and limbs. Small disruptions in these hormone signals can significantly alter the way these body parts and systems will develop, and these altered effects will last a lifetime.

This does not diminish their (EDCs) importance (in adults), but contrasts with their effects in the fetus and neonate where a hormone can have permanent effects in triggering early developmental events such as cell proliferation or differentiation. Hormones acting during embryonic development can, cause some structures to develop (e.g. male reproductive tract) or cause others to diminish (e.g. some sex-related brain regions). Once hormone action has taken place, at these critical times during development, the changes produced will last a lifetime [16].

The actions of EDCs on the development of endocrine and physiological system in fetuses are considered to be programming events. They set how these systems will function in adults. The WHO and UNEP study found that up to 40% of young men in some countries have low semen quality as well as an increase in genital malformations in baby boys such as non-descending testes and penile malformations. There is an increase in adverse pregnancy outcomes such as preterm birth and low birth weight. There is an increase in neurobehavioral disorders in children that is associated with thyroid disruption. The age of breast development in girls is decreasing, and this is considered a risk factor in developing breast cancer later in life. Breast, endometrial, ovarian, prostate, testicular, and thyroid cancers are increasing. These are endocrine system–related cancers [16–19].

6.10 PROTECTING OUR CHILDREN

The current regulatory systems have failed to protect children and the wider population from diseases caused by pesticides. It is important to remember that the majority of people get their exposure to pesticides from food. Most people, including children, carry a body burden of cocktail of these toxic chemicals with no scientific evidence that they are safe. However, there is ample evidence that these chemicals are harming our children [5].

Currently, for consumers the only way to avoid synthetic pesticides is to eat organically grown food. Most children get their pesticide exposure either directly by consuming food with pesticide residues or though the placenta and breast milk due to the pesticides in their mothers' food. Several scientific studies show that eating organic food is the best way to protect children as most pesticide exposure comes from eating food from conventional farming systems.

A study published in Environmental Health Perspectives found that children who eat organic fruits, vegetables, and juices can significantly lower the levels of organophosphate pesticides in their bodies. The University of Washington researchers who conducted the study concluded,

The dose estimates suggest that consumption of organic fruits, vegetables, and juice can reduce children's exposure levels from above to below the U.S. Environmental Protection Agency's current guidelines, thereby shifting exposures from a range of uncertain risk to a range of negligible risk. Consumption of organic produce appears to provide a relatively simple way for parents to reduce their children's exposure to OP (organophosphate) pesticides [20].

Researchers in a 2006 study found that the urinary concentrations of the specific metabolites for malathion and chlorpyrifos decreased to undetectable levels immediately after the introduction of organic diets and remained undetectable until the conventional diets were reintroduced. The researchers from Emory University, Atlanta, Georgia; the University of Washington, Seattle, Washington; and the Centers for Disease Control and Prevention, Atlanta, Georgia, stated,

> In conclusion, we were able to demonstrate that an organic diet provides a dramatic and immediate protective effect against exposures to organophosphorus pesticides that are commonly used in agricultural production. We also concluded that these children were most likely exposed to these organophosphorus pesticides exclusively through their diet [21].

6.11 CONCLUSION

The fact is that the current testing protocols can only tell us if a pesticide causes a disease. It cannot tell us if a pesticide is safe. Finding no evidence that a pesticide does not cause cancer, autism, or other diseases in a study is not the same as saying that the chemical in question does not cause these diseases.

In my opinion, it is a gross misrepresentation to say that any of the current published toxicology studies can be used to say that any of the thousands of pesticide products used in the world do not cause cancer and other diseases. The fact is there is no evidence that they are safe.

Very concerning is that when they can cause diseases, the number of people who can be affected by the chemical is extraordinarily high.

REFERENCES

1. World Health Organization, The Global Health Observatory, Noncommunicable Diseases, https://www.who.int/data/gho/data/themes/noncommunicable-diseases, accessed March 30, 2021.
2. Organisation for Economic Co-operation and Development (OECD) Guidelines for the Testing of Chemicals, Section 4, Health Effects, ISSN: 20745788 (online) https://doi.org/10.1787/20745788, accessed May 12, 2020.
3. U.S. Cancer Statistics Working Group. U.S. Cancer Statistics Data Visualizations Tool, based on November 2017 submission data (1999–2015): U.S. Department of Health and Human Services, Centers for Disease Control and Prevention and National Cancer Institute; www.cdc.gov/cancer/dataviz, June 2018.
4. Chronic Disease Data, Centers for Disease Control and Prevention, https://www.cdc.gov/chronicdisease/data/statistics.htm, accessed Oct 31, 2019.
5. Leu, Andre, Poisoning our children, the parent's guide to the myths of safe pesticides, Acres U.S.A. Greely, Colorado, USA 2018, ISBN 978-1-601-73140-1.
6. Data & Statistics on Autism Spectrum Disorder, Content source: National Center on Birth Defects and Developmental Disabilities, Centers for Disease Control and Prevention, https://www.cdc.gov/ncbddd/autism/data.html, Page last reviewed: September 3, 2019.
7. "U.S. President's Cancer Panel 2008–2009 Annual Report; Reducing Environmental Cancer Risk: What We Can Do Now." Suzanne H. Reuben for the President's Cancer Panel, U.S. Department Of Health And Human Services, National Institutes of Health, National Cancer Institute, April 2010.
8. Qiao, Dan, Frederic Seidler, and Theodore Slotkin. "Developmental neurotoxicity of chlorpyrifos modeled in vitro: Comparative effects of metabolites and other cholinesterase inhibitors on DNA synthesis in PC12 and C6 cells." *Environmental Health Perspectives* 109, no. 9 (September 2001): 909–913.
9. Rauh, Virginia, Srikesh Arunajadai, Megan Horton, Frederica Perera, Lori Hoepner, Dana B. Barr, and Robin Whyatt. "7-year neurodevelopmental scores and prenatal exposure to chlorpyrifos, a common agricultural insecticide." *Environmental Health Perspectives*, 119 (2011): 1196–1201. Published online April 21, 2011.
10. Pastor, Patricia N. and Cynthia A. Reuben, "Diagnosed attention deficit hyperactivity disorder and learning disability: United States, 2004–2006." *National Center for Health Statistics, Vital and Health Statistics*, 10, no. 237 (July 2008): 1–14.
11. Engel, Stephanie M., James Wetmur, Jia Chen, Chenbo Zhu, Dana Boyd Barr, Richard L. Canfield, and Mary S. Wolff. "Prenatal exposure to organophosphates, paraoxonase 1, and cognitive development in children." *Environmental Health Perspectives* 119 (2011): 1182–1188. Published online April 21, 2011, http://ehp.niehs.nih.gov/1003183/.

12. Rauh, Virginia, Frederica P. Perera, Megan K. Horton, Robin M. Whyatt, Ravi Bansal, Xuejun Hao, Jun Liu, Dana Boyd Barr, Theodore A. Slotkin, and Bradley S. Peterson. "Brain anomalies in children exposed prenatally to a common organophosphate pesticide." *Proceedings of the National Academy of Sciences of the United States of America* 109, no. 20 (May 2012): 7871–7876.

13. Manikkam, Mohan, Carlos Guerrero-Bosagna, Rebecca Tracey, Md. M. Haque, and Michael K. Skinner. "Transgenerational actions of environmental compounds on reproductive disease and identification of epigenetic biomarkers of ancestral exposures." *PLoS ONE* 7, no. 2 (February 2012): e31901.

14. Manikkam, Mohan, Rebecca Tracey, Carlos Guerrero-Bosagna, and Michael K. Skinner. "Pesticide and insect repellent mixture permethrin and DEET induces epigenetic transgenerational inheritance of disease and sperm epimutations." *Journal of Reproductive Toxicology* 34, no. 4 (December 2012): 708–719.

15. Guerrero-Bosagna, Carlos, Trevor R. Covert, Matthew Settles, Matthew D. Anway, and Michael K. Skinner. "Epigenetic transgenerational inheritance of vinclozolin induced mouse adult onset disease and associated sperm epigenome biomarkers." *Reproductive Toxicology* 34, no. 4 (December 2012): 694–707.

16. Bergman, Åke, Jerrold J. Heindel, Susan Jobling, Karen A. Kidd, and R. Thomas Zoeller, (eds.) State of the Science of Endocrine Disrupting Chemicals 2012. United Nations Environment Programme and the World Health Organization, 2013.

17. Cadbury, Deborah. *The Feminization of Nature: Our Future at Risk*. Middlesex, England: Penguin Books, 1998.

18. Colborn, Theo, Dianne Dumanoski, and John Peterson Myers. *Our Stolen Future: Are We Threatening Our Fertility, Intelligence, and Survival? A Scientific Detective Story*. New York: Dutton, 1996.

19. Vandenberg, Laura N., Theo Colborn, Tyrone B. Hayes, Jerrold J. Heindel, David R. Jacobs Jr., Duk-Hee Lee, Toshi Shioda, Ana M. Soto, Frederick S. vom Saal, Wade V. Welshons, R. Thomas Zoeller, and John Peterson Myers. "Hormones and endocrine-disrupting chemicals: Low-dose effects and nonmonotonic dose responses." *Endocrine Reviews* 33, no. 3 (June 2012): 378–455. First published ahead of print March 14, 2012, as doi:10.1210/er.2011-1050 (Endocrine Reviews 33: 0000-0000, 2012).

20. Curl, Cynthia, Richard A. Fenske, and Kai Elgethun. "Organophosphorus pesticide exposure of urban and suburban preschool children with organic and conventional diets." *Environmental Health Perspectives* 111, no. 3 (March 2003): 377–382.

21. Lu, Chensheng, Kathryn Toepel, Rene Irish, Richard A. Fenske, Dana B. Barr, and Roberto Bravo. "Organic diets significantly lower children's dietary exposure to organophosphorus pesticides." *Environmental Health Perspectives* 114, no. 2 (February 2006): 260–263.

7 Pesticides and the Crisis in Children's Health

Michelle Perro

CONTENTS

We owe it to ourselves and to the next generation to conserve the environment so that we can bequeath our children a sustainable world that benefits all.

Dr. Wangari Maathai

7.1 INTRODUCTION

The role of pesticides as a major factor and prime cause in the downward trajectory of our children's health has been known for decades.[1] There are innumerable studies that have documented how pesticides work and why they are so dangerous for our children. Yet, agribusinesses and government officials ignore both the well-vetted scientific and clinical information available and continue to promote pesticide usage despite these alarming negative trends in our children's health globally.

The goal of this chapter is to provide an understanding of the present health crisis facing our children via an overview and update of the status of their health. While there are many disorders and diseases now facing our children, several disorders have been selected due to either their prevalence (obesity), lack of awareness of their existence (intestinal/liver), and because of their rapidly escalating incidence (neurocognitive dysfunctional disorders such as autism spectrum disorder or ASD). This small representation of the vast number of challenges facing our children are pivotal examples of health consequences and the cost of convenience from decades of pesticide applications. New studies and clinical perspectives will be presented highlighting the links between these diseases and the most commonly applied pesticides.

The narrative of this chapter will hone in on the particular challenges of pesticide usage and their effects on African children. Research demonstrating the health benefits of food free from chemical inputs is discussed to conclude with practical and applicable solutions away from chemical farming in Africa that would benefit and promote well-being in children's lives.

7.2 DEFINITIONS

Pesticides are a broad term and encompass various classes of killing agents including herbicides, insecticides, fungicides, rodenticides, and bactericides (disinfectants). It is unusual to be exposed to only one of these chemicals individually and, more commonly, multiple chemicals are applied concomitantly. The toxic effects of this "chemical soup" have been inadequately studied.[2] It should be clearly understood that studies are often reporting on one chemical only. Additionally, an entire pesticide product with the "inert" adjuvants is not studied along with the "active" ingredient. The true toxic picture on children's health from multiple chemical exposures simultaneously is unknown.

7.3 TOXIC SOUP

Presently, over 85,000 chemicals are used in the United States causing exposures to both known and potential health hazards. These chemicals are detected in various body fluids of almost all Americans and in the global population as well.[3] They are categorized based on the type of usage, route of exposure, toxicologic effects, and their longevity/half-lives in biologic tissue.[4] The scope of the problem of trying to monitor blended effects from chemical mixtures is unwieldy. However, one can narrow down the largest groups employed and that cause some of the greatest harm. Those chemicals are pesticides.

7.4 PUBLIC HEALTH – A PRECARIOUS BALANCE

A particular problem faced by African children that confounds the issue with pesticide application is the usage of pesticides in public health. There are serious diseases spread by animal and insect vectors (malaria, dengue fever, Zika virus, Lyme disease, etc.), and these organisms can contaminate food crops/processing facilities, homes, and schools. In order to protect children from these health risks, pesticides have played a vital role.

A cogent example of the employment of pesticides in the treatment of vector-borne illness, such as malaria, is the usage of insecticide-treated mosquito nets as promoted by the World Health Organization (WHO). In 2015, 438,000 people were killed by malaria and 70% of those killed were children under the age of 5 years.[5] However, even as this measure saved so many lives, a study in 2002 demonstrated mosquito resistance to pyrethroids, the class of insecticide used in the netting, and recommended the addition of a carbamate insecticide to enhance the efficacy of the nets.[6] However, children would have significant chemical exposure via respiratory and dermal routes with the use of a dual pesticide combination (childhood exposure will be discussed later in this chapter), and both of these insecticides have been linked to neurologic disorders.[7,8]

The delicate balance between children's health and pesticides from a public health perspective elucidates the need for vector control. However, it begs the question as to whether we can do better in terms of health protection by simultaneously decreasing toxicant exposures for the sake of protection while preserving our children's health. It is important to note that this public health issue should not be seen as an open invitation for the biotechnology industry to put pressure on African nations to accept novel forms of transgenic technology such as gene drives and genome editing, with the goal of manipulating biologic ecosystems (such as The Target Malaria Project).[9]

In this regard, it is worth referencing a recent report to best appreciate the African perspective, *Profiteering From Health and Ecological Crises in Africa*, published by *The African Centre for Biodiversity;*

> ...A historical review of malaria eradication success stories in Africa shows that traditional methods can be successful in eradicating this killer disease. If, in the light of new discoveries related to how transmission is evolving, novel approaches to control malaria are warranted, applications that are based on shifting ecosystems and that entail unprecedented environmental and human health risks should be unequivocally rejected by African governments and society. African ecosystems have already reached tipping points; what we need is to protect our biodiversity and support domestic strategies to combat malaria that are focused on bolstering health care systems and fixing public infrastructure.[10]

7.5   THE STATE OF OUR CHILDREN'S HEALTH

7.5.1   Children's Health Landscape

Just observationally looking at children playing in a park or school in the United States, stark changes are revealed that one might not have noted just a few decades ago. Presently, one might expect to see one in five to one in three of the kids with obesity; excessive behavioral disorders such as overt aggression or social isolation with ASD affecting 1 in 23 boys; one in six to one in eight children with asthma inhalers in their back pockets and cell phones in many of their hands, spending more time on their phones than in other physical activities. (As an aside, the epidemic of cell phone usage at schools has created flocks of kids looking downward consistently, so much so that a newly diagnosed condition called "text neck" has emerged.)[11] Additionally, mental health challenges are dominating adolescence, for example, one in six US youth aged 6–17 years experiences a mental health disorder each year.[12]

7.5.2   What Do the Statistics Say about African Children's Health?

A look at obesity: The African region is experiencing similar obesity rates as many parts of the world.[13] A common misconception is that obesity and other chronic diseases are affecting primarily the wealthy, and the poorer populations are immune. This has been shown to be a fallacy; the upward trend in obesity is associated with a change in diet to sugary/highly processed foods, decreased consumption of legumes and vegetables, the navigation away from traditional diets, and an increase in sedentary lifestyles. Obesity in children is not a benign disorder and is associated with the development of metabolic syndrome, hypertension, mental health issues such as low self-esteem, bullying, and risks of future chronic diseases in adulthood.[14]

While Africa comprises 54 countries with a vast array of different languages and cultures, there is a continental trend of countries undergoing a nutritional transition, with nutrition-related diseases such as nutrient deficiencies (zinc, iron, vitamin A) and growth retardation coexisting with obesity in children causing a significant health burden.[15] As noted globally, the pattern of changing diets to more "western diets" and decreased physical activity are contributing to this dilemma.[16] Upon a review of the literature on African children and diet, the role of infectious diseases is noted and reported, as well as the aforementioned factors; however, there is little mention of the role of contamination by pesticides as another root cause of nutritional disease. While there are contradictory studies and an element of industry bias in research, most of the data supports the thesis that organically grown crops contain more iron, magnesium, phosphorus, and vitamin C, as well as fewer nitrates than conventional crops.[17] In addition, there are also lower levels of some heavy metals in organic crops compared to those that are grown conventionally.[18] In other studies, contamination of glyphosate-based herbicides with heavy metals has been reported.[19] Indeed, providing the most nutrient-dense food free of contaminants and toxicants such as lead, aluminum, and mercury should be a major consideration in childhood nutrient analyses, especially in lieu of the fact that small children do not eat large quantities. Therefore, one of the goals is to maximize nutrient density in smaller amounts of food.

As discussed above, there are many chronic health disorders affecting children. Of concern, are the escalating rates of endocrinologic issues, atopic disorders (asthma/allergies/eczema), behavioral and mental health issues, and serious neurocognitive dysfunction. Using US data for comparison, the prevalence of ASD can be seen as a canary in the coal mine or the flagship of the deterioration of our children's health and will be presented in comparison to African children. As we begin to dissect the causes of these issues in children, a brief review on the physiologic differences between children and adults will be presented as well as an explanation as to why this is so important in understanding the role of pesticide toxicity on children's health.

7.6   HEALTH LINKS TO PESTICIDES – WHAT DO THE STUDIES SAY?

7.6.1   UNDERSTANDING PESTICIDE EXPOSURE IN CHILDREN

In relation to their body weight, children have a larger skin surface area than adults, breathe in more air, and drink and eat more, hence they experience a much higher exposure rate than adults. Common routes of exposure for many children occur via the dermal and inhalation routes. Children are closer to the ground, often play barefoot in pesticide-sprayed areas, and explore much of their world via hand-to-mouth behaviors. Pesticides can cross the epithelium of the skin and mucous membranes that exchange gases (alveoli) or via the gastrointestinal mucosa. The rate of absorption depends on the chemical properties, amount of the chemical, length of exposure, and the physical state of the actual pesticide molecules. There are also other factors that may contribute to increased absorption, such as skin absorption that is higher when there is vasodilatation (as in summer or with heating), which would be more of a problem on the African continent. Respiratory absorption is many times higher when respiration is more rapid, e.g., when playing or running. Children also have a much higher respiratory rate than adults and thus have a greater exposure via inhalation.[20]

Another important consideration is how children detoxify xenobiologics, particularly in comparison to adults. The first line of defense in processing foreign chemicals resides in the intestinal microbiota. The collection of organisms that are found in the large intestine are vital to the health of the child and one of the most important areas of study and treatment in modern medicine. This vast community of organisms (consisting of bacteria, archaea, fungi, and viruses) have been shown to be impacted by pesticides that act as antibiotics.[21] For example, glyphosate, the main ingredient in Roundup®, was patented by Monsanto (now Bayer) as an antibiotic in 2010.[22] Mechanisms on how glyphosate affects the intestinal microbiota have been demonstrated as well as potentially elucidating the link between glyphosate and cancer.[23] (In 2015, the WHO, via its specialized cancer agency, the *International Agency for Research on Cancer* (IARC), classified glyphosate as a class IIa carcinogen.[24,25]) Glyphosate has been found in both breast milk[26] and infant formula[27], so either method of nutrition will expose infants to glyphosate from the very first stages of life.

From a clinical perspective, again, there is tremendous importance in the role of the microbiota in terms of children's health. Children acquire their microbial communities from their mothers during vaginal births and breastfeeding. (Of note, progress is being made in Africa to support mothers to breastfeed their infants for the first 6 months, which is a source of nutrients and immune development as well as infant immune support via the breast milk microbiota.)[28] If the mother has been exposed to antibiotics either via food or pharmacologics, her microbiome will be altered, which will impact the baby's microbiota inheritance. One of the roles of the infant's microbiota is the establishment of the innate immune system.[29] The emergence of childhood health issues from obesity, immune dysregulation (allergies/asthma), and ASD is linked to early microbiome disturbances.[30] Based on the importance of the acquisition of the most robust microbiota possible from the mother, antibiotic usage in all forms for mother and infant should be used judiciously. Based on our current understanding of the antimicrobial effects of the ubiquitous glyphosate herbicide, it should be banned outright.

Of note, in addition to glyphosate, other pesticides also have been shown to affect the microbiota. There are many studies that have documented how organophosphate (OP) insecticides, for example, alter the microbiota.[31] Of interest, many of the studies propose techniques to modify the microbiota to reduce toxicity rather than elimination of OPs; and thus, do not acknowledge or address the true treatment, which is removal of the root cause. OP insecticides will be addressed further below.

In trying to understand how much pesticide exposure such as glyphosate (the most commonly applied pesticides globally) African children receive, a quick review will be presented regarding glyphosate application.[32] The amount of glyphosate applied in continental Africa depends on the individual country. For example, South Africa accounts for 2% of global pesticide usage and there are 96 glyphosate-based herbicides registered.[33] Its adoption of genetically modified organisms (GMOs) tolerant of glyphosate and other pesticides will greatly increase usage and dietary exposure to the chemicals applied. Kenya imports 2.6 million kilograms (kgs) of glyphosate-based herbicides annually and is present in 70 of their 1,540 pesticide products approved for crop production.[34] Uganda reports that 42 of 300 herbicides are glyphosate-based.[35] It is important to note that the usage of glyphosate-based herbicides in Africa has created significant controversy. Organizations, such as *The African Centre for Biosafety*,[36] are running a campaign to ban glyphosate. Additionally, there are other African countries looking at the health ramifications of glyphosate. In 2019, the Cancer Association of South Africa (CANSA) published a paper supporting the findings of IARC and its position on glyphosate to be carcinogenic, as well as an endocrine disruptor, cytotoxic agent, genotoxic agent, and teratogen.[37]

What should be alarming is that despite the vigorous research and number of organizations and countries that are engaged in either restricting or banning glyphosate-based herbicides, little attention has been paid to the ongoing exposure, from a health perspective, of children, our most vulnerable population. As mentioned previously, South Africa grows GMO crops and, according to the *African Center of Biodiversity*, half of South Africa's maize crop and 100% of the soya crop are GMO, which means that they are grown with glyphosate.[38] Maize and soya are main staples of the South African diet.

Further analyzing the question of how much African children are being exposed to glyphosate-based herbicides, let's look quantitatively at the actual amounts of glyphosate that children consume in their diets. In a South African study of 81 off-the-shelf maize and soybean food products tested for glyphosate, 67% contained 27–2,357 parts per billion (ppb) glyphosate. Thirty of 57 maize products contained from 27 to 95 ppb glyphosate. All 11 soya products contained anywhere from 27 to 142 ppb glyphosate. All six corn–soy blends tested positive for glyphosate in the range of 43–65 ppb. Seven texturized soy protein products tested positive for glyphosate in the range of 41–2,257 ppb. Clearly, the main diet of South African children, heavily comprised of maize and soya, is contaminated with very high levels of glyphosate, which are much higher than acceptable levels determined to be safe in children.[39] The significance of these findings will be discussed below.

7.7 REMOVE THE BLINDERS TO GLYPHOSATE-BASED HARM

7.7.1 DOES DOSE MAKE THE POISON?

In a study by Mesnage et al., rats administered 0.1 ppb of Roundup® showed signs of enhanced liver injury via several biological systems and oxidative stress and demonstrated that **ultralow doses** of glyphosate-based herbicides can cause fatty liver (NAFLD – nonalcoholic fatty liver disease).[40]

This issue has now surfaced as a major epidemic. In America, one in three adults is presently diagnosed with NAFLD.[41] This disorder can progress to more serious forms of liver disease and cirrhosis. Children are not faring better than adults in respect to NAFLD. The prevalence of NAFLD in the pediatric population is estimated between 3% and 12% and found in a shocking 70%–80% of children with obesity.[42] Acknowledging the reported obesity in African children, the obvious question is whether the impact of pesticide exposure on this serious health problem is even being addressed or considered. With the high levels of glyphosate reported in the foods children consume, one can surmise that these children are overloaded with pesticides, and their ability to detoxify may be significantly impaired. The reality exists that a significant number of children may have undiagnosed/unrecognized NAFLD with the potential for further advancement to even more serious liver disease. A further look at the detoxification capacity of children will follow in the next section.

Research continues to focus on genetics and "environmental factors" as causative agents in the epidemic of liver disease with some researchers pointing the finger at high-fructose corn syrup (HFCS) as an etiology for the rise in NAFLD as well. Equally disturbing is that a review of the literature, reports, or further discussion of the link between ultralow doses of Roundup causing NAFLD in either children and adults could not be found despite the research previously published by Mesnage, and the American National Liver Foundation ringing the bell regarding the epidemic of NAFLD. Clearly, we must also address the derivatives of HFCS as well as the contaminating GMOs and their associated pesticide components as being potential causal factors for NAFLD.

How could we have ever believed that it was a good idea to grow our food with poisons?

Jane Goodall

7.7.2 CAN CHILDREN CLEAR THE TOXIC LOAD?

The second line of defense in clearing xenobiologics after the frontline defenders in the microbiota is the liver, which has many housekeeping responsibilities in health maintenance. The impact of NAFLD and its sequelae of worsening liver disease could potentially impair detoxification. One of the ways detoxification occurs is via the cytochrome p450 pathway. The cytochrome p450 pathway and the paraoxonase1 (PON1) pathway play a major role in the metabolism and detoxification of OP insecticides as well as the metabolism of oxidized lipids. The role of OPs in children's health such as neurodevelopment disorders is well known.[43]

Children's detoxification mechanisms are not well developed until several years after birth.[44] The role of the PON1 plasma enzyme in the detoxification scheme is an undervalued variable. In a study from 1953,[45] PON1 was shown to hydrolyze the toxic metabolites of OP insecticides. Hence, the information regarding this enzyme has been available for decades. Studies have shown that the PON1 levels of newborns are one-third to one-fourth the levels of adults.[46] These levels have an impact on OP exposures in small children, which can effect developmental neurotoxicity.[47]

OPs can reach children by various routes of exposure such as via fetal exposure by crossing the placenta and in breast milk. The blood–brain barrier is not developed until after age 1 and does not protect the infant from unwanted chemicals crossing into the brain.[48] There is significant genetic variability of the PON1 gene, which impacts susceptibility to OPs. PON1 can be seen as a biomarker of susceptibility to neurodevelopment effects.[49]

Due to its toxicity, OPs are being phased out in many parts of the world; however, the usage forecast in Africa has been estimated to continue to grow at a rate of 4.8% between 2017 and 2022, with South Africa and Egypt requiring the highest demands.[50]

7.8 PESTICIDES AND THE LINK TO NEURODEVELOPMENT DISORDERS

ASD is now a global epidemic facing children according to the transcontinental data presented in Figure 7.1.

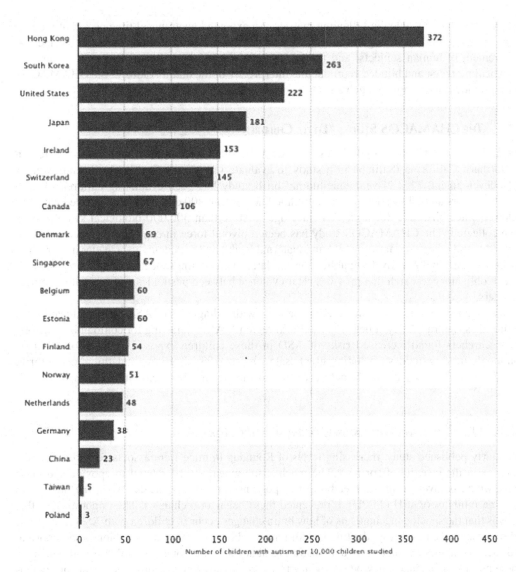

FIGURE 7.1 Prevalence of autism spectrum disorder among children in select countries worldwide as of 2020.

However, rates of ASD in Africa are not clearly defined nor are any African countries included in global reporting. One research paper laid out all of the peer-reviewed studies on autism from the 46 countries in sub-Saharan Africa. Eighty percent of the studies focused on South Africa and Nigeria, two of the wealthier countries in Africa.[51] This study showed that there were inadequate methodologies utilized and a lack of standardized diagnostic tools. Existing studies were generally small-scale in terms of the assessment of ASD in Africa. However, unpublished data shows that 1% of children from Kenya are autistic, in-line with global reports.[52] Of concern, are the acknowledged lack of resources, educational limitations, diagnostic challenges such as late-onset of diagnoses and a paucity of specific programs to assist ASD children and their families. What is not being addressed, both in Africa and in countries such as the United States, is determining the root causes of the autism epidemic. Specifically, what are the links between pesticides and the autism epidemic?

One might argue that in order to show a clear causation between pesticides and neurocognitive disorders, there are specific research requirements that demonstrate causation and not just correlation, which include the following metrics: An extended meaningful time period of the participants being studied, a large number of participants, exclusion of other confounding factors, involvement of human subjects, and the use of multiple monitoring markers (such as surveys, biological markers, and blinded reproducible interpreters of the data). A look at the CHAMACOS study provides an example of such a study to be followed.

7.8.1 THE CHAMACOS STUDY: "LITTLE CHILDREN"[53]

The University of California (Berkeley, CA, USA) enrolled women from an agricultural region of Northern California (Salinas) in a study to evaluate the effect of pesticides on farmworker neurodevelopment. This 21-year longitudinal birth study has been evaluating children's health every 1–2 years and followed them into adulthood. More than 800 children were enrolled with 536 followed since birth and 305 followed since age 9. More than 300,000 biological samples have been collected. The CHAMACOS study has been a pivotal force in exposing the critical role of pesticides and other toxicants on child and maternal health and is undeniably a leader in providing impervious data with nearly 150 publications to date. Their findings meet the requirements of the type of obligatory research that provides clear evidence between pesticides and neurodevelopment disorders.

In another groundbreaking study, 2,961 patients with a diagnosis of ASD were evaluated from California's Central Valley, a heavy agricultural region.[54] After adjusting for confounding factors, the researchers found increased risks of ASD in those children exposed to seven of the most commonly used pesticides (including glyphosate, chlorpyrifos, diazinon, malathion, permethrin, bifenthrin, and methyl bromide) before birth and during the first year of life compared with controls.[55]

7.8.2 UNRAVELING THE MECHANISMS OF NEUROLOGIC TOXICITY

In a newly published study, increasing levels of Roundup in mice were associated with ASD-like behavior in the juvenile offspring.[56] The mechanism appears to be related to epoxide hydrolase (EH), which is involved in the metabolism of polyunsaturated fatty acids. When the mice were given an inhibitor of sEH (TPPU), it prevented these behavioral changes. The importance of this study is that the specific mechanisms of how brain changes occur in children with ASD from pesticide exposures are now being identified. Additionally, changes in the gut microbiota were reported, and there is significant data now demonstrating abnormal microbiota in children with autism.[57] These microbiota changes may be secondary from the already-known antibiotic effects of glyphosate as previously discussed. The communication between the gut and the brain via the enteric nervous system is now accepted, with the microbiota producing a vast array of metabolic-modifying compounds affecting the brain.[58] The gut–brain axis appeared to be affected by the administration of increasing doses of Roundup as noted in this study with the subsequent development of neurocognitive changes seen in ASD and death in some of the animals. An even stronger study would have identified a control group that was fed an organic chow diet and given Roundup-free water (since Roundup is commonly present in animal feed and the water supply). Further clarification of whether the effects reported are from the glyphosate or its toxic adjuvant, polyethoxylated tallow amine (POEA), is clinically relevant.

From a clinical perspective, the removal of pesticides and chemical inputs is the first step in my treatment plan (as well as being practiced by colleagues from meetings, organizations, and personal correspondence) to reverse neurobehavioral disorders. But let it be clear that although individuals bear responsibility for their own actions, it is equally true that social determinants of health (such as access to quality food, water, air, education, racial equality) create the fabric in which we live. These other factors must be taken into consideration in order to nurture and create well-being for children.

7.8.3 AN END TO PESTICIDE PROFITEERS

It is indisputable that children are victims to ubiquitous herbicide usage, such as glyphosate-based herbicides. Exposure may be from the food they consume, indirectly from drift, contamination from air and water, living in proximity to farms and farm workers, or access via inappropriate storage of dangerous chemicals. Africa has a large agricultural base, and many of those farming are children. Occupational hazards from sprayers with leaky hoses or workers without appropriate personal protection result in a high dermal exposure. Often, there are inadequate safeguards in place for worker protection, including medical access. In 2016, the European Union called for a ban on the toxic adjuvant component of Roundup, POEA. Will Africa be subject to "overstocked" chemicals and receive these unwanted rejects?[59]

African organizations, such as The Alliance for Food Sovereignty (AFSA), are taking a strong position against not only glyphosate, but also other toxic pesticides and have created a call to action away from pesticides and toward agroecology via the creation of networks of support with farmers, consumers, NGOs, and international partners.[60] It is clear that Africans should decide what is best for their communities without political and profit pressure. A movement is underfoot to promote agroecology, diverting production away from unsustainable chemical farming, which undermines self-determination and resilience. Evidence exists that agroecology is capable of feeding the world's population.[61]

> If governments won't solve the climate, hunger, health and democracy crises, then the people will. Regenerative agriculture provides answers to the soil crisis, the food crisis, the health crisis, the climate crisis and the crisis of democracy.

Dr. Vandana Shiva

(per 10,000 children)[62]

REFERENCES

1. Perro M and Adams V. 2017. *What's Making our Children Sick? How Industrialized Food is Causing an Epidemic of Chronic Illness, and What Parents (and Doctors) Can Do About It; Exploring the Links Between GM Foods, Glyphosate, and Gut Health.* White River Junction, VT: Chelsea Green Publisher.
2. Henn BC, et al. 2019. Chemical mixtures and children's health. *Current Opinion in Pediatrics* 26(2) (Apr): 223–229.
3. *Fourth National Report on Human Exposure to Environmental Chemicals Updated Tables*, January 2019, Volume One; https://cfpub.epa.gov/ncea/risk/hhra/recordisplay.cfm?deid=23995.
4. Braun JM, Gray K. 2017. Challenges to studying the health effects of early life environmental chemical exposures on children's health. *Plos Biology* Dec. 2017; doi: 10.1371/journal.pbio.2002800.
5. United Nations Children's Fund, 'Malaria', 2016. *UNICEF*, New York, 22 July.
6. Corbel V, et al. 2002. Insecticide mixtures for mosquito net impregnation against malaria vectors. *Parasite* 9(3) (September): 255–259. doi: 10.1051/parasite/2002093255.
7. Viel J-F, et al. 2015. Pyrethroid insecticide exposure and cognitive developmental disabilities in children: The PELAGIE mother–child cohort. *Environment International* 82 (September): 69–75. doi: 10.1016/j.envint.2015.05.009.
8. Rowe C, et al. 2016. Residential proximity to organophosphate and carbamate pesticide use during pregnancy, poverty during childhood, and cognitive functioning in 10-year-old children. *Environmental Research* 150 (October): 128–137. doi: 10.1016/j.envres.2016.05.048.
9. https://www.acbio.org.za/sites/default/files/documents/202006/profiteering-health-and-ecological-crises-africathe-target-malaria-project-and-new-risky-ge_0.pdf.
10. Ibid.
11. Fares J, et al. 2017. Musculoskeletal neck pain in children and adolescents: Risk factors and complications. *Surgical Neurology International* 8: 72. doi: 10.4103/sni.sni_445_16.
12. https://www.nami.org/mhstats.
13. https://www.afro.who.int/health-topics/obesity.

14. Weiss R, et al. 2005. The metabolic consequences of childhood obesity. *Best Practice & Research Clinical Endocrinology & Metabolism* 19(3) (September): 405–419. doi: 10.1016/j.beem.2005.04.009.
15. Mbogori T, and Mucherah W. 2019. "Nutrition transition in Africa: Consequences and opportunities". *Global Journal of Transformative Education* 1(1): 5–10. doi: 10.14434/gjte.v1i1.26141.
16. Ibid.
17. Worthington V. 2001. Nutritional quality of organic versus conventional fruits, vegetables, and grains. *The Journal of Alternative and Complementary Medicine* 7(2): 161–173.
18. Ibid.
19. Defarge N. 2018. Toxicity of formulants and heavy metals in glyphosate-based herbicides and other pesticides. *Toxic Rep* 5: 156–163.
20. Health Council of the Netherlands. 2004. Pesticides in food: assessing the risk to children. The Hague: Health Council of the Netherlands.
21. Mesnage R, et al. 2019. Shotgun metagenomics and metabolomics reveal glyphosate alters the gut microbiome of Sprague-Dawley rats by inhibiting the shikimate pathway. *BioRxiv.* doi: 10.1101/870105.
22. https://patents.google.com/patent/US7771736B2/en.
23. Ibid 14.
24. https://www.iarc.fr/wp-content/uploads/2018/07/MonographVolume112-1.pdf.
25. Guyton KZ, et al. 2015. Carcinogenicity of tetrachlorvinphos, parathion, malathion, diazinon, and glyphosate. *The Lancet Oncology* 16(5): 490–491. doi: 10.1016/S1470-2045(15)70134-8. Epub 2015 Mar 20.
26. https://www.telesurenglish.net/news/Brazil-Poisonous-Agrotoxin-Found-Over-80-of-Breast-Milk-Samples-in-Urucui-20180809-0008.html.
27. https://www.reuters.com/article/us-food-agriculture-glyphosate/fears-over-roundup-herbicide-residues-prompt-private-testing-idUSKBN0N029H20150410.
28. https://www.dw.com/en/many-african-countries-urged-to-support-breastfeeding-mothers/a-39966423.
29. Al-Shehri S, et al. 2015. Breastmilk-saliva interactions boost innate immunity by regulating the oral microbiome in early infancy. *PLoS One* 10(9): e0135047.
30. Eshraghi RS, et al. 2018. Early disruption of the microbiome leading to decreased antioxidant capacity and epigenetic changes: Implications for the rise in autism. *Cell Neuroscience* 15 (August). doi: 10.3389/fncel.2018.00256.
31. Roman P, et al. 2019. Microbiota and organophosphates. *NeuroToxicology* 75 (December): 200–208.
32. Vlavanidis A. 2018. Glyphosate, the most widely used herbicide, health and safety issues: Why scientists differ in their evaluation of its adverse health effects. *Research Gate* (March).
33. https://www.agribusinessglobal.com/markets/africa-middle-east/the-tenuous-future-of-glyphosate-in-africa/; accessed 27 May 2020.
34. Ibid.
35. Ibid.
36. https://www.greenafricadirectory.org/listing/the-african-centre-for-biosafety-acb/.
37. https://www.cansa.org.za/files/2019/04/Fact-Sheet-and-Position-Statement-on-Glyphosate-web-April-2019.pdf. The original paper is from this source, but could not be accessed: https://scholar.ufs.ac.za:8080/xmlui/bitstream/handle/11660/6536/KoortzenBJ.pdf.
38. Ibid.
39. Ibid.
40. Mesnage R., et al. 2017. Multiomics reveal non-alcoholic fatty liver disease in rats following chronic exposure to an ultra-low dose of Roundup herbicide. *Scientific Reports* 7, Article number: 39328.
41. https://liverfoundation.org/?s=NAFLD.
42. Bush H. et al. 2017. Pediatric non-alcoholic fatty liver disease. *Children (Basel)* 4(6): 48.
43. Rowe C, et al. 2016. Residential proximity to organophosphate and carbamate pesticide use during pregnancy, poverty during childhood, and cognitive functioning in 10-year-old children. *Environmental Research* 150 (October): 128–137. doi: 10.1016/j.envres.2016.05.048. Epub 2016 Jun 6.
44. Freeman K. 2009. Capability to detoxify pesticides. *Environmental Health Perspectives* 117, No. 10: 1. doi: 10.1289/ehp.117-a454b.
45. Aldridge WN. 1953. Serum esterases. I. Two types of esterase (A and B) hydrolysing p-nitrophenyl acetate, propionate and butyrate, and a method for their determination. Biochemical Journal 53: 110–117.
46. Huen K. 2009. Developmental changes in PON1 enzyme activity in young children and effects of PON1 polymorphisms. *Environmental Health Perspectives* 117(10): 1632–1638.

47. Furlong C, et al. 2006. PON1 status of farmworker mothers and children as a predictor of organophosphate sensitivity. *Pharmacogenet Genomics* 16(3) (March): 183–190._doi: 10.1097/01.fpc.0000189796. 21770.d3.

48. Eaton DL, et al. 2008. Review of the toxicology of chlorpyrifos with an emphasis on human exposure and neurodevelopment". *Critical Reviews in Toxicology* 38 (suppl 2): 1–125.

49. Marsillach J, et al. 2016. Paraoxonase-1 and early-life environmental exposures. *Annals of Global Health* 82(1) (January–February): 100–110.

50. https://www.mordorintelligence.com/industry-reports/africa-organophosphate-pesticides-market.

51. Franz L, et al. 2017. Autism spectrum disorder in sub-Saharan Africa: A comprehensive scoping review. *Autism Research* 10(5) (May):723–749. doi: 10.1002/aur.1766. Epub 2017 Mar 7.

52. https://www.spectrumnews.org/features/deep-dive/autism-remains-hidden-africa/.

53. https://cerch.berkeley.edu/research-programs/chamacos-study.

54. Ehrenstein OS, et al. 1962. Prenatal and infant exposure to ambient pesticides and autism spectrum disorder in children: population based case-control study. *BMJ.* 364. doi: 10.1136/bmj.1962.

55. https://www.gmoscience.org/early-exposure-to-pesticides-linked-to-increased-risk-of-autism/.

56. Pu, Y, et al. 2020. Maternal glyphosate exposure causes autism-like behaviors in offspring through increased expression of soluble epoxide hydrolase. *PNAS* 117 (21) (May 26): 11753–11759; first published May 12, 2020. doi: 10.1073/pnas.1922287117.

57. Eshragi R, et al. 2018. Early disruption of the microbiome leading to decreased antioxidant capacity and epigenetic changes: Implications for the rise in autism. *Frontiers in Cellular Neuroscience* (15 August). doi: 10.3389/fncel.2018.00256.

58. Carabotti M. 2015. The gut-brain axis: Interactions between enteric microbiota, central and enteric nervous systems. *Annals of Gastroenterology* 28(2) (April–June): 203–209.

59. EU agrees to ban on glyphosate co-formulant. *Euractiv* (Published 12 July 2016). https://www.euractiv. com/ section/agriculture-food/news/eu-agrees-ban-on-glyphosate-co-formulant/.

60. https://www.acbio.org.za/sites/default/files/documents/Africa_must_ban_glyphosate_now_Aug2019. pdf.

61. https://www.agroecology-pool.org/showcases/.

63. https://www.statista.com/statistics/676354/autism-rate-among-children-select-countries-worldwide/.

62. https://www.statista.com/statistics/676354/autism-rate-among-children-select-countries-worldwide/.

8 Animal Health Issues with Increased Risk from Exposure to Glyphosate-Based Herbicides

Arthur Dunham
Ryan Veterinary Service

CONTENTS

8.1 INTRODUCTION

My large animal veterinary medicine practice in northeast Iowa has serviced a mixture of swine, dairy, cow-calf, and feedlot operations. Since graduating from Iowa State University in 1974, the practice has also included some sheep, goat, and horse farms. Because nutrition is important to health, considerable experience analyzing the diets of all the species mentioned has been obtained. Over the past 46 years, I have observed clinically significant differences when comparing the health of farm animals prior to and after the introduction of Roundup Ready® genetically engineered technology. Differences in soil, plant, and animal health on farms that do not use glyphosate, compared to those farms that use this chemical, also have been observed. Over time, I have found agronomists, plant pathologists, other health professionals, and researchers who share similar observations and apprehensions. It seems clear that the herbicide glyphosate has negatively impacted animal health within our large animal veterinary practice. Similar impacts occur on a national and global scale.

Prior to its use as a broad-spectrum herbicide, glyphosate had other uses. In 1964, it was patented as a broad-spectrum chelator of minerals. Since a chelator binds to metal ions, glyphosate was used to descale metal pipes and boilers. In 1969, it was patented as an herbicide and was approved for such use in the early 1970s. After the introduction of Roundup Ready® soybeans in 1996 and Roundup Ready corn in 1998 tolerant to glyphosate, the herbicide became widely used all over the world. In 2010, it was patented for human parasite control and as an antimicrobial agent [1]. All three of these patents depend at least partially on glyphosate's ability to bind with cations (minerals) that are necessary cofactors in biological enzyme systems.

To understand how glyphosate negatively affects animal nutrition, one must have some understanding of its chelating ability. Glyphosate chelates divalent cations in the following order, based on the log of chelate formation constants or K values (shown in parentheses): copper (11.93), zinc (8.74), nickel (8.10), cadmium (7.29), cobalt (7.23), iron (6.87), manganese (5.47), magnesium (3.31), and calcium (3.25) [2]. When a molecule "chelates" an ion, it attaches like a claw. These K value numerical constants are in logarithmic scale. They are read in factors of ten (such as pH values) so a six on this scale will chelate a mineral ion with a force ten times stronger than a five. The trace mineral cations are essential cofactors in many metabolic reactions and are needed, and provided, in tiny ppb and ppm levels. This makes residues in feed from the 0.5ppm standard spray dose of glyphosate (with four possible chelation sites per molecule) potentially dangerous. We do not know the disassociation values for the various cations chelated with glyphosate when this endocrine-disrupting chemical is present in any mammal's body. We know that glyphosate bound to cations in the soil can be desorbed and made active again by phosphate fertilizers. Phosphate anions present in mammalian metabolism can more than likely do the same. Therefore, at any point in time with glyphosate residues in the living animal, it is impossible, with current technology, to know what cations are bound and what are available for metabolism. This is true even if we analyze glyphosate levels and trace mineral cation levels at the same time.

Unfortunately, most scientists working with glyphosate-based herbicides have not paid attention to the power of this nutrient immobilizing "claw" nor to the first or last patents mentioned above. They also have ignored the difficulty in breaking glyphosate's man-made carbon to phosphorus bond. Few soil bacteria have enzymes that can break this C–P bond, so that glyphosate's half-life in soil varies from a few months to 20+ years depending on the type of soil and microbial activity therein. In the field, we observe a half-life that is much longer than that quoted by industry. The half-life increases with increased glyphosate use since soil biological activity goes down in response to the presence of the glyphosate and its degradation products (such as AMPA).

These oversights on the part of the majority of the science community have led to an underestimation of the potential harm glyphosate can cause within living systems. These biological systems rely on essential cations to function properly. What we know about living systems also has changed. Epigenetics and fetal programming are emerging new sciences. The scientific community now recognizes the importance of the gut microbiome. Forty percent of a mammal's genome and the metabolic processes instructed by that genome are similar to a plant's genome. Thus, can we accurately say that residues of herbicides in diets are nontoxic to mammals? Current government pesticide regulation has resulted in a lack of concern for chronic toxicity since it is largely based on determining the one-time dose that kills 50% of the lab rats fed that dose and then arbitrarily using one-thousandth to one-hundredth of that dose as a regulatory limit. When glyphosate was first used as an herbicide, there were very few scientists that considered its impact on the nontarget ecology. Even fewer saw a need to include these impacts when establishing regulatory criteria for its use.

Since there is only limited metabolism to break down glyphosate in plants, including those that are Roundup Ready, the functional importance of the microbiome alone should curtail the indiscriminate use of glyphosate as an antimicrobial herbicide and prompt society to get glyphosate out of our food supply. Food animals do not have a long life. If a veterinarian can see health issues related to glyphosate use in livestock, humans with exposure over the relative longer term of their lives are undoubtedly suffering health effects as well. The rest of this chapter will describe clinical observations that have become common with increased glyphosate use. An argument can then be made that an in-depth evaluation of glyphosate's effects on animal and human health should be a high priority.

8.2   INFLUENCE OF GLYPHOSATE ON TRACE MINERALS

Let us start with some trace mineral deficiencies. Cropping systems that rely on glyphosate often have damaged root systems and impaired activity of the rhizosphere microbial community responsible for plant nutrient uptake. Keep in mind that different soils have different levels of trace minerals as well. Continuous removal of crops with little to no trace mineral replenishment through fertilization also has led to lower levels of trace minerals in the soil and, subsequently, in the plants produced on these soils. Many nutritionists, therefore, add trace minerals to diets as if the diets being fed have next to no trace mineral content to begin with. Because of this, many veterinarians think that there is no need to worry about crops that are mineral-deficient. I disagree. If the crop itself is providing much of the mineral needs, animal feed cost is lowered. There also is less chance of interference between trace minerals when they are not all added together at high levels in the mineral premix. Nevertheless, what I have observed will not be seen everywhere. In addition, glyphosate in the diet might affect different, less plentiful, cations in another region more severely.

8.2.1   MANGANESE (MN++) DEFICIENCY

Manganese is the cofactor for the enzyme EPSPS in the Shikimate pathway that plants and many bacteria possess to make the aromatic amino acids phenylalanine, tyrosine, and tryptophan. Besides being building blocks for protein, these amino acids are precursors for plant defense substances such as lignin and pyrrolnitrin, a potent antifungal metabolite. Unlike plants, mammals do not have the Shikimate pathway. However, their gut bacteria certainly use this pathway. In mammals, these three aromatic amino acids are also building blocks for protein as well as precursors for neurohormones that include dopamine and serotonin.

My veterinary partner and I started to see and recognize manganese deficiency in swine, dairy, and cow-calf herds in 2005. Even though glyphosate was approved as an herbicide in the mid-1970s, its use did not skyrocket until the introduction of Roundup Ready® crops. According to a US geological survey, 200 million pounds was used solely on US corn and soybean acres in 2016 [3]. The National Research Council's nutrient requirements of domestic animals for both dairy and swine say that it is hard to create manganese deficiency without having too much calcium, iron, or zinc in the feed or water. Balance is always important, and excessive amounts of any or all three of these minerals can be a risk factor; however, we have confirmed manganese deficiency in the absence of excesses of calcium, iron, or zinc.

According to the National Research Council, manganese deficiency can result in non-cycling sows and cows even when they are in positive energy and protein balance. Manganese-deficient cows often have small pea-sized ovaries confirmed on rectal palpation, and these cows do not come into heat regularly. Nonbreeding sows and gilts sent to local butchers have the same small ovaries when examined. The divalent cation manganese is a cofactor in an enzyme system essential for immune system function (National Research Council [4]), and calves born from deficient cows can suffer from recurring scours or navel infection.

Attention to manganese deficiency increased in the spring of 2006, when four calves with recurring scours along with potential nervous system involvement were born in an excellently managed purebred beef herd. The nervous signs were not due to a thiamine deficiency (commonly known as polio) nor were they related to acidosis that is common with diarrhea in young calves. These signs were probably due to deformed bones of the middle ear. Two of the four calves that could barely stand would, after assistance getting up, along with loud clapping, wake out of a stupor and go nurse. All four of these calves required multiple treatments with intravenous fluids on top of adequate oral electrolyte therapy. One calf died and we submitted a liver sample to the Iowa State University Diagnostic Laboratory. The sample showed low manganese levels (Figure 8.1). Some calves were born with chondrodysplasia—short legs and big joints—like a bulldog (Figure 8.2). This occurrence with manganese deficiency is why I felt the bones in the middle ear may have been deformed as well contributing to the lack of balance in these calves. Also, some cows had stillborn calves that had low liver manganese levels when analyzed by the Veterinary Diagnostic Lab at Iowa State University. Unfortunately, Iowa State's lab did not have a plasma mass spectrograph in 2006. To keep costs down, it was not possible to analyze for most of the cations like we can do

Veterinary Diagnostic Laboratory
Iowa State University
College of Veterinary Medicine
Ames, Iowa 50011-1134
Phone: 515-294-1950
Fax: 515-294-3564

Accession: **2010010357**

Final Report
Accessed Date: 05/24/2020 11:28 am

DR. ARTHUR G. DUNHAM
RYAN VETERINARY SERVICE
PO BOX 48
RYAN , IA 52330

Site   :

Premises ID#   :

Lot/Group ID   :
Source/Flow ID :
Reference   :
Case Tags   :

Owner
Diagnostician: ENSLEY, STEVEN

Client Phone: 563-922-2094	Species: Bovine	Age:
Client Fax: 563-932-2996	Breed: Unknown	Weight:
Client Account#: 000401331	Sex:	Received: Fresh tissue
Date Received: 04/13/2010	Previous Case:	
Sample Taken: 04/12/2010	Farm Type:	Reason:
Accompanying Cases:	Animal ID(s): Liver	
Final Report(s): 4/16/2010 2:08 pm		

Reference ranges for bovine liver

	Reference range ppm	Liver
Na (ppm)	900-1800	1146
Mg (ppm)	100-250	127
P (ppm)	2000-4000	2123
K (ppm)	1400-4000	2715
Ca (ppm)	30-200	79
Cr (ppm)	.04-3.8	0.050
Mn (ppm)	2.5-6	0.9
Fe (ppm)	45-300	367
Cu (ppm)	25-100	71
Co (ppm)	.02-.085	0.019
Zn (ppm)	25-100	153
Se (ppm)	.25-.50	0.28
Mo (ppm)	.14-1.4	0.38
Cd (ppm)	.02-1	0.003
Se (ppm) fetus	0.3-1.2	

FIGURE 8.1 Iowa State University Veterinary Diagnostic Laboratory Report (This farmer not only had a still born calf but had five "bulldog" calves from a group of 15 heifers that conceived out of the 25 heads he tried to breed. (Normal calves are sometimes born with a lower level than the range given, but they should still be at least 1 ppm Mn on a wet liver basis.))

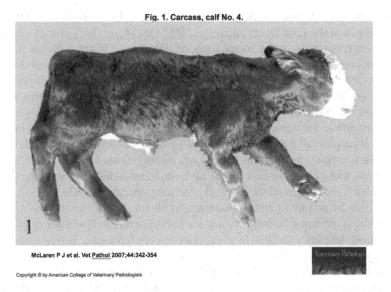

Fig. 1. Carcass, calf No. 4.

McLaren P J et al. Vet Pathol 2007;44:342-354

Veterinary Pathology

FIGURE 8.2   Carcass calf No. 4.

today, so only Mn was added until Iowa State University purchased their plasma mass spectrograph in 2008.

National Research Council data indicates that newborns are born with low levels of manganese compared to adults. Therefore, unless the stillborn and young calf livers came back under 1 ppm on a wet liver basis, they were considered normal. Some came back under the 0.5 ppm detection limit at that time when a normal adult needed to be over 2.5 ppm. Growing animals, such as first calf heifers and gilts, have a higher manganese requirement than mature animals and will often show deficiency first. Heifers in manganese-deficient herds have more chondrodysplasia calves and much lower pregnancy rates than older cows. In many herds, younger animals become energy and protein-deficient because of competition for feed. This deficiency of energy and protein can also result in small ovaries with no cyclicity even in the absence of manganese deficiency. However, the herds in our practice that had symptoms of manganese deficiency were different in that all animals (including some of the middle-aged mature cows and sows) exhibited pea-sized ovaries. In these herds, all animals, including younger animals, were in positive energy and protein balance.

After Iowa State University purchased a plasma mass spectrograph in 2008, our practice could economically check for more liver cations that included calcium, iron, and zinc. While high levels of calcium can occur because of excellent alfalfa hay or from excessive feeding of limestone added to the ration, this was not observed in our manganese-deficient herds. Similarly, too much ferrous iron in well water and/or zinc methionine added to rations can increase the risk of manganese deficiency, especially if the zinc methionine is left in the ration too long for foot rot control. Again, several of our cases did not have elevated levels of calcium, iron, or zinc. The common factor in our confirmed manganese deficiency cases was the presence of glyphosate residues in Roundup Ready feeds and supplements. By the process of elimination, glyphosate was recognized as a risk factor for manganese deficiency and related herd health problems.

These observations were brought to the attention of Dr. Jeremy Schefers, diagnostician at the Minnesota Veterinary Diagnostic Laboratory, at the Minnesota Dairy Health Conference that was held May 17–19, 2011. Dr. Schefers presented a seminar discussing manganese deficiency where he stated that "it had always been around" [5]. He reported that it was first documented in Australia with crops raised in dry soil following an extended drought. I was amused by the dry weather comment since the Midwest was drenched with heavy rains during the time he was collecting tissues and documenting the problem. While reflecting on this information, I remembered the article by Dr Don Huber that I had seen in the John Deere Furrow in 2007. One of our dairy clients with a manganese deficiency problem brought the article to my attention. He was not happy with our veterinary work. "My cows do not come into heat, my calves get sick, and your vaccine programs do not work—so read this!" In the article, Dr. Huber described the importance of foliar feeding soybeans with manganese and zinc about 10–14 days after spraying with glyphosate for weed control. Dr. Huber said that this was most important on light sandy calcareous soils with high pH. This description fit the soils on the farms where we were having problems. Sandy soils leach trace minerals more easily and high pH alkaline soils oxidize the cations rapidly—so not as many cations remain available for plant uptake. The second spray pass was needed when applying these trace minerals since mixing the divalent cations with glyphosate in the spray tank would have chelated them with the glyphosate. This would result in rendering both the herbicide and the minerals less effective.

A few years later, Monsanto figured out how to add the traces in a slow-release complexed form so clients using the brand name herbicide would not see the severe yellow flashing of the leaves that is due to manganese deficiency induced by the application of glyphosate. Yellow flashing of soybean leaves was easily seen on dry hillsides when clients used the cheaper generic glyphosate.

After reading Dr. Huber's article, I wanted the experts at Iowa State University to contact Dr. Huber; however, in central Iowa with heavier black dirt and lower soil pH than that which we have in the northeastern part of the state, they were not seeing the problem. When they refused, I phoned Dr. Huber. I got a response—and a friend and mentor for life.

We have a client feeding beef cows that have had three chondrodysplasia calves out of 60 born this spring (Figure 8.3). We are generally seeing fewer cases of manganese deficiency and a lower incidence of chondrodysplasia in our area that is probably due to awareness and much higher supplementation of manganese in the feed. Much of that supplementation is via a manganese amino acid complex versus the inorganic salts such as manganese sulfate that used to be used. This complexing of the trace mineral with amino acids makes it harder for any of the four chelation sites on glyphosate to "grab" the cations compared to when they are not complexed.

Our practice has observed manganese deficiency since 2005. On March 2, 2020, while eating a quick lunch, a local nutritionist told me that our neighboring veterinary practice had recently diagnosed and confirmed a case of manganese deficiency in a beef herd that was losing calves with chondrodysplasia. This herd also had poor pregnancy rates in heifers. This herd is in an area with heavier black soils of lower pH and generally higher manganese availability in forage. This difference in soil might be why it took more years of glyphosate use for the problem to become obvious to them. They, like so many other operations in our Midwest, were probably less profitable in previous years due to the influence of subclinical manganese deficiency. None of our clients are feeding Roundup Ready® alfalfa. I was sought out and complimented by a dairy nutritional

FIGURE 8.3 Picture of manganese deficient calf spring 2020.

consultant at the AABP annual meeting in Omaha in 2017 who had heard me comment about the increased risk of manganese deficiency due to glyphosate at some earlier meetings. Because he remembered those comments, he easily recognized manganese deficiency in a large group of custom-fed dairy heifers being fed Roundup Ready alfalfa in addition to Roundup Ready corn and corn silage.

8.2.2 COBALT (Co++) DEFICIENCY

Dr. Mike Sheridan, a swine consultant from Steinbach, Manitoba, saw what he called "squatter" hogs in a large farrow to finish operation that was feeding small grains to both sows and their offspring where about 2% of the finisher hogs, weighing 150 pounds and up, became paralyzed in their rear quarters. Spinal cord tissue sent to two veterinary labs and to a human health lab all confirmed demyelination of the spinal cord due to vitamin B12 deficiency; however, increasing folate and vitamin B12 in the feed for both the sows and the pigs, at 5-10X the NRCS requirement, did not stop the problem [6]. In responding to Dr. Sheridan's request for help on the American Association of Swine Veterinarians international e-mail, I asked him if the grain in the diets that were fed to the pigs was spray dried with glyphosate. His affirmative response prompted a recommendation to switch the source of the purchased grain to small grains not desiccated with glyphosate as well as discontinuing the practice on the farm. These changes appeared to stop the "squatter" problem. Swine consultants in the US Midwest see "squatters" sporadically but no extensive diagnostic workups have been done to rule in or out demyelination. "Squatters" are usually attributed to bacterial abscesses such as those caused by *Strep suis* or to injury.

Spray drying of non–Roundup Ready crops with glyphosate prior to harvest, as was done in Dr. Sheridan's "squatter" case, is probably a greater risk for demyelination than glyphosate residue in Roundup Ready corn and Roundup Ready beans and bean meal. The glyphosate present from spray drying has had less time to chelate with Ca++, Mg++, etc., and more of its four sites are still available to physiologically immobilize Co++ in the animals that are consuming this food.

Glyphosate chelated to cations in the soil can be desorbed by phosphate fertilizers to release it from its relatively inactive state to become an active chelator, herbicide, and antimicrobial agent again. It is reasonable to think that glyphosate residues remaining after spring weed spraying that are absorbed in the intestines may also be desorbed by phosphates during metabolism.

Internationally known dairy veterinarian and consultant, Dr. Earl Aalseth, recommended that dairies give 7.5 cc of 1,000 mcg/cc vitamin B12 intramuscularly to all fresh cows twice a day for 2–3 weeks post freshening. This treatment gained up to 5 pounds of milk/cow/day throughout the lactation on lower feed intake with better reproductive and overall health [7]. Fiber digesters in the rumen have a high cobalt requirement. Glyphosate's K value for cobalt (7.23) shows that the force of chelation for this cation is about 100× more than for the manganese cation (5.47). Glyphosate's "cousin," glufosinate, the herbicide used with Liberty Link seed, chelates cobalt as well. Would Dr. Aalseth's advice work in an organic herd with, hopefully, much less glyphosate or glufosinate in the feed?

Again, there are four chelation sites per glyphosate molecule. Based on glyphosate's molecular weight, 1 μg (1 mcg/kg is 1ppb.) of glyphosate has 3.561×10 to the twelfth molecules in it. The pig is a model for B12 (cyanocobalamin) in humans. Dr. Mike Sheridan supplemented his "squatter" pigs with 15 ppb B12, and that level would also be realistic in a human diet. A cobalt atom is in the center of every one of those B12 molecules and makes up about 5% of its mass. Is it any wonder that we are seeing B12 deficiency and demyelination of nerves in humans when the US Federal Register allows up to 30 ppm glyphosate in bread and cereal products? [8] In total, 30 ppm glyphosate would be over 30,000 × more potential chelator than the not quite 1 ppb cobalt needed in daily B12. The common residue level of 1 ppm that occurs because of the permitted preharvest desiccation of the wheat that goes into bread would still be 1,000 times greater than the trace mineral present. Bread is not our only food contaminated with glyphosate. About 70 human edible crops in the United States and Canada are currently being spray dried with this herbicide. This includes other small grains, legumes, and peanuts.

Since it is common practice to supplement human diets with folate, demyelination of nerves becomes the first symptom of B12 deficiency instead of pernicious anemia [9,10]. Demyelination can involve any nerves and does not have to involve the spinal cord as experienced with Dr. Sheridan's "squatters." Nurse Sally Pacholok, author of *Could it Be B12? An Epidemic of Misdiagnoses*, estimates that 40% of Americans are now low in B12.

8.2.3 COPPER DEFICIENCY IN SHEEP AND GOATS

Prior to 1996 and 1998, when Roundup Ready soybeans and Roundup Ready corn were commercialized, copper poisoning was a common cause of death in sheep. Although sheep and goats need copper, they are much less tolerant of it than other livestock. Because of this, most sheep and goat minerals are not supplemented with copper. The copper needed is supposed to come from the forage and grain in the diet. With hogs raised on solid floors in the 1970s and 1980s, there were a lot of cases of enteric Salmonellosis as the pigs recycled their manure. Copper sulfate, added at 2 pounds/ton of feed for 3 weeks, was a common follow-up treatment for enteric Salmonellosis and worked well if the client did not leave the copper sulfate in too long and cause copper poisoning in the hogs.

It is important to maintain balance in the copper–molybdenum–sulfur "triangle"; however, today, it is not uncommon to find many soils that are low in both sulfur and molybdenum. Some of these same soils have been heavily fertilized with hog manure laden with copper sulfate or with dairy manure contaminated with discarded copper sulfate footbath water. It seems somewhat surprising that when we pasture and grow small ruminant hay on these soils that should be high in copper while low in molybdenum and sulfur, we have had sheep and goats dying from copper deficiency. The deficiency in copper was confirmed by liver and feed submissions to Iowa State University (Figure 8.4).

Copper deficiency looks similar to copper poisoning. Both cause severe anemia and high death loss with copper deficiency also causing little to no wool growth. Clients that used non–Roundup Ready corn, resulting in less glyphosate in their sheep and goat rations, had less severe issues with low copper. Instead of death loss and lack of wool growth, their herds had poor immune function related to the lack of copper, resulting in extra trouble with foot rot, pneumonia, and worms. When feeding free choice mixed minerals to sheep or dairy goats, Amish clients needed to mix one bag of calf mineral containing 1,150 ppm copper with one bag of sheep and goat mineral (with no additional copper added) to make their mineral mix. In contrast, other clients using Roundup Ready corn had to add three bags of the calf 1,150 ppm copper mineral to one bag of the no added copper mineral to "fix" the problem. In the fall of 2016, I visited the Netherlands and met Australian organic farmer Steve Marsh. Steve had had 3,000 head of ewes on his ranch. After his land was contaminated by glyphosate, his ewes suffered from copper deficiency with many dying of anemia and many growing no wool. He did not see swayback lambs that the literature associates with copper deficiency and neither have we.

8.2.4 MOLYBDENUM (MO) DEFICIENCY

Molybdenum levels in most Midwest soils have been low for years. This spring, several speakers at a meeting hosted by Central Iowa Agronomy and Supply stated that 95% of Iowa soils are low in molybdenum [11]. This seems to make my recent low molybdenum swine liver submissions more understandable since very few Midwest corn producers are fertilizing with molybdenum. The National Research Council's dietary guidelines for dairy and swine state that it is hard to create a molybdenum deficiency. According to their guidelines, most feedstuffs have the amount required and toxicity has often been the worry. Cattle on pastures with 20 ppm or

Client Phone: Species: Ovine   Age:

Client Fax: Breed: Unknown   Weight:

Client Account#: 000401331   Sex: Received: Feed

Date Received: 09/19/2016   Previous Case:

Sample Taken: 09/15/2016   Farm Type: Other   Reason: General Diagnostics

Accompanying Cases: Animal ID(s):

Final Report(s): 9/27/2016 2:18 pm

Chemistry:
Results are below.

Sheep Diet DM basis

	Reference range ppm	Maximum tolerable level	Feed
Cd (ppm)			0.022
Ca (ppm)	2000-8200		4686
Cr (ppm)			4.100
Co (ppm)	0.1-0.2	10	0.756
Cu (ppm)	7-11	25	4
Fe (ppm)	30-50	500	120
Mg (ppm)	1200-1800		1331
Mn (ppm)	20-40	1000	51.8
Mo (ppm)	0.5	10	3.72
P (ppm)	1600-3800		3491
K (ppm)	5000-8000		6029
Se (ppm)	0.1-0.2	2	0.38
Na (ppm)	900-1800	35000	1540
Zn (ppm)	20-33	750	115

FIGURE 8.4 Sheep diet mineral ranges (dry matter basis). (ISU lamb liver submission in this case showed 9 ppm copper, when the reference range is 25–100 ppm.)

greater molybdenum will quickly develop diarrhea. However, the National Research Council does list molybdenum as an essential dietary ingredient. Molybdenum is a cofactor in four essential mammalian enzyme systems. Recently, scientists have realized that it is also a cofactor in a fifth enzyme called nitrate reductase that is used by bacteria in a mammal's oral cavity and saliva to convert nitrate into nitrite.

In the soil, molybdenum is found as an anion and, as such, has a negative charge. In the enzymes where it acts as a cofactor, it is oxidized to positive valences. In all five of these enzyme systems, it becomes part of a pterin protein and is found as a cation. In the redox reactions in which molybdenum takes part, it is commonly shifting between valences of four and six [12,13].

Molybdenum, as a cofactor for nitrate reductase, is needed by the bacteria in root nodules of legumes to fix nitrogen. It is needed by all plants, including corn, to synthesize protein. Bacteria in saliva and the oral cavity of animals use nitrate reductase in the same process to convert nitrate to nitrite. In pigs and humans, the oral nitric oxide produced, initiated with the nitrate reductase pathway, provides around 10% of the total nitric oxide needed to close the pylorus. If the pylorus does not shut completely or for long enough, stomach acids and pepsin do not do a good job of breaking down whole proteins. These whole proteins include both infectious viruses and proteins in foods. A virus is going to have a much better chance of causing clinical disease if many virions go through the stomach and into the intestinal intact. If whole proteins in foods are not broken down into amino acid pieces in the stomach, they will go through fermentations in the intestine, often leading to excessive gas, leaky gut, and malnutrition.

Dr. Mike Sheridan gave advice to fellow Canadian veterinarians when, via contaminated feed, the porcine epidemic diarrhea virus (PEDV) was introduced into Canada in 2015. Porcine epidemic diarrhea virus is a coronavirus that causes a severe diarrhea and high mortality in piglets that are nursing sows as well as in pigs post weaning. Apple cider vinegar is a good source of molybdenum as well as *Acetobacter*. These bacteria produce nitrate reductase by synthesizing the pterin protein where molybdenum is a structural component. After making a stock solution by adding 2 oz. of apple cider vinegar to a gallon of water, Dr. Sheridan had the veterinarians and managers of the hog units with PEDV meter this apple cider vinegar stock solution into drinking water at a rate of one English ounce (30 ml) per gallon. All sows and pigs on virus positive farms were placed on the apple cider vinegar water treatment. The treatment rapidly dropped morbidity and mortality where the pigs and sows had clinical signs. The treatment worked as a preventative as well since nursery rooms that were still negative on positive farms stayed that way. This simple intervention was successful on one site where there were almost 40,000 pigs.

My veterinary partner and I have used some organic apple cider vinegar in practice as well. We have used it to treat bloody bowel in fat hogs and as an aid in treating porcine reproductive and respiratory syndrome virus (PRRS) in positive wean-to-finish operations to hold down both virus and secondary bacterial infections.

More recently, I am adding one-half pound of sodium molybdate (0.2% molybdenum) to 1 ton of hog feed in a couple of operations. This product is primarily manufactured for sheep producers to use in cases of copper poisoning. As discussed earlier, demand for this particular use is now low, so the addition of this product to swine diets is currently not costly. Too much molybdenum is a worry for good reasons. The product must be added to the premix, and then that molybdenum-amended premix is carefully and thoroughly incorporated into the complete diet. I hope our pork producers, as well as the farmers growing grain for them, will include molybdenum in their crop fertility program. Then, it will not have to be added to the feed, and crop production will benefit as well.

Does glyphosate truly increase the risk of molybdenum deficiency? Molybdenum is part of the anion molybdate in the soil profile. When added to feed, a sodium molybdate salt is used. Once bound to its pterin cofactor as an enzyme, its valences commonly shift between positive four and positive six. Even if chelation does not happen, the antimicrobial action of glyphosate could diminish bacteria in the oral cavity that normally create the pterin cofactor and provide nitrate reductase. If the normal flora of the oral cavity is killed, either by antimicrobials such as glyphosate or by other factors, including low humidity, the molybdenum–pterin enzyme system cannot work even if adequate molybdenum is present [12,13].

8.2.5 OTHER POSSIBLE MINERAL ISSUES

Because glyphosate has different K values for the different cations, glyphosate can cause toxicities as well as deficiencies. This can potentially occur in the same animal, and every deficiency sets the ground for a potential toxicity!

Manganese miners in Australia suffer from a prion brain disease where manganese takes the place of copper in a brain protein [14]. Any of these miners would not want glyphosate in their diet since the K value for copper and glyphosate chelation is over 1 million times stronger than the chelation force for manganese and glyphosate. With glyphosate in their diet, there would be a chance that glyphosate bound to excessive manganese would release the manganese to preferentially chelate copper. Instead of having copper incorporated into brain proteins, there would be a better chance of having an occasional manganese used in its place to result in a bent protein prion.

We need much more research to answer questions about possible mineral interactions with glyphosate, and its metabolite aminomethylphosphonic acid (AMPA). Unbiased science is now hard to find and hard to do, since Ag business provides much of the funding. Ag business has convinced farmers and the public that one weed in a field is one too many. Because of blind acceptance of substantial equivalence, important questions are not being asked or answered. Substantial equivalence for the Roundup Ready gene alteration was based on pooled historical data of crop analysis and not on replicated side-by-side studies that follow accepted science. For example, if the gene-altered crop containing its herbicide fell within a range of previously recorded mineral values for that crop, the genetically altered crop was declared substantially equivalent to the natural crop.

Many nutritionists want researchers to update the required mineral values for our food animals since today's dairy cows and sows have larger demands than those of the past. The goal will be worthy if an attempt is made to codify the deficiencies and excesses that have resulted from the actions of glyphosate and its effects on the microbiome. Older texts like Puls Toxicology may need to be used for guidance since available scientific procedures may have trouble in consistently measuring the amount of a mineral cation that is truly available for metabolism in glyphosate-contaminated feeds. This will be true even if glyphosate and AMPA levels are measured at the same time as the minerals.

Here are a few of the questions that need to be answered by unbiased scientific research:

1. Dr. Don Huber has shown how glyphosate increases the toxicity of boron in sugar cane [15].

    Some swine veterinarians are adding boron to sow diets to strengthen bones. Will glyphosate in the sow's regular diet increase the chance of boron being toxic to those animals?

2. If the heavy and dangerous metal cadmium is in the environment and in feedstuffs, will glyphosate in the soil and diet help carry this toxic cation into animals?

3. In areas with iron poor soils, will glyphosate residues in feed increase the need for iron supplementation?

4. Many areas have soils deficient in selenium. Both selenium and vitamin E are important antioxidants to inhibit viral infections [16]. Does glyphosate chelate this divalent cation, and what is its K value?

5. What are the dissociation values for the different cations when all of them are available in the feed mix? Can this question be answered in a living mammal?

6. A copper-dependent enzyme, thiol oxidase, is responsible for building disulfide bonds in keratin. Are the Amish in our area correct when they think that feeding non–Roundup Ready feeds to their horses results in better hoof walls? Is better hoof wall health why dairy herds that do not use glyphosate have no or minimal trouble with hairy heel warts?

7. Is this why those same herds have less trouble with heel horn erosion triggered by zinc deficiency?

8. Today's veterinary profession is convinced that chelated organic trace minerals perform better than the inorganic salts and that they are worth the extra expense. When discussing manganese, I mentioned that most of our dairy clients use a manganese amino acid complex in their mineral premix. Are complexed trace minerals necessary in diets without residual glyphosate?

8.3 TOXICOINFECTIOUS BOTULISM

I was first introduced to this disease term by Professor Monika Krueger from the University of Leipzig in Germany. Dr. Krueger holds PhDs in mycology, microbiology, and veterinary pathology. She spent much of her career at The Human Health Center at the University of Leipzig studying sudden infant death syndrome and crib death. After meeting Dr. Don Huber, she became even more interested in glyphosate. In her lab, she showed how glyphosate, at a common residue rate of 0.1 ppm, inhibits some beneficial bacteria. *Enterococcus faecalis* and *Enterococcus faecium* are two of the bacteria that keep the Clostridial bacteria in check and, in the case of *Clostridium botulinum*, reduce the production of the nerve toxin (BoNt) that causes botulism [17,18]. Glyphosate does not inhibit bacteria in the Clostridial family unless it is at a concentration of 5,000–10,000 ppm. This concentration of glyphosate does not occur outside of a laboratory. I met Dr. Krueger at the World Food Safety Conference in China in 2014. Previously, she had coached me via e-mail as I was trying to help one of our dairy clients whose herd was experiencing a problem. I will discuss his situation later. Before Dr. Krueger did animal studies with cattle and chickens, she was approached by the German brewing industry. Farmers raising barley for malt started to dry their crop by spraying glyphosate preharvest. Just as it is hard to make good cheese out of antibiotic contaminated milk, it is hard to ferment beer with the antimicrobial herbicide glyphosate in the mix.

Monsanto started the preharvest spraying practice of grains worldwide in 2002. Mainly, because of Dr. Krueger's work, Germany became the first major economy to outlaw all desiccation or spray-drying of non–Roundup Ready human edible crops with glyphosate. The United States and Canada still allow the spraying of about 70 human edible crops immediately ahead of harvest, and Monsanto-Bayer promotes this use as a "harvest aid." No government has directly approved the use of glyphosate as a drying agent or harvest aid. Recommending weed control after the weeds are mature and have largely gone to seed is questionable. Additionally, spraying glyphosate ahead of winter assures harm is done to the beneficial mycorrhizal community and other soil organisms that should be maintained for next year's crop. A longer half-life for the herbicide is also assured with late season applications because there is little or no bacterial breakdown of the product through the cold winter months.

The US CDC considers the two Enterococcus bacteria that Dr. Krueger found useful as inhibitors of Clostridial toxin production, to be "bad bugs" and not treatable (ESKAPE pathogens) [19]. In veterinary medicine, these two organisms are routinely found in probiotic products for scouring foals and calves. They have become important treatments. Quoting pediatrician Michelle Perro from a recent e-mail, "The human microbiome is made up of a microbiome, mycobiome, and microvirome. There are really no such things as 'good' and ;bad' bacteria. They live in balance and keep each other in check. They work collaboratively via quorum sensing and EMF transmissions to optimize their own situation…."

It seems logical to assume that a broad-spectrum chelator-antimicrobial hanging onto minerals while in the gastrointestinal tract would alter this quorum sensing and EMF (electromagnetic frequency).

Eventually I found out that toxicoinfectious botulism was described years earlier in the United States by Dr. Tom Swerczek. He wrote an AVMA Journal article about this disease in racehorse foals and adults in February of 1980 [20]. Dr. Swerczek had to come up with his own culture media and technique to prove that the disease existed. Later, the US CDC used his culture media and technique to diagnose botulism in humans.

Although the clinical picture of toxicoinfectious botulism in horses is slightly different than what is seen in cows, there are many similarities. Dysphagia, resulting in horses and cattle occasionally biting off their tongues, is one of the similarities, along with aspiration pneumonia. Dr. Swerczek's cases were dependent on wet feed, as were our cattle cases. In his opinion, cases went up dramatically when glass-lined silos were used to store forage in the early 1970s. Horses, as lower gut fermenters, cannot handle fermented feeds as well as ruminants can. Even with Dr. Swerczek's great reputation, he had trouble convincing other equine veterinarians that he knew what he was talking about. This remained true even after he was able to culture *Clostridium botulinum* from case tissues. Medical doctors that believe in the involvement of *Clostridium botulinum* neurotoxins (BoNts) in chronic fatigue syndrome have the same problem proving that toxicoinfectious botulism might contribute to this syndrome.

Regrettably, the mouse inoculation test used to diagnose classical botulism does not work on the toxicoinfectious form of the disease. Less toxin is elaborated in the gut with toxicoinfectious botulism compared to the quantity of toxin present in classical botulism. In classical botulism, the toxin is produced in putrid feed before ingestion, and there is enough remaining preformed toxin present in the gastrointestinal (GI) contents of dead animals that diagnostic mice die when injected interperitoneally with material from these specimens.

Since mice are not as sensitive to the BoNts as are horses, cows, pigs, birds, humans, and many other species, diagnostic mice will not die when injected with specimens from the GI tracts of any of these animals that have died from toxicoinfectiuos botulism.

Using PCR to detect the gene that enables a bacterium to produce BoNt, the test used at the Veterinary Botulism Lab at New Bolton, Pennsylvania, is not a reliable test either. Even if PCR is positive for the bacterial gene, there is no proof that the toxin was produced. Culture is still not easy. Clostridia that produce BoNt are obligate anaerobes and hard to grow. This is true even if they are put in the right media at collection. It is now known that *Clostridium novyi* and *Clostridium hemolyticum* can both produce BoNts similar to that produced by *Clostridium botulinum* [21]. *Clostridium butyricum* can also produce BoNt. The latter species is named for the smelly butyric acid that it makes when growing in silages that have been made too wet. The cases of toxicoinfectious botulism we have seen with dairy cows have also involved haylage made too wet. Haylage made over 70% moisture favors the growth of *Clostridium butyricum*, not only in the silage, but also in the GI tracts of the animals eating the silage. After 9/11 in 2001, botulism was labeled a select agent in the United States. Therefore, no land grant college laboratory can do research with BoNts since their labs currently lack the level of biosecurity required to handle agents labeled as potential weapons of terrorism or war.

Dr. Monika Krueger developed ELISA test kits for the seven BoNts (A–G) before she retired. These ELISA tests can detect the actual small protein toxins that have no DNA or RNA. In contrast, PCR detects DNA or RNA and cannot detect the small protein BoNts. Access to her Elisa tests in Canada or if someone else could develop similar assays in Canada would be a tremendous diagnostic tool to have available. Canada did not have the same regulatory response to 9/11 as the United States and, therefore, BoNt research is not as restricted. In the advent of such services available outside of the United States, diagnostic samples could then be sent across the border for confirmation. In my opinion a better option would be lifting the ban on studying botulism at college institutions and other research laboratories.

The initial cases of toxicoinfectious botulism we saw involved haylage made too wet that was fed with Roundup Ready corn grain and corn silage. These Roundup Ready crops were also grown on ground where glyphosate had been used continually for many years. We still work for a dairyman who has never used glyphosate. Ten to fifteen years ago, shortly before the time of our first cases of toxicoinfectious botulism, he was guilty of consistently making haylages too wet. This resulted in excessive growth of *Clostridium butyricum*, and the smell of butyric acid was apparent before turning into his driveway. When he fed this appetite-suppressing haylage to prefresh cows, dry matter intake was too low for these animals' needs. This resulted in a great many cows having a displaced abomasum after calving because the lack of a full rumen along with space created by an empty uterus allowed that organ to shift position. Without glyphosate in the picture, the client never experienced a problem with botulism.

My veterinary partner and I recognized the toxicoinfectious botulism herd problems more easily than many would have since we had seen a classical botulism case. In this herd, many cows became recumbent and died after they ate a total mixed ration that had a dead decomposing animal picked up at harvest inadvertently mixed into it.

Botulism is a "cousin" to lockjaw, which is caused by *Clostridium tetani*. The tetanus toxin works on sensory neurons while the botulism toxin works on motor neurons. In the case of tetanus, no clinician tries to diagnose it through lab submissions. The diagnosis is based on the clinician's recognition of the clinical signs. Clinical tetanus results in a sawhorse stiff stance. The third eyelids of an animal suffering from tetanus flip across the eyes when an examiner claps in front of the animal's face. Lab data may be used to confirm a diagnosis, but it is rarely used to make a diagnosis of tetanus. At this time, without confirmatory ELISA tests for the seven BoNts, clinical signs must also be relied upon for making a diagnosis of toxicoinfectious botulism.

In our first dairy herd case, the entire herd backed down on feed. Milk production went down and one cow in the 70-cow dairy died. This dairy rarely had death loss, so the dairyman thought the death was a fluke and did not request a postmortem. A few days later, many cows were completely off feed and some had constipation and others had diarrhea. BoNt works on smooth muscles of the GI tract as well as on skeletal muscles. Some of the cows got what appeared to be allergic reactions with edematous, puffy eyelids, swollen skin, and edema around the anal sphincter that was probably due to toxin buildup following gut stasis caused by BoNt. In addition, some cows had their tongues hanging out of their mouths so far that they had a great deal of difficulty eating and drinking. Some of these cows dropped from over 100 pounds of milk per day down to almost nothing, along with loosing over 150 pounds of body weight in 4–5 days (Figure 8.5). A few cows bit their tongues off. At first, we thought they were getting their tongues caught on something sharp. Then more did it without anything sharp around. Intramuscular administration of the third-generation penicillin Excenel seemed to help keep the cows alive but did not help them come back into production. First and second-generation penicillins given intramuscularly also helped alleviate clinical signs; however, none of these penicillins did anything for the BoNt that was already elaborated, but they knocked the Clostridium overgrowth down enough to stop production of more BoNt. The use of Excenel on cows in the US diaries that are failing in production and condition, and who are subsequently sent to slaughter, is more than likely covering up more than a few cases of toxicoinfectious botulism. When I asked for help on the international American Association of Bovine Practitioners list-serv at the time of this case, I got no immediate response. Dr. Monika Krueger told me to give sauerkraut juice orally since cabbage is the only common vegetable with acetylcholine in it. That neurohormone is the one blocked at the motor neuron junction by BoNt. This recommendation was followed with later cases when I found a source for the sauerkraut juice. She also mentioned the

FIGURE 8.5   Cow with paralyzed tongue from chronic botulism.

ability of the nonabsorbable sugar lactulose, given orally, to bind to the BoNt while it is still in the intestinal tract. We tried some of it without being able to tell if it helped. The client discarded the rest of the haylage. Four cows died, and about 25 were culled because of this break. No glyphosate has been used on this farm since, and we have not experienced another outbreak of the disease even though haylage has occasionally been made too wet.

In our second dairy case, a new client moved to an acreage with a 400-head free-stall barn. Since no crop ground was purchased, all feed for the cows was purchased. The purchased corn silage was Roundup Ready that had been grown on light-textured ground that had been producing only Roundup Ready corn and beans for at least 12 years. Cows at this dairy could not consume enough water and could sometimes be observed drinking somewhat like cats, with their tongues lapping and moving side to side. Even in cool weather, abnormal numbers of cows would be standing around the waterers, lapping. Some of them would have their front feet in the water trough. This helped water consumption by letting water run down the esophagus by gravity. It did not seem to matter what nutritionist was trying in an effort to help. This included my second guessing! Without adequate water intake, feed intake remained low. Nothing could get dry matter intake above 47 pounds per head per day when 52–53 pounds or more would be much more typical consumption for big Holsteins like this client had. Since his forages were considered better than average for feed quality, that high quality should have helped daily intakes as well.

The client was constantly hauling cows to slaughter who had persistent right-sided gas. Abomasums were often not motile and contained gas, but they were not true right-sided displacements. Surgeries for right-sided displaced abomasum on these cows did not help. On those that I operated on, the abomasum was not twisted, but just dilated. Eventually I did some exploratory surgeries immediately ahead of slaughter on two right-sided gassy cows. We sent abomasal, intestinal, or colon contents to Iowa State University for culture, depending on which parts of the digestive tract were gas filled. One of the spiral colon samples came back with no growth. This still makes my head spin along with the heads of some clinicians at Iowa State University. We know we cannot culture more than 80% of the bacteria in any mammalian gastrointestinal tract. We also know we will never be able to culture all of them since they are so dependent on each other. Even with this knowledge, it is still shocking that nothing grew from a 5 mL inoculum from a cow's spiral colon (Figure 8.6).

On two feedlot farms with unexplained sudden deaths in feeder calves, we had wild pigeons that were eating the same feed as the cattle. These birds were losing the ability to fly and then dropping over dead with their necks flopped to one side. We sent some of these dead birds to Dr. Darrell Trampel, long-time poultry pathologist at Iowa State University, for postmortem. He concurred that it was limberneck caused by botulism. He was interested in Dr. Krueger's ELISA tests and was willing to try to go "through the hoops" to get them to the United States and to help standardize them for use. Regrettably, he died suddenly on August 31, 2014, before he could bring Dr. Krueger's BoNt ELISA tests to the United States.

In conjunction with a trip to the Netherlands in the fall of 2016 to testify at the Monsanto Tribunal at The Hague, I went on a farm tour organized by Dr. Don Huber and my host, crop consultant, Roelf Havinga. On the first day of the tour, we went to two dairies that were both having herd issues with toxicoinfectious botulism. At the first 600-cow rotary parlor dairy, many cows were around the water tanks drinking like cats and coughing, with bits and pieces of cud flying out their mouths. Some of the cows were stumbling away from the water tank with a lack of good motor control and coming down too hard on their feet.

The second dairy toured had two cows that had become extremely lame a few days before. One of the two cows had fractured a pedal bone in a front foot, likely due to lack of motor control of her legs. Both dairies were purchasing wet-brewers' grains from a Danish brewery across the strait and feeding it with their wet grass haylages. The Danish brewery just happened to be in the hometown of Danish pork producer Ib Pedersen. Ib had also testified at the Tribunal and then stayed for this tour where he was able to provide valuable information. The Danes had voluntarily quit spray-drying their barley with glyphosate in previous years, but this particular fall had decided to let their farmers spray-dry (desiccate with glyphosate) again if they so desired. Dr. Krueger had helped Ib Pedersen test the barley and the by-product from this plant for glyphosate before the voluntary ban was in place. About 2–3 weeks before my trip, a new crop of barley was being harvested, and some of it was spray-dried again. The two dairymen, as well as others, had used wet brewers from this plant with no apparent issues that summer until they got this new spray-dried glyphosate-contaminated feed.

A neighboring dairyman to the first dairy toured had lost over 30 cows in the previous week. Postmortems were performed at the veterinary school in Utrecht. The cause of death was ruled aspiration pneumonia. The next day, we went back to the first dairy to talk to farmers, vets, and

---

Veterinary Diagnostic Laboratory
Iowa State University
College of Veterinary Medicine
Ames, Iowa 50011-1134
Phone: 515-294-1950
Fax: 515-294-3564

Accession: **2017070105**

Final Report
Accessed Date: 05/24/2020 06:28 pm

DR. ARTHUR G. DUNHAM
RYAN VETERINARY SERVICE
PO BOX 48
RYAN , IA 52330

Site

Premises ID# :

Owner
Diagnostician: SAHIN, ORHAN

Lot/Group ID :
Source/Flow ID :
Reference :
Case Tags :

Client Phone	Species: Bovine	Age:
Client Fax	Breed: Holstein	Weight:
Client Account#: 000401331	Sex:	Received: 5 Stomach Content
Date Received: 10/19/2017	Previous Case:	
Sample Taken:	Farm Type: Other	Reason: General Diagnostics
Accompanying Cases:	Animal ID(s): A1, A2, A3, B1, B2	
Final Report(s): 11/9/2017 3:07 pm		

History:
- Intestinal and stomach contents (obtained via exploratory biopsy) submitted from 2 lactating cows that have not been eating well or producing milk well.

Bacteriology:
- Routine culture: See detailed **Final** results below under **Culture Summary**.

Molecular:
- See results below.
- Please note that the *Clostridium novyi* and *Clostridium haemolyticum* are non-validated tests and results should be interpreted with caution. Please call if there are questions.

Comments:
- Please contact the laboratory if you have any questions. (11/9/17 os/clm)

Orhan Sahin, DVM/MS, PhD, DACVM
Assistant Professor
Bacteriology Section Leader, Microbiologist

FIGURE 8.6 Iowa State University Veterinary Diagnostic Lab Report. (Iowa State University Veterinary Medical Diagnostic report, Bacteriology.)

(Continued)

Bacteriology

Culture Summary

Animal ID	Specimen	Enrichment	Growth	Organism	Comment
A1, Tube #1	Intestinal contents		Few	B.CER --> Bacillus cereus	
A2, Tube #2	Intestinal contents		Couple	E.COLR --> Rough Escherichia coli	
A2, Tube #2	Intestinal contents		Couple	NTG --> Noted Growth	likely alpha hemolytic Strep
A2, Tube #2	Intestinal contents		Single	E.COLH --> Escherichia coli haemolytic	
A3, Tube #3	Intestinal contents			NG --> No Growth	
B1, Tube #4	Intestinal contents		Single	NTG --> Noted Growth	Klebsiella-like
B1, Tube #4	Intestinal contents		Few	B.CER --> Bacillus cereus	
B1, Tube #4	Intestinal contents		Few	NTG --> Noted Growth	likely alpha hemolytic Strep
B2, Tube #5	Intestinal contents		Moderate	E.COLR --> Rough Escherichia coli	
B2, Tube #5	Intestinal contents		Few	E.COI --> Escherichia coli intermediate	
B2, Tube #5	Intestinal contents		Moderate	NTG --> Noted Growth	likely alpha hemolytic Strep
B2, Tube #5	Intestinal contents		Few	BAC.SP --> Bacillus species	

Research & Development

QA/QC testing

Animal ID	Specimen	Test	Target Agent	Result	Comment
A1, Tube #1	Abdominal contents	PCR	Other	Negative	Negative for C. Novyii and C. Haemolyticum
A2, Tube #2	Abdominal contents	PCR	Other	Negative	Negative for C.Novyii and C. Haemolyticum
A3, Tube #3	Abdominal contents	PCR	Other	Negative	Negative for C. Novyii and C. Haemolyticum
B1, Tube #4	Abdominal contents	PCR	Other	Negative	Negative for C. Novyii and C. Haemolyticum
B2, Tube #5	Abdominal contents	PCR	Other	Negative	Negative for C. Novyii and C. Haemolyticum

FIGURE 8.6 (CONTINUED) Iowa State University Veterinary Diagnostic Lab Report. (Iowa State University Veterinary Medical Diagnostic report, Bacteriology.)

agricultural officials about toxicoinfectious botulism. I argued that adult cows should never have a herd case of aspiration pneumonia. If they cannot swallow normally and must lift their heads for the help of gravity when the epiglottis is partially paralyzed, some water and ingesta will run down their windpipes.

While speaking in a reception area above the rotary parlor, I could look out a window and see a wheat field that was yellow brown from recently being sprayed with glyphosate. Not everyone liked the presentation since many that were present were also using glyphosate to dry their crops. Those not owning cows were not worried about keeping cows alive. It rains a lot in northwest Holland and Denmark in the fall. Finding some dry days to dry windrowed small grains is not easy; however, windrowing was used until 2002 when the agricultural industry promoted spray-drying worldwide. Holland outlawed all glyphosate spray-drying of human edible crops shortly after 2016. I like to think that I was a part of the effort to solidify that decision.

In December 2017, I participated in a glyphosate roundtable at the University of Calgary Veterinary School. Veterinary clinician Dr. Eugene Janzen made me stop my power point to empha-size that cattle with toxicoinfectious botulism will bite their tongues off. Canadian veterinary consultant, Dr. Ted Dupmeier, and Dr. Janzen are seeing this damage along with considerable death loss in feeders that are eating the glyphosate spray-dried pea and lentil foliage in their total mixed rations after the human edible peas and lentils are harvested. In this case, the glyphosate used for spray-drying does not have weeks and months to bind to micro and macro mineral cations in the soil and plant, as it does when used in the spring. Because of this, it acts as an active antimicrobial with many of its chelation sites open and does not need desorption to make it active. *Enterococci* are inhibited, and Clostridia proliferate. Remember that 0.1 ppm of glyphosate in the GI tract can kill Clostridia-suppressing *Enterococci*.

Due to experience using oral antibiotics, swine practitioners realize that 0.1 ppm does not have to be 0.1 mg kg^{-1} body weight but can be 0.1 mg kg^{-1} solute in the GI tract. An effective dose for an enteric problem can be much lower than that for a systemic problem. Many critics forget this and think that the amount of residual glyphosate in feed, when compared to body weight, is too low to cause antimicrobial issues. They forget (or ignore) that the glyphosate is having local—rather than systemic—action. Dr. Jorn Erri reported that he also has seen numerous cases of toxicoinfectious botulism through his work for a semen supplier in Denmark. He has supported me in my concerns via e-mail. Dr. Erri has good experience improving the performance and stopping clinical signs of toxicoinfectious botulism in both calves and cows using a vaccine for botulism; however, this vac-cine is not available in the United States [22]. He agreed to meet me at the World Dairy Expo in Madison in 2018. Immediately before our visit, he toured four Midwest dairies. Three of the four dairies he visited had cows drinking like cats. In his clinical judgment, they were suffering from toxicoinfectious botulism.

8.4 MYCOTOXIN ISSUES

Dr. Bob Kremer, long-time Agricultural Research Scientist and now at the University of Missouri, has done multiple studies showing how glyphosate use at real-world rates can favor the *Fusarium* mold family [23]. Just because more mold growth is encouraged does not mean that there will necessarily be more mycotoxin production, but increased risk of that happening is certainly there. The USDA did recent studies to contradict Kremer's work. The USDA studies "stacked the deck" in industry's favor by selecting a Midwest farm that still practiced crop rotation with a small grain crop ahead of hay, before going back to corn and beans. Glyphosate was not used every year. Molds like *Fusarium* are ubiquitous in soil and multiply rapidly in the crop residues of minimum and no-till fields. Alternating crops and turning residue under in a crop rotation certainly help control

Fusarium problems. Regrettably, most Iowa farms are not rotating anymore. Many growers are just alternating between beans and corn, or they are just planting continuous corn. Herbicides used in these latter scenarios often contain significant rates of glyphosate every year.

8.4.1 Fusarium Mycotoxins

Another misconception that was set straight at a Fusarium Head Blight Conference in Florida in 2009 is the idea that you have to have insect or hail damage to the seed head, along with the right weather conditions at flowering, to get *Fusarium* mycotoxins in the grain. I saw how this is not the case after a severe hailstorm in 2010 when I helped collect samples in our practice area for an Iowa State University hail damage study. Many fields of Roundup Ready corn that did not get hit by hail had much higher levels of mycotoxins than some of the non-Roundup varieties that were severely stripped by the hail.

Canadian scientist Miriam Fernandez has shown that weather is not as important a predictor of head blight in wheat as is glyphosate use in previous years [24]. Weeds gaining resistance to glyphosate tend to be the ones that are shade and moisture-tolerant. These weeds are more tolerant of molds that proliferate under these conditions. Johal et al. showed how plants grown in sterile soil are not killed by glyphosate but just momentarily stunted [25]. When glyphosate shuts down the Shikimate pathway and prevents the subsequent synthesis of aromatic amino acids, there are fewer plant defense chemicals manufactured. *Fusarium* proliferation occurs in the root zone and mycotoxins produced there can be translocated to the grain. Fungicides applied to the crop may help with certain fungal leaf diseases, but they have little value in killing *Fusarium* at ground level. The four *Fusarium graminearum* mycotoxin families we most commonly deal with in swine and cattle diets are: vomitoxin (DON), zearalenone, fumonisin, and T2. The 2009 and 2018 crop-growing seasons in northeast Iowa, USA, were very conducive to the growth of *Fusarium* and the mycotoxins it produces. 2019, with a prolonged wet Fall and gas shortages for drying harvested grain, also resulted in higher levels of mycotoxins in grains. Even with these peak years, companies that make binders for these toxins know that the trend in levels has kept going up since the late 1990s. Levels have certainly increased recently (Figure 8.7). Cattle are usually more tolerant of low levels of mycotoxins than pigs, but the levels produced in 2009, 2018, and 2019 were high enough to trigger health problems in cattle as well. Rumen and intestinal bacteria can sometimes unmask toxins that do not show up in lab tests [26].

Corn by-products, like the modified distillers dried grains being fed as a protein source to hogs, or the wet or dry gluten by-product fed to cattle, are three to four times higher in toxin levels than the corn that was delivered to the plant to make them. The toxins are concentrated in these products when the relatively clean germ and endosperm are used up producing ethanol. Combinations of the four toxins, along with their many first cousins, are usually present if any toxin is elaborated. That makes it difficult to predict clinical effects, even if using more elaborate tests like Alltech's 37+® or Olmix's Myco'Screen Report®. These tests check for multiple toxins rather than checking for the most common two to three (Figure 8.8).

8.4.2 Signs of Mycotoxin Intoxication

One common clinical sign in dairy cattle suffering from T2 combinations is porphyria. The toxin causes liver damage, which results in the animal not being able to handle the porphyrin break-down products from the normal metabolic breakdown of hemoglobin. Cattle affected will have a reddish or pinkish color to the inside of the mouth and nostrils. Usually, in these herds, more than one animal is affected. Typically, all sorts of other diseases crop up due to poor immune function,

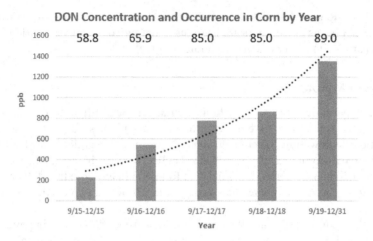

Zearalenone concentration and occurrence in corn by year.

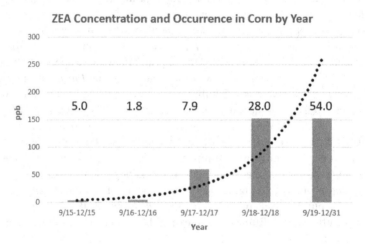

FIGURE 8.7   Two graphs from Max Hawkins. Samples analyzed by Alltech.

including mastitis, pinkeye, foot rot, pneumonia, and poor reproductive performance. The incidence of a right displaced abomasum also sometimes increases.

I saw these patterns clearly on three dairy farms in our practice area in 2010 when the cows were eating the previous year's corn and corn silage. Since each cow has its own individual microbiome, some handled the toxins relatively well. Cows that developed porphyria crashed in dry matter intake and milk production in just a few days. They would die if not taken off the mycotoxin-laden feed. If they did survive, they never came back into full production. These animals were culled despite their massive weight loss.

Some cows had swollen vulvas and swollen limbs due to the presence of zearalenone and fumonisin. Virgin heifers on the same feed would sometimes bag up and squirt milk due to the estrogenic zearalenone, the same mycotoxin that contributed to the swollen vulvas in the cows (Figure 8.9). Fumonisin was more than likely responsible for some of the vascular damage and swollen limbs. Feedlots feeding the same type of feed can have more than an occasional calf display rectal and/or vaginal prolapse.

8.5   MORE FUNGAL ISSUES

Roundup Ready corn stalk bedding has also led to some health issues. One beef client using corn stalk bedding had a couple of his 30 head of early spring 300-pound calves die with an apparent nonresponsive pneumonia. The corn stalks were made correctly when they were dry. They had also been stored inside before use. When the calves were examined on postmortem, the calves looked like they had tumors. These masses were not only in the lungs, but also in the kidneys, liver, and throat-latch area. Tissue submissions to Iowa State University diagnostic laboratory confirmed

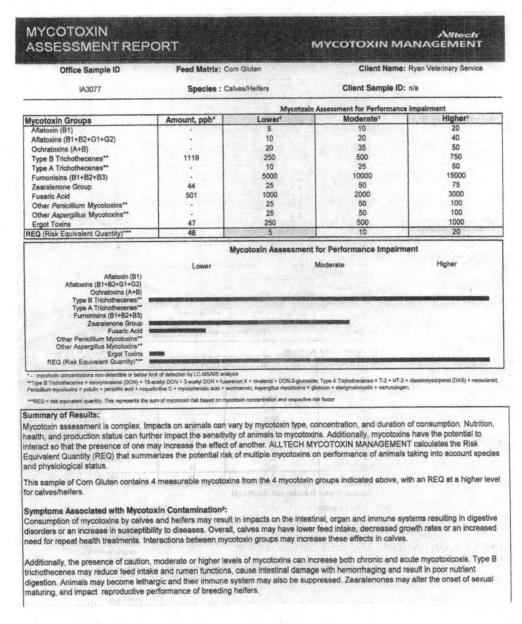

MYCOTOXIN ASSESSMENT REPORT   *Alltech* **MYCOTOXIN MANAGEMENT**

Office Sample ID	Feed Matrix: Corn Gluten	Client Name: Ryan Veterinary Service
IA3077	Species : Calves/Heifers	Client Sample ID: n/a

Mycotoxin Assessment for Performance Impairment

Mycotoxin Groups	Amount, ppb*	Lower†	Moderate†	Higher†
Aflatoxin (B1)	-	5	10	20
Aflatoxins (B1+B2+G1+G2)	-	10	20	40
Ochratoxins (A+B)	-	20	35	50
Type B Trichothecenes**	1119	250	500	750
Type A Trichothecenes**	-	10	25	50
Fumonisins (B1+B2+B3)	-	5000	10000	15000
Zearalenone Group	44	25	50	75
Fusaric Acid	501	1000	2000	3000
Other *Penicillium* Mycotoxins**	-	25	50	100
Other *Aspergillus* Mycotoxins**	-	25	50	100
Ergot Toxins	47	250	500	1000
REQ (Risk Equivalent Quantity)***	46	5	10	20

Mycotoxin Assessment for Performance Impairment

	Lower	Moderate	Higher
Aflatoxin (B1)			
Aflatoxins (B1+B2+G1+G2)			
Ochratoxins (A+B)			
Type B Trichothecenes**			
Type A Trichothecenes**			
Fumonisins (B1+B2+B3)			
Zearalenone Group			
Fusaric Acid			
Other Penicillium Mycotoxins**			
Other Aspergillus Mycotoxins**			
Ergot Toxins			
REQ (Risk Equivalent Quantity)***			

* - : mycotoxin concentrations non-detectible or below limit of detection by LC-MS/MS analysis
**Type B Trichothecenes = deoxynivalenol (DON) + 15-acetyl DON + 3-acetyl DON + fusarenon X + nivalenol + DON-3-glucoside; Type A Trichothecenes = T-2 + HT-2 + diacetoxyscirpenol (DAS) + neosolaniol; Penicillium mycotoxins = patulin + penicillic acid + roquefortine C + mycophenolic acid + wortmannin; Aspergillus mycotoxins = gliotoxin + sterigmatocystin + verruculogen;

***REQ = risk equivalent quantity. This represents the sum of mycotoxin risk based on mycotoxin concentration and respective risk factor

Summary of Results:
Mycotoxin assessment is complex. Impacts on animals can vary by mycotoxin type, concentration, and duration of consumption. Nutrition, health, and production status can further impact the sensitivity of animals to mycotoxins. Additionally, mycotoxins have the potential to interact so that the presence of one may increase the effect of another. ALLTECH MYCOTOXIN MANAGEMENT calculates the Risk Equivalent Quantity (REQ) that summarizes the potential risk of multiple mycotoxins on performance of animals taking into account species and physiological status.

This sample of Corn Gluten contains 4 measurable mycotoxins from the 4 mycotoxin groups indicated above, with an REQ at a higher level for calves/heifers.

Symptoms Associated with Mycotoxin Contamination²:
Consumption of mycotoxins by calves and heifers may result in impacts on the intestinal, organ and immune systems resulting in digestive disorders or an increase in susceptibility to diseases. Overall, calves may have lower feed intake, decreased growth rates or an increased need for repeat health treatments. Interactions between mycotoxin groups may increase these effects in calves.

Additionally, the presence of caution, moderate or higher levels of mycotoxins can increase both chronic and acute mycotoxicosis. Type B trichothecenes may reduce feed intake and rumen functions, cause intestinal damage with hemorrhaging and result in poor nutrient digestion. Animals may become lethargic and their immune system may also be suppressed. Zearalenones may alter the onset of sexual maturing, and impact reproductive performance of breeding heifers.

FIGURE 8.8   Mycotoxin assessment report.

(Continued)

Alltech 37+   RESULTS: MYCOTOXINS LEVELS MEASURED AT _71.09_% DRY MATTER

Sample ID #:	IA3077		Customer Sample ID:	n/a	
Origin:	Iowa		**Feed Matrix:**	Corn Gluten	
Species:	Beef				

Internal Ref # 115-020-11866	Mycotoxins	Levels Detected (ppb)	± Stdev (ppb)	Detection Limit (ppb)	Lower Quantification Limit (ppb)
1	Aflatoxin B1	ND	ND	0.129	0.429
2	Aflatoxin B2	ND	ND	0.684	2.281
3	Aflatoxin G1	ND	ND	0.449	1.495
4	Aflatoxin G2	ND	ND	0.422	1.408
5	Ochratoxin A	ND	ND	0.362	1.208
6	Ochratoxin B	ND	ND	0.302	1.008
7	Deoxynivalenol	ND	ND	5.713	19.044
8	3-AcDon	ND	ND	4.058	13.526
9	15-AcDon	1119.47	516.11	7.442	24.806
10	DON-3-Glucoside	ND	ND	16.651	55.500
11	Nivalenol	ND	ND	53.988	179.960
12	Fusarenon X	ND	ND	2.489	8.295
13	Fusaric Acid	501.42	232.77	0.017	0.055
14	T2 Toxin	ND	ND	0.744	2.481
15	HT2 Toxin	ND	ND	2.296	7.655
16	Diacetoxyscirpenol	ND	ND	1.505	5.017
17	Neosolaniol	ND	ND	0.946	3.154
18	Fumonisin B1	ND	ND	20.426	68.086
19	Fumonisin B2	ND	ND	1.804	6.012
20	Fumonisin B3	ND	ND	2.918	16.493
21	Zearalenone	ND	ND	2.545	8.482
22	α Zearalanol	ND	ND	12.964	43.213
23	β-Zearalanol	ND	ND	10.910	36.467
24	Zearalanone	44.09	16.89	3.427	11.424
25	Patulin	ND	ND	16.669	55.562
26	Mycophenolic Acid	ND	ND	2.496	8.319
27	Roquefortine C	ND	ND	0.196	0.653
28	Penicillic Acid	ND	ND	11.693	38.978
29	Wortmannin	ND	ND	0.764	2.545
30	Gliotoxin	ND	ND	5.608	18.692
31	Sterigmatocystin	ND	ND	0.184	0.612
32	Verruculogen	ND	ND	0.331	1.104
33	2-Bromo-Alpha-Ergocryptine	ND	ND	0.838	2.794
34	Ergometrine/Ergonovine	ND	ND	0.573	1.911
35	Ergotamine	46.75	8.98	0.502	1.673
36	Lysergol	ND	ND	0.457	1.522
37	Methylergonovine	ND	ND	0.048	0.161

FIGURE 8.8 (CONTINUED)   Mycotoxin assessment report.

Lichtheimia ramosa (Figure 8.10). This fungal infection is showing up more often than it has in the past. Even with the use of ultrasound in human medicine, it is easy for doctors to confuse this fungal infection in the lungs with cancer or severe bacterial pneumonia. Antimicrobials and chemotherapy are certainly not the thing to use with fungal infections.

Unbred heifer – June 10, 2010

Zearalenone

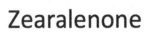

Crashing Cows –Same Dairy

FIGURE 8.9 Udder development in unbred heifers from consuming feed containing zearalenone mycotoxin.

Veterinary Diagnostic Laboratory
Iowa State University
College of Veterinary Medicine
Ames, Iowa 50011-1134
Phone: 515-294-1950
Fax: 515-294-3564

Accession: **2017033115**

Final Report
Accessed Date: 05/24/2020 06:22 pm

DR. ARTHUR G. DUNHAM
RYAN VETERINARY SERVICE
PO BOX 48
RYAN , IA 52330

Site

Premises ID#   :

Lot/Group ID   :
Source/Flow ID   :
Reference   :
Case Tags   :

Owner
Diagnostician: ARRUDA, BAILEY

Client Phone:	Species: Bovine	Age: 3 Months
Client Fax:	Breed: Simmental	Weight: 250 Pounds
Client Account#: 000401331	Sex: Female	Received: Fresh
Date Received: 05/12/2017	Previous Case:	
Sample Taken: 05/10/2017	Farm Type: Other	Reason: General Diagnostics
Accompanying Cases:	Animal ID(s): A	

Final Report(s): 6/9/2017 3:56 pm, 3/22/2018 1:49 pm

History: 3-month-old bovine with enteric and respiratory signs reported.

Gross Pathology:
Laryngitis, fibrinonecrotic

Hemorrhage surrounding the large arteries (no laceration noted)

Renal necrosis and hemorrhage, multifocal

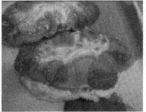

FIGURE 8.10   Iowa State University Veterinary Diagnostic Laboratory report.

(*Continued*)

Hepatic infarcts with fibrinous serositis

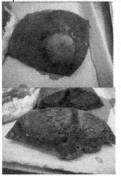

Histopathology:
Liver: Arteries commonly contain fibrin thrombi with fungal hyphae. Vessel walls are disrupted by degenerate neutrophils. Hepatocytes in these areas are commonly necrotic or degenerate.

Kidney: Arteries commonly contain fibrin thrombi. Renal tubules are occasionally lined by degenerate or necrotic epithelium.

Ancillary Diagnostic Tests:
Completed results appear below

Laboratory Diagnosis:
Liver: Thrombosis and vasculitis with fungal hyphae and hepatocyte necrosis
Kidney: Thrombosis and renal necrosis
Laryngitis, fibrinonecrotic
Hemorrhage surrounding the large arteries (no laceration noted)

Comments:
Histologic lesions are consistent with the systemic dissemination of a fungal etiology. The pathogenesis usually involves a breach of epithelial barriers with secondary colonization, local invasion and vasculitis which can lead to systemic dissemination. (6/5/17 ba/clm)

SUPPLEMENTAL REPORT
Fungal culture and identification was unsuccessful. (6/9/17 ba/clm)

Bailey Arruda, DVM, PhD
Assistant Professor
Diagnostic Pathologist

KEY: Tests: FA = Fluorescent Antibody, IHC = Immunohistochemistry, MALDI = Matrix-assisted laser desorption/ionization, MLV = Modified Live Virus, ORF = Open Reading Frame, PCR = Polymerase Chain Reaction, RFLP = Restriction Fragment Length Polymorphism, SS = Special Stain, VI = Virus Isolation. Agents: BCV = Bovine Coronavirus, BRSV = Bovine Respiratory Syncytial Virus, BVDV = Bovine Viral Diarrhea Virus, CSF = Classical Swine Fever, HPS = *Haemophilus parasuis*, IAV = Influenza A Virus, IBRV = Infectious Bovine Rhinotracheitis Virus, MHP = *Mycoplasma hyopneumoniae*, MHR = *Mycoplasma hyorhinis*, MHS = *Mycoplasma hyosynoviae*, PCV = Porcine Circovirus, PDCV = Porcine Deltacoronavirus, PEDV = Porcine Epidemic Diarrhea Virus, PPV = Porcine Parvovirus, PRCV = Porcine Respiratory Coronavirus, PRRSV = Porcine Reproductive & Respiratory Syndrome Virus, PRV = Pseudorabies Virus, SVA = Senecavirus A, TGEV = Transmissible Gastroenteritis Virus.

Test Ordered	Order Date	Current Status	Complete Date
Culture Summary	05/15/2017	Result Released	05/15/2017
Routine Culture - Tissue	05/12/2017	Result Released	05/15/2017
Fungal Culture Results	05/19/2017	Result Released	06/06/2017
Hematoxylin and Eosin Slides	05/12/2017	Result Released	05/12/2017
^PCR - Fungal 18s gene - R&D	06/14/2017	Result Released	06/14/2017
^Fungal 18s gene sequencing - R&D	06/14/2017	Result Released	06/14/2017

^ Testing performed in part or in total at a Referral Laboratory.

FIGURE 8.10 (CONTINUED) Iowa State University Veterinary Diagnostic Laboratory report.

Histopathology

Hematoxylin and Eosin Slides

Animal ID	Specimen	Slides	Comments
A	Assorted	2	

Molecular Diagnostic

PCR - Fungal 18s gene - R&D

Animal ID	Specimen	Result	Comment
A, Tube #1	Culture	Positive	

Fungal 18s gene sequencing - R&D

Animal ID	Specimen	Comment
A, Tube #1	Isolate	Direct sequencing from the liver

Genbank Reference	Query coverage	Max identity	Organism ID
HQ285700	100	99	Lichtheimia ramosa

Nucleotide

Bacteriology

Culture Summary

Animal ID	Specimen	Enrichment	Growth	Organism	Comment
A, Tube #1	Assorted			NSG --> No Significant Growth	

Fungal Culture Results

Animal ID	Specimen	Growth	Result	Comment
A	Liver	Couple	No significant growth	fungal didn't grow up

FIGURE 8.10 (CONTINUED)   Iowa State University Veterinary Diagnostic Laboratory report.

The postmortem of a fat finished steer that died suddenly showed liver damage typical of mycotoxicosis. The Iowa State University diagnostic lab checked our tissue submission for aflatoxin, since it is the only mycotoxin that can be confirmed using animal specimens known to cause liver damage (Figure 8.11). Aflatoxin is produced by *Aspergillus flavus* in drought conditions. Cool, wet weather, and "not drought," would describe that year's crop season. As we expected, the aflatoxin test was negative. There is a good chance that a toxin in the T2 family was involved. The client was using moldy corn stalk bedding made during that wet fall's harvest. He also was feeding the animals corn grain and corn by-product that may have been contaminated with *Fusarium* mycotoxins. The rest of the fat steers in this group were slaughtered, so the feed was not checked.

Veterinary Diagnostic Laboratory
Iowa State University
College of Veterinary Medicine
Ames, Iowa 50011-1134
Phone: 515-294-1950
Fax: 515-294-3564

Accession: **2020013965**

Final Report
Accessed Date: 05/24/2020 08:07 pm

DR. DONALD A. COOK
RYAN VETERINARY SERVICE
PO BOX 48
RYAN , IA 52330

Site

Premises ID#   :

Lot/Group ID   :
Source/Flow ID   :
Reference   :
Case Tags   :

Owner
Diagnostician: BURROUGH, ERIC

Client Phone: 563-922-2094	Species: Bovine	Age: 14 Months
Client Fax: 563-932-2996	Breed: Holstein	Weight:
Client Account#: 000401331	Sex: Castrate	Received: Fresh & Fixed
Date Received: 02/20/2020	Previous Case:	
Sample Taken:	Farm Type: Other	Reason: General Diagnostics
Accompanying Cases:	Animal ID(s): A	
Final Report(s): 3/5/2020 9:39 am		

History:
This steer had looked a little slow the last couple days and was found dead in the afternoon.
Concerned about monensin toxicity or mycotoxins.

Gross Pathology:
Per submitter, the heart was greatly enlarged and the walls appeared thickened and appeared dark and hemorrhagic.
The valves appeared normal. The liver was also markedly enlarged and was very congested and oozed blood. The
lungs appeared normal colored but felt firm to the touch. There were no other lesions.

Histopathology:
Liver: Zone 3 hepatocytes are diffusely atrophied or absent and there is concurrent congestion of sinusoids.
Hepatocytes in zone 1 are often expanded by one or more lipid-type vacuoles and there are increased numbers of bile
duct profiles. The interlobular connective tissue is variably and often severely expanded by fibrosis.
Lung and heart: Evaluated sections from these tissues are unremarkable.

Laboratory Diagnosis:
Toxic hepatopathy

Comments:
Lesions are consistent with a toxic hepatopathy. Aflatoxin M1 testing was negative on liver tissue; however, a feed
sample is required for other mycotoxins. Is there potential for other toxin exposure in the environment? Please contact
the laboratory if testing for other toxic compounds on the liver tissue is desired.

Histopathology

Histopathology - With Interpretation

Animal ID	Specimen	Slides	Comments
A	Assorted	2	

Necropsy

Diagnostic Pathology Interpretation - Routine

Animal ID	Specimen	Slides	Comment
A	Assorted	2	

FIGURE 8.11 Iowa State University Veterinary Diagnostic Laboratory Final Report.

(Continued)

Toxicology

Aflatoxin-QUAL M1 Liver

Animal ID	Specimen	Analysis	Result
A	Liver	Aflatoxin M1	< 1 ppb Same RT by HPLC

Comments:

Bacteriology

Culture Summary

Animal ID	Specimen	Enrichment	Growth	Organism	Comment
A, Tube #1	Liver			NSG --> No Significant Growth	

FIGURE 8.11 (CONTINUED)   Iowa State University Veterinary Diagnostic Laboratory Final Report.

In 2009–2010, 2018–2019, and 2019–2020, we saw cases of pneumonia in feeder calves after clients have had them on feed for 2–3 months. This has happened even when there was no shipping fever pneumonia or death loss after delivery. These cases were more than likely triggered by immunosuppression from mycotoxins delivered via bedding, corn grain, silage, and/or corn by-product feeds.

8.5.1 MYCOTOXICOSIS OF PIGS

Pigs are much more susceptible to the effects of mycotoxins in the feed than are cattle. This past fall was very wet, and harvest was delayed. Farmers were not able to get gas to dry their grain, and many of them harvested their crops anyway. They placed the crop in a bin and thought it would be okay during a cold winter, with plans for drying later; however, the *Fusarium* mold that was present in the field kept growing in the 15%–20% moisture grain. Usually, vomitoxin over 1 ppm does not make pigs vomit. It certainly causes appetite suppression at those levels since pigs do not like its smell or taste. The pigs root the feed out into the pit instead of eating it. At first, too many clients think that this behavior is due to a feeder adjustment issue. Some knowledgeable pork producers have "cranked" the feeders down, limiting intake even more. We have had a few clients with feed as high as 25 ppm vomitoxin. In these cases, the clients had to borrow a grain vacuum from a coop to empty feeders and start over with clean feed.

Fumonisin levels over 1 ppm can damage the heart. Heart damage causes pulmonary congestion and swollen livers, which compounds the effects of poor immune function. We had to have a client sell light hogs suffering from this heart damage to keep many of them from dying on the farm before they reached their desired market weight. If the ratio of the width of the left ventricular wall plus the width of the septum between the two ventricles in the numerator, over the width of the right ventricular wall in the denominator comes out over ten (when under four is normal), you know that these pigs are in trouble (Figure 8.12).

If zearalenone is present, gilts (no matter at what age and weight) look like they are going to pig (deliver) within a week because of their swollen vulvas and teats (Figures 8.13 and 8.14). As is the case in cattle, we can sometimes have a spike in rectal and vaginal prolapses as well. When diets contain T2 or its cousins in the mix, we see the worst death losses with PRRS, influenza, circovirus, *H. parasuis*, *Strep suis*, *Pasteurella*, etc.

Veterinary Diagnostic Laboratory
Iowa State University
College of Veterinary Medicine
Ames, Iowa 50011-1250
Phone: 515-294-1950
Fax: 515-294-3564

Accession: **2010011961**

Final Report
Report Date: 4/29/2010

Dr. Arthur G. Dunham
Ryan Veterinary Service
PO Box 48
201 Main St
Ryan, IA 52330

Owner:
Unknown
Hopkinton, IA 52237

Diagnostician: V.L. Cooper

Client Phone:
Client Fax:
Client Account#: 401331
Date Received: 4/27/2010
Preliminary Report: 4/27/2010, 4/28/2010

Species: Porcine
Breed: Unknown
Sex: Mixed
Previous Case:

Age: 5.5 Months
Weight: 200 Pounds
Received:
Fresh tissue

Macroscopic evaluation.
A: RV = 0.6 cm IVS = 3.0 cm LV = 3.3 cm for a ratio of 10.5 heart valves are thickened by bluish white areas on leaflets
B: RV = 1.0 cm IVS= 2.0 cm LV = 2.4 cm for a ratio of 4.4
Normal range is 2.38 - 3.84 for swine with a mean of 2.94.

Histopathology: Myocardial fibers are hypertrophic. Valve sections are characterized by expansion of leaflets by amorphous amphophilic material.

Diagnosis: A: Hypertrophic cardiomyopathy in association with valvular endocardiosis

Comments: Lesions are consistent with hypertrophic cardiomyopathy associated with valvular lesions. Lesions likely are representative of degenerative change or remnants of a previous inflammatory insult. [Results Emailed: 04-27-2010/vlc] (4/29/10 vlc/kw)

FIGURE 8.12   Iowa State University Veterinary Diagnostic Laboratory Final Report 4/29/10.

TRIANGLE AGRI SERVICES
19799 HWY 151
MONTICELLO IA   52310

ACCOUNT NO: 25659

DESCRIPTION   w.th distillers
HOG FIN DIST RYAN VET

SAMPLE NO.
971407

DATE
12/07/11

ANALYSIS

INTERPRETATION

ZEARALENONE	49	PPB	MARGINALLY SAFE	0 - 200	PPB
VOMITOXIN	1.200	PPM	POTENTIALLY UNSAFE	> 0.400	PPM
T-2	31	PPB	MARGINALLY SAFE	0 - 100	PPB
FUMONISIN	1.700	PPM	POTENTIALLY UNSAFE	> 1.000	PPM

* MARGINALLY SAFE   - THESE LEVELS OF TOXINS ARE NOT EXPECTED TO CAUSE
                     ANIMAL PERFORMANCE PROBLEMS IN MOST SITUATIONS,
                     HOWEVER IT SHOULD BE EMPHASIZED THAT THE ONLY
                     COMPLETELY SAFE LEVEL OF MYCOTOXINS IS ZERO.

* POTENTIALLY UNSAFE - THESE LEVELS OF MYCOTOXINS CAN POTENTIALLY CAUSE
                      ANIMAL PERFORMANCE PROBLEMS.

** GROWING AND FINISHING CATTLE AND SWINE MAY BE ABLE TO TOLERATE UP TO
100 PPB OF AFLATOXIN.

FIGURE 8.13   Triangle Agri Services mycotoxin report.

(Continued)

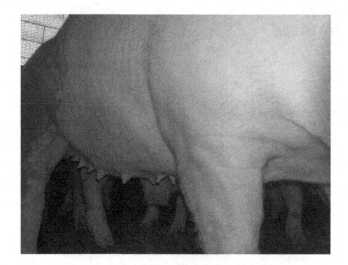

FIGURE 8.13 (CONTINUED)   Triangle Agri Services mycotoxin report.

Most of our hog clients are no longer grinding their own feed. Now, most feed is prepared at elevators where much of the corn and other feedstuffs are being sourced from non-livestock farmers. If any of the pooled corn, corn distillers, or other feedstuffs are contaminated with toxins, it is essential to have an effective binder in the feed. This is especially critical with the first feed delivered. If it is contaminated and fed when the pigs have just been weaned and do not have much fat for reserves, look out. Pharmaceuticals in feed or water cannot remediate immune stress and starvation. The binders we use in our practice for *Fusarium* mycotoxins were all developed in Europe. The northern Europeans have never had the corn-growing weather we have been blessed with. Because of their damper, colder weather, their governments have required them to approve mycotoxin binders in the same way we approve antimicrobials in the

United States. Lot of testing is required. The *Fusarium* mycotoxins are bigger molecules than the flat, polar aflatoxin produced by *Aspergillus flavus* during droughts. The aluminum silicates, charcoals, sodium bentonites, and clays that bind the flat polar aflatoxin do not work well on the bigger, three-dimensional *Fusarium* toxins. Since aflatoxin comes through milk, and since it is one of the most teratogenic (birth defect causing) agents known to man, our government has enforceable action levels. No more than 20 ppb aflatoxin in dairy feed and no more than 0.5 ppb aflatoxin in milk are allowed.

The T2 family of *Fusarium* toxins are extremely dangerous as well, but in their case, we have non-enforceable advisory levels of 100 ppb. In 2009–2010 and in 2018–2020, our clients could not

Veterinary Diagnostic Laboratory
Iowa State University
College of Veterinary Medicine
Ames, Iowa 50011-1134
Phone: 515-294-1950
Fax: 515-294-3564

Accession: 2016039006

Web Report
Report Date: 09/16/2016 10:26 AM

Dr. Arthur G. Dunham
Ryan Veterinary Service
PO Box 48
Ryan, IA 52330

Site   : Home Farm
               Unknown
               Unknown , Unknown 00000

Premises ID#   : 005MELP
Lot/Group ID   :
Source/Flow ID :

Owner   Reference   :
Division :   Diagnostician : Pineyro, Pablo

Client Phone: Species: Bovine   Age: 10
Client Fax: Breed: Crossbreed   Weight: 500
Client Account#: 000401331   Sex: Female   Received: 1 Fresh - Rabies & Others NHE
Date Received: 07/18/2016   Previous Case: Reason: General Diagnostics
Sample Taken: 07/15/2016   Farm Type: Other
Preliminary Report: 07/20/2016, 07/22/2016

History: 10-month-old bovine. CNS signs reported

Histopathology:
Brain: In section of brainstem there is multifocal suppurative encephalitis. Multifocally vessels are also congested and cuffed by dense aggregates of neutrophils and lymphocytes. In the subjacent gray matter, there are increased numbers of glial cells (gliosis) with small clusters of neutrophils and the neuropil is multifocally pale and minimally vacuolated (spongiosis).

Morphologic diagnoses:
Encephalitis, subacute, multifocal, moderate, with microabscesses and perivasculitis.

Virology: Rabies testing was negative.

Bacteriology:
    Routine culture: No significant growth

    Listeria monocytogenes IHC Positive

Laboratory diagnosis:
Encephalitis à Listeria monocytogenes

Comments:
    • Histological lesions are consistent with Listeria monocytogenes and supported by bacterial identification by IHC.
    • Please correlate clinically and contact the laboratory for further testing or if questions arise. (7/26/16 pp/clm)

Pablo Pineyro, DVM, MVSc, DVSc, PhD.

KEY: Tests: FA = Fluorescent Antibody, IHC = Immunohistochemistry, MLV = Modified Live Virus, ORF = Open Reading Frame, PCR = Polymerase Chain Reaction, RFLP = Restriction Fragment Length Polymorphism, VI = Virus Isolation. Agents: BCV = Bovine Coronavirus, BRSV = Bovine Respiratory Syncytial Virus, BVDV = Bovine Viral Diarrhea Virus, CSF = Classical Swine Fever, HPS = Haemophilus parasuis, IAV = Influenza A Virus, IBRV = Infectious Bovine Rhinotracheitis Virus, MHP = Mycoplasma hyopneumoniae, MHR = Mycoplasma hyorhinis, MHS =

FIGURE 8.14   Veterinary Diagnostic Laboratory Web Report 9/16/2016.

HISTOPATHOLOGY

Hematoxylin and Eosin Slides

Animal ID	Specimen	Slides	Comments
A	Assorted	15	

Immunohistochemistry Listeria

Animal ID	Specimen	Result	Comment
A	Brain	Positive	

VIROLOGY

Test:	FA - Rabies virus
Result:	Negative
Technician:	SFK
Human Exposure Information:	
Exposed Person's Name:	No human exposure
Exposed Person's Town:	Delhi
Comments:	

BACTERIOLOGY

Culture Summary

Animal ID	Specimen	Enrichment	Growth	Organism	Comment
A, Tube #1	Brain			No Significant Growth	

Listeria Enrichment - Result Pending

Animal ID	Specimen	Result
A	Brain	Listeria enrichment result is pending. Please allow 12-14 weeks for final result

FIGURE 8.14 (CONTINUED)   Veterinary Diagnostic Laboratory Web Report 9/16/2016.

risk using distillers (protein left after alcohol fermentation) in place of bean meal to save a few dollars per hog marketed. This was especially the case when their own corn was contaminated since the distillers are bound to be even more highly contaminated. (As stated before, distillers' grain is typically three times higher in mycotoxins than the corn hauled into the plant to make it since ethanol comes from the germ while the by-product comes from the hull where the mycotoxin is present.) This is true even if they are using a European binder such as Olmix's Algonite®, Biomin's Biofix®, or Alltech's Integral®. The European binders have higher levels of toxins to contend within the US Midwest than in Europe, and using them, even though they are tested, is not a guarantee of success. In my opinion, the Europeans can be relieved that they did not adopt the Roundup Ready GMO program full-scale as was done in North America.

8.6   ANTIMICROBIAL ACTION OF GLYPHOSATE CAN ALTER MICROBIOME FUNCTION

The Clostridia discussed earlier that produce BoNts are not the only Clostridial species favored by the antimicrobial action of glyphosate. Before Roundup Ready was introduced in the late 1990s, we had only a few cases of *Clostridium perfringens* enterotoxemia in calves since our sandy calcareous soils are not favorable for the growth of these anaerobic bacteria. Now enterotoxemia in both dairy and beef calves is common. The use of probiotics that contain the two glyphosate-sensitive *Enterococci* species mentioned before has helped treat this condition. Our favorite calf scour boluses are now gel capsules filled with a clinoptilolite called zeolite [27]. The veterinary profession is realizing that avoiding antimicrobial treatment of the GI microbiome is a good thing not only for treatment of the GI symptoms, but also for lowering the risk of pneumonia and other systemic infections later.

The diagnostic labs at both the University of Minnesota and Iowa State University admit they are culturing more *Listeria monocytogenes* in mastitis cases, even though it is easy to overlook since it grows slowly and is often obscured by the overgrowth of other organisms [28]. Listeriosis used to be associated with poorly made dry corn silage that does not have enough lactic acid to stop the bacteria's slow growth at low temperatures. Now we see cases of listeriosis in cattle on dry hay if they are also being fed Roundup Ready corn (Figure 8.15).

Researcher Megan Niederwerder at Kansas State University has shown that pigs with a more diverse gut microbiome have lower morbidity and mortality from porcine reproductive and respiratory syndrome virus (PRRS) and circa virus than those with a microbiome that is less diverse. Since her original work in 2015, she has more recent work documenting the same correlation [29].

One of our swine clients has used non-Roundup corn for over 10 years. I had him address the Iowa State University Swine Peer Group (of which I am a member) this past Fall. This swine client is in a hog-dense area and is still able to farrow and finish hogs on the same site. This practice is almost unheard of in this day. Typically, farrowing barns are separate from finishing facilities because of the constant threat of infection with PRRS virus. PRRS switches back and forth from a respiratory problem in weanlings and finishers to an abortion and weak pig problem in the sow herd. His area is involved in a PRRS surveillance and control program where neighbors share clinical problems and lab results. In the Winter and Spring of 2018, neighboring herds were decimated

www.fhr1.com

Crop Science

2510 HWY 63 NE
PO BOX 157
STEWARTVILLE, MN 55976

877-907-1444

HERBICIDE - PESTICIDE TESTING

CLIENT: RYAN VETERINARY   8/7/2013

Elisa Testing Results Results

TEST	SAMPLE #	DESCRIPTION	ELISA RESULT	mg/kg	CONCENTRATION/ACRE*
Glyphosate 1:500	072613-4	SOYHULL FEED *pellets*	1.708	8.625	

*Acre Furrow Slice = 7.5 inches deep ≡ 2x10⁶ lb soil; example 1 mg/kg = 2lbs/acre is equal to 0.5 gal Glyphosate (AI)

Manure 10,000 lb/acre (For top 3 inches) = 0.0025 lbs Glyphosate/acre

(4lb Glyphosate (AI) per gallon. AI = Active Ingredient)

Laboratory Director: George Kindness BS, MS, PhD MiBol

Agronomist: John Oolman, BS, MS, CCA

Agronomist: John Mayernak BS, MS

FIGURE 8.15 Crop Science Glyphosate Investigation.

by a 1-7-4 cut pattern PRRS that was RNA mapped at Orf 5 (Open reading frame 5) and which caused abortions as well as respiratory and wasting issues in the pigs. (This diagnostic work does not have to be totally understood by nonveterinarians. The take-home message is that the strain our client had was genotyped and shown to be nearly identical to the other viruses isolated from the neighboring herds) (Figure 8.16).

Email Link to Report   Submit Additional Test Requests   Email Attachment of Report

Veterinary Diagnostic Laboratory   Accession: 2018049286
Iowa State University
College of Veterinary Medicine   **Final Report**
Ames, Iowa 50011-1134   Accessed Date: 09/24/2019 01:07 pm
Phone: 515-294-1950
Fax: 515-294-3564

                                                      Site   : HOMESTEAD

                                                                         UNKNOWN , IA 00000

                                                      Premises ID#   :

                                                      Lot/Group ID   : Homestead
                              Owner   Source/Flow ID   :
                              Diagnostician: GAUGER, PHILLIP   Reference   :
                                                      Case Tags   :

Client Phone: ▮▮▮   Species: Porcine   Age: NA
Client Fax: ▮▮▮   Breed: Unknown   Weight:
Client Account#: ▮▮▮   Sex: Received: 2 Oral Fluid
Date Received: 07/11/2018   Previous Case:
Sample Taken: 07/10/2018   Farm Type: Other   Reason: Other
Accompanying Cases: Animal ID(s): north, south
Final Report(s): 7/16/2018 11:22 am

The PRRSV PCR retest on Oral Fluid tube #1 repeated positive. Please call if there are questions. (7/12/18 pcg/clm)

Phil Gauger

Phil Gauger, DVM, MS, PhD
Associate Professor
Molecular and Viral Diagnostics Section Leader; Diagnostic Pathologist

SUPPLEMENTAL REPORT
PRRSV sequencing results are reported below. Please contact the laboratory if you would like any further interpretation or have other sequences for comparison to this one. (07/13/18 kmh/dd)

Karen M. Harmon

Karen Harmon, PhD
Clinical Associate Professor
Molecular Diagnostician

KEY: Tests: FA = Fluorescent Antibody, IHC = Immunohistochemistry, ISH = *in situ* hybridization, MALDI = Matrix-assisted laser desorption/ionization, MLV = Modified Live Virus, ORF = Open Reading Frame, PCR = Polymerase Chain Reaction, RFLP = Restriction Fragment Length Polymorphism, VI = Virus Isolation. Agents: BCV = Bovine Coronavirus, BRSV = Bovine Respiratory Syncytial Virus, BVDV = Bovine Viral Diarrhea Virus, CSF = Classical Swine Fever, HPS = *Haemophilus parasuis*, IAV = Influenza A Virus, IBRV = Infectious Bovine Rhinotracheitis Virus, MHP = *Mycoplasma hyopneumoniae*, MHR = *Mycoplasma hyorhinis*, MHS = *Mycoplasma hyosynoviae*, PCV = Porcine Circovirus, PDCV = Porcine Deltacoronavirus, PEDV = Porcine Epidemic Diarrhea Virus, PPV = Porcine Parvovirus, PRCV = Porcine Respiratory Coronavirus, PRRSV = Porcine Reproductive & Respiratory Syndrome Virus, PRV = Pseudorabies Virus, SVA = Senecavirus A, TGEV = Transmissible Gastroenteritis Virus.

Test Ordered	Order Date	Current Status	Complete Date
PCR Applied Biosystems - PRRSV	07/11/2018	Result Released	07/11/2018
PCR - PRRSV ORF5	07/11/2018	Result Released	07/13/2018

FIGURE 8.16   ISU PRRS Rope testing report.

(Continued)

PRRSV X3 Oral Fluid Antibody	07/11/2018	Result Released	07/12/2018
^Sequencing and Analysis - PRRSV	07/11/2018	Result Released	07/13/2018
PCR - PRRSV Applied Biosystems - N/C RETEST	07/11/2018	Result Released	07/12/2018

^ Testing performed in part or in total at a Referral Laboratory.

Molecular Diagnostic

PCR Applied Biosystems - PRRSV

Animal ID	Specimen	US Ct / Result	EU Ct / Result	Comment
north, Tube #1	Oral fluid	29.1 / Positive	>=37 / Negative	Will retest to confirm
south, Tube #2	Oral fluid	>=37 / Negative	>=37 / Negative	

PCR - PRRSV ORF5

Animal ID	Specimen	Result	Comment
north, Tube #1	Oral fluid	Positive	

Sequencing and Analysis - PRRSV

Animal ID	Specimen	Target Gene	RFLP	Comment
north, Tube #1	Oral fluid	ORF5	1-7-4	Wild type

Sequence Homology

Reference Virus	Inglevac ATP	Lelystad	Prime Pac	Inglevac MLV	Fostera
Percent Identity	86.8 %	63 %	87.6 %	87.7 %	86.8 %

Nucleotide

ATGTTGGGGAAATGCTTGACCGCGGGCTGTTGCTCGCAATTGCCTTTTTTGTGGTGTATCGTGCCGTTCTGTTTTGTTGC
GCTCGTCAACGCCAACAACAACGACAGCTCCCATTTACAGTTGATTTATAACCTGACGATATGTGAGCTGAATGGCACAG
ATTGGCTAAATAAAAGTTTTGATTGGGCGGTGGAGACCTTTGTTATTTTTCCTGTGTTGACTCATATTGTCTCCTATGGA
GCCCTCACCACCAGCCATTTCCTTGACACAGTCGGCCTGATCACCGTGTCTGCCGCCGGATATTACCACAGGCGGTATGT
CTTGAGTAGCATTTACGCCGTCTGCGCCCTGGCTGCGTTAACTTGCTTCGTCATCAGGCTAACAAAAAATTGTATGTCCT
GGCGTTACTCATGCACCAGATACACTAACTTTCTTCTGGACACCAAGGGCAAACTCTATCGTTGGCGGTCTCCTGTCATC
ATAGAGAAAGGGGGNNAAATTGAGGTCGAAGGTCACCTGATCGACCTCAAGAGAGTTGTGCTTGACGGTTCCGCGGCAAC
CCCTGTAACCAAAGTTTCAGCGGAACAATGGGGTCGTCCTTAG

PCR - PRRSV Applied Biosystems - N/C RETEST

Animal ID	Specimen	US Ct / Result	EU Ct / Result	Comment
north, Tube #1	Oral fluid	29.1 / Positive	>=40 / Negative	

Serology

TITER RESULT INTERPRETATION

Samples that generate titers with a numerical value are considered positive for antibody detection to the stated agent at the reported dilution of the tested sample.
The (>) symbol would indicate the sample is positive for antibody detection at the highest sample dilution tested.
The (<) symbol would indicate the sample antibody level is below the detection sensitivity of this assay at the beginning sample dilution, therefore negative at this dilution.

PRRSV X3 Oral Fluid Antibody - Caution: The PRRS X3 oral fluid antibody assay may detect antibodies against PRRSV in samples collected from pigs consuming diets containing spray dried plasma of porcine origin.

Animal ID	Tube#	S/P / Result
north	1	3,078 / Pos
south	2	1,533 / Pos

26252

View in Excel format | Download the CSV file

FIGURE 8.16 (CONTINUED)   ISU PRRS Rope testing report.

Even though our client's pigs had no clinical signs of PRRS, he hung ropes in his two finishing houses to monitor for the disease. (The pigs chew on cotton rope strands from mop heads and then their saliva is squeezed out and checked via PCR for the nucleic acid of PRRS virus). His finishing pigs became PRRS-positive and stayed that way for about 4 months, as documented by results of repeated rope testing. Unlike his neighbors' pigs, these pigs never showed clinical signs of being sick even though they were not vaccinated for the disease and even in the presence of shedding virus. The virus never spread from the finishing barns to the sow herd. The local veterinarian thought that perhaps the client's success was because his genetics were immune to PRRS. The PRRS virus attaches to alveolar macrophages, thus infecting the cells. Attempts are being made to genetically engineer a pig that lacks the viral attachment site on its macrophages. In the spring, we found out that his genetic line was not immune to PRRS when another client that contract-farrows for this same breeder had a herd break out with a severe case of a similar Orf 5 mapped strain of PRRS. However, this contract client fed Roundup Ready corn. Abortions and respiratory deaths were numerous enough that this contract herd was depopulated.

The novel influenza A virus that moved into Midwest poultry flocks in 2015 failed to infect most of the small outdoor organically fed flocks. Most experts thought that the small outdoor flocks with little to no biosecurity to keep wild carrier birds out would be the first to become infected. When a few small flocks did break, it was attributed to neighborhood spread. This spread was thought to be caused by managers of nearby large infected flocks hauling dead birds to landfills. In the small flocks that became infected, morbidity and mortality of market-size birds were not any higher than those of young birds.

In the large turkey flocks feeding glyphosate-contaminated feeds, young birds did not die fast if they were infected early in life. If the infection started in older birds of market size, mortality approached 95% in just 2–3 days. Does this make sense for a supposedly novel, just introduced, influenza A virus? In pigs, a new strain of influenza A, with no immunity gained from their mothers' colostrum, should kill young pigs just as fast as older ones. Iowa State University poultry veterinarian, Dr. Yuko Sato, was puzzled by this difference in death rate between the young poults and older birds as well [30]. Did the diet play a role? Is there a possibility that the higher morbidity and mortality in the older birds on a glyphosate-contaminated diet may have resulted from these birds having an altered and less diverse GI microbiome, and therefore, compromised immunity, for a longer time compared to the young birds?

My veterinary partner and I have seen cases of nursing beef calves on pasture breaking (becoming deathly ill) with *Mycoplasma bovis* pneumonia if the cows and calves are supplemented with Roundup Ready corn silage. This has also happened when the calves are supplemented while nursing with the use of creep feeders. The feeders are usually filled with Roundup Ready grains and by-products. Chelation of trace minerals, the presence of mycotoxins, and/or changes in the microbiome could all influence the immune function of these calves, thus increasing their susceptibility to Mycoplasma. All are possibilities with glyphosate in the diet. Shiva Ayyadurai's metabolic model for human metabolism predicts less glutathione (part of the vitamin E system necessary for good immunity) and more formaldehyde with glyphosate in the diet [31]. The antimicrobial action of glyphosate in the GI tract appears to increase the chance of hemorrhage after castration. Many of today's veterinarians are using banders on larger bull calves, instead of castration, because of this issue. Alteration of the microbiome could decrease vitamin K production, and vitamin K is an important factor in blood clotting. We witnessed this tendency for excessive blood loss in the 1970s and 1980s with castration of weanling pigs that were consuming sulfonamides in their feed. The antimicrobial action of the sulfa drugs decreases the gut bacteria that manufacture vitamin K. It appears that residual glyphosate in the diet may do the same.

8.6.1 Increased Risk of Antimicrobial Resistance

Swine veterinarians, including my partner and I, occasionally run into *Clostridium difficile* scours in baby piglets. Danish pork producer, Ib Pedersen, saw clostridial scours within days when his sow herd was switched from non–Roundup Ready bean meal to Roundup Ready bean meal. Similar scours often occur after sows or pigs are treated aggressively with antibiotics for another disease, causing dysbiosis (a severe alteration in the normal gut microbiome) in the pigs. A similar situation sometimes causes *Clostridium difficile* to flourish in people that are on extended or aggressive antibiotic treatment. Often, fecal transplantation is the only and best therapy for this disease. Dr. Larry Weiss, CEO of Persona Biome, speaking at the 2019 Bard Symposium on Reimagining Human Health, explained how an indigenous hunter and gatherer tribe in Venezuela showed no *Clostridium difficile* in the infant fecal samples cultured. In contrast, this organism is present in the fecal cultures of nearly all babies in the United States and Canada [32]. It seems that changes to our human microbiome over time have also resulted in modern populations having much more trouble with noncommunicable diseases.

We can now use genomics to identify GI bacteria, fungi, and viruses. We may never be able to culture over 80% of them since they are so interrelated and dependent on each other. Another speaker at the Bard Symposium, Dr. Rodney Dietert, PhD emeritus immunologist from Cornell University, does an excellent job in his book, *The Human Super-Organism*, explaining the importance of protecting our microbiome [33].

Glyphosate in feed can sometimes cause a dysbiosis. Without concurrent pneumonia, we have seen bloating when calves are weaned and put on dry, pelleted feed with glyphosate-contaminated soy hulls in the pellets. The soy hulls helped bring the total glyphosate level in the diet to over 8 ppm. As discussed previously, 0.1 ppm glyphosate in the bowel contents can kill some beneficial bacteria. Glyphosate's antimicrobial patent 7771736 says that most generally a dose of 1–2 ppm is adequate to achieve bacterial control [34]. Dr. Monika Krueger, Veterinary Pathologist at Leipzig University in Germany, indicated that she was hired by German breweries before she did animal studies. The breweries could tell when the barley delivered for malt was spray-dried with glyphosate preharvest because fermentation was inhibited. Dr. Krueger documented this in her lab. This is proof again that the US Patent No. 7771736 antimicrobial patent for glyphosate is accurate. Doctors tell us not to put any unused human antibiotics down the drain to avoid environmental contamination. Why worry about this route of environmental contamination by antibiotics when we used over 200 million pounds of antimicrobial glyphosate on just US corn and beans in 2017? (Figure 8.17).

Heinemann et al. have shown how commercial rates of glyphosate influence antibiotic resistance patterns in pathogens to important commonly used antibiotics [35]. Starting January 1, 2017, US veterinarians are required to write veterinary feed directives (VFDs) for any legal antibiotic put in livestock feed. This regulatory oversight was deemed necessary to cut down on the indiscriminate, and inappropriate, use of antibiotics and to thus lower the induction of antimicrobial resistance. If agriculture is causing issues with human antimicrobial resistance, then environmental contamination with antibiotics should be a main area of concern. In my opinion, leaving the antimicrobial herbicide glyphosate, a broad-spectrum antimicrobial that works on malaria, gonorrhea, and many other pathogens, out of the discussion, is not science.

I attended the National Institute of Animal Agriculture's Second Annual Meeting on Antimicrobial Resistance in Columbus, Ohio, with an agronomist in 2012. The whole meeting was devoted to environmental contamination and even presented research showing that soil contamination with copper compounds could change resistance patterns [36]. The two of us spent considerable time trying to inform the speakers of patent No. 7771736 and glyphosate's antimicrobial action. The USDA estimated that the total antibiotic use in animal agriculture was around 30 million pounds in 2016 [37]. Can we say we believe in science and the data it creates when we totally ignore well over 200 million pounds of an antimicrobial herbicide that has been shown to alter resistance patterns to antibiotics used in human medicine?

8.7   MORE ON EARLY-TERM PREGNANCY LOSS

A Hoard's Dairyman article, as well as an article in Beef Magazine entitled "Going No Where Fast" [38], talked about how less than 85% of US beef cows and heifers exposed to artificial insemination and bulls for breeding ever have a calf. This conception rate is worse than in the 1990s. One of our long-term clients with a 130-cow beef herd had only 1 year out of the 43 we worked for him where he had fewer animals sold than the number of heifers and cows exposed to artificial insemination

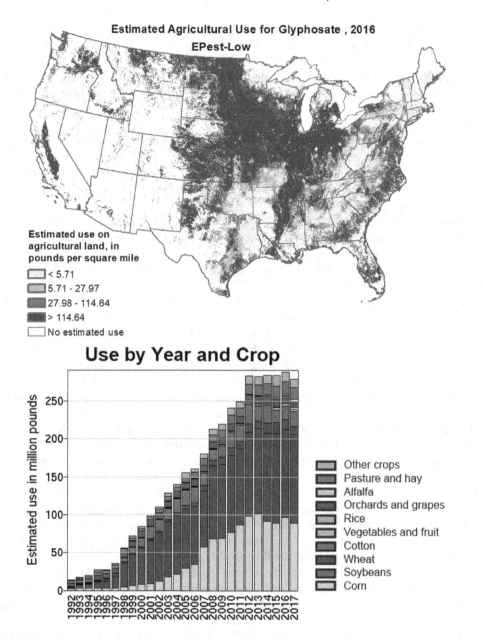

FIGURE 8.17   https://water.usgs.gov/nawqa/pnsp/usage/maps/show_map.php?year=2016&map=GLYPHOS ATE&hilo=L&disp=Glyphosate.

and bulls the previous year. Keeping his twin calves alive kept his calving rate above 100% for most years. This client was a respected and well-known purebred Simmental breeder that did not grow Roundup Ready crops except for the 1 year that he followed the advice of his coop crop advisor. That crop year was the only year he fed Roundup Ready corn and Roundup Ready corn silage to his cows since he always fed his own home-raised feeds. The following year was the 1 year out of 43 when he had chondrodysplasia calves and a much lower than 100% calving rate.

A swine client who was very profitable and who was questioned by extension experts on how he used so few boars switched from non–Roundup Ready corn to Roundup Ready corn. Within a year of the switch, fertility in the herd was miserably low. We documented manganese deficiency, as mentioned earlier, by sending a few open sows and gilts to a local locker and then submitting their livers for mineral analysis. This herd also had many abortions in a pattern that looked like a herd breaking with PRRS. However, the abortion storms were in the summer, rather than the more typical winter pattern of PRRS abortions. In the winter months, while in early pregnancy, the sows and gilts were bedded with Roundup Ready corn stalks. Many of these animals aborted near term, in the summer months. Sows and gilts who were in early pregnancy in the summer, and who received no bedding in their dirt floor hoop houses in the summer, had healthy litters in the winter. Two dairy client neighbors, one using Roundup Ready corn and Roundup Ready corn silage and the other not, had huge differences in their pregnancy check rates. The conception rate of the Roundup user's cattle was much poorer, even though he was considered a good manager.

Is there any additional science to support these clinical observations besides glyphosate's link to manganese deficiency and extra mycotoxins like the estrogenic zearalenone? Dr. Tom Swerczek, then a veterinary pathologist working at the Glock Research Center at the University of Kentucky, described "an agent" he'd observed with electron microscopy when he was investigating the cause of mare infertility syndrome (MIS) in 2001. His work, described below, determined that this agent was causative to MIS. Before his retirement in 2019, he published a paper describing this agent. A second paper was just published in June 2020. The agent Dr Swerczek described has remained largely ignored by the scientific establishment and has not yet been completely characterized [39,40].

I sent split samples of edematous swine placenta from aborting sows in the swine herd described above to both Iowa State University and Dr. Swerczek. Iowa State University could not make a diagnosis but eliminated PRRS as a cause. With the use of electron microscopy, Dr. Swerczek found large quantities of the vascular-damaging agent in the swollen, edematous placentas. He had identified the same agent in placental tissue from mares that aborted with MIS in the spring of 2001. He put some of the agent collected from the tissue I submitted, plus agent from other tissues submitted to him, into chicken embryos, experimental horses, and experimental goats. The presence of this agent caused some of these animals to abort. The two dairymen mentioned above collected amniotic fluid from cows at the start of calving by withdrawing a small sample of the fluid before the feet of the calves punctured through the amnion. Our office manager deliberately intermingled the numbered samples from both farms before sending them to Dr. Swerczek. Dr. Swerczek was able to sort the samples into the two herds by the amount of agent present in each sample. Samples from the Roundup user's herd, with poor fertility, had the most agent.

Dr. Swerczek first saw the agent during his investigation of MIS while he was examining semen from stallions fed bean meal. Dr. Swerczek also isolated what appeared to be his agent from Goss's wilt corn samples and from Sudden Death Syndrome soybean samples, which were submitted via Dr. Huber, emeritus professor from Purdue University.

Anthony Samsel and Dr. Stephanie Seneff have a plausible description of this agent, suggesting that it is a biomatrix in which the "synthetic glycine with a side chain" (that is, the glyphosate molecule), along with some minerals chelated to the glyphosate, attempts to replace glycine during protein formation, and thus bioaccumulates into the protein matrixes. Monsanto described this peptoid in the 1980s when they looked at the distribution dynamics of glyphosate by using C-14-labeled glyphosate. Not all the labeled glyphosate was excreted in the urine and feces. It took several years of hard chemistry to learn how to extract this approximately 1% out of these tissues [41]. The mammalian metabolic enzyme system is not going to be able to do this. Dr. Swerczek determined that the agent was particularly enhanced by high-protein diets, low salt, and fermented feed. He completed a lot of studies, under tremendous pressure, and on his own time, using experimental horses and goats. He held up millions in insurance payments when he said his agent was responsible for MIS in 2001. A majority of equine specialists decided that a web-worm poisoning was the cause of MIS. Pregnant mares that aborted were thought to have eaten dead webworms after a late killing frost in 2001. Cutting down the wild cherry trees to rid the area of webworms did not stop the problem in subsequent years. Dr. Swerczek's conclusions earned him the support of the ex-governor of Kentucky. The advice he provided to the ex-governor on how to handle this agent (and other issues) nutritionally kept the ex-governor in the racehorse business. This client's newborn foals not only survived, but some turned into winners of multiple national filly races.

I want readers of this chapter to read Dr. Swerczek's papers mentioned in the previous foot note. In my opinion, veterinary medicine needs some researchers that are willing to continue Dr. Swerczek's work. Dr. Swerczek has proven that the EM agent is not an artifact. Now it needs identi-fication. There is more than one theory describing what this agent is. As I stated previously, Samsel and Seneff have a plausible description of this agent, suggesting that it might be a biomatrix in which the "synthetic glycine with a side chain" along with minerals chelated to the glyphosate attempts to or does replace the glycine during protein formation. If this happens, glyphosate can bioaccumulate into protein matrixes.

In the last 10 years, the US dairy industry has dramatically dropped recommended protein levels for lactating cows. Dairymen have followed the recommendation, and reproductive performance has improved. Too much protein in relation to carbohydrate in the diet (an Atkin's diet) results in great metabolic expense so that is one reason why higher protein diets result in poorer reproductive success. If glyphosate is a risk factor with Dr. Swerczek's vascular damaging agent, and if glypho-sate is occasionally incorporated into protein, then because of this, a lower-protein diet might be of less risk.

Dr. Swerczek is certain that high protein can ruin horse hooves and that the vascular damaging agent is more than likely a contributor to this as well. I have shared this information with dairy vet-erinarians. Both Dr. Swerczek and I think excessive protein is also bad for dairy cow feet. Laminitis in cows on lush pasture, or laminitis when the cows are eating high-protein total mixed rations, does not have to be caused by high sugar content alone.

8.8   SOIL HEALTH RISKS, INCLUDING DECLINING ORGANIC MATTER AND DECLINING PHOSPHORUS LEVELS

After a typical 1 pound glyphosate/acre spray dose, the population of oxidizing rhizosphere organ-isms goes up by about threefold, while the population of beneficial reducing organisms goes down about threefold. It is these reducing organisms that are essential in the transformation of trace min-erals into their available positive two valences for uptake by the plants. Beneficial Pseudomonads that produce the antifungal compound, pyrrolnitrin, are also harmed, along with beneficial fungi that make the glue that aggregates the soil [42] for porosity and water penetration.

Ag business claims we have "saved the soil" by adopting glyphosate. Although we have less visible erosion of particulate matter after drastically cutting back on use of the mold-board plow, and the shift to no-till and low-till practices, we have largely gone backward on maintaining (and increasing) soil organic matter. Increasing organic matter means more carbon dioxide storage. An increase in organic matter increases water retention, important in both floods and droughts. Higher soil carbon can hold the nitrogen and other nutrients applied and, therefore, would improve both crop and water quality.

Glyphosate use is playing a role in declining soil organic matter. It also is contributing to more leaching of soluble phosphorus. That leached phosphorus contributes to algae blooms (Figure 8.18). More soluble phosphate anions can result in greater desorption of glyphosate from the soils. The desorbed glyphosate then contributes to the bloom via its antimicrobial action.

There are many pork producers growing crops on soils with declining soluble phosphorus levels even with their annual application of liquid hog manure.

We have some clients whose pigs are raised on solid floors with corn stalk-bedded, open front buildings while other pigs on the same farm are raised in fully enclosed buildings and housed on cement slats over full liquid manure pits. Even though these pigs of similar size are raised in

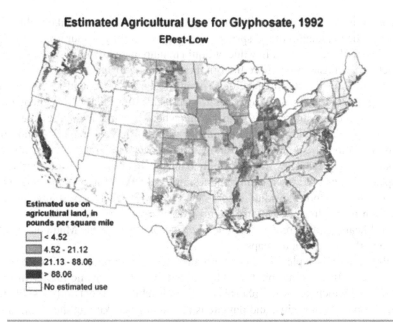

Estimated Agricultural Use for Glyphosate, 1992

EPest-Low

Estimated use on
agricultural land, in
pounds per square mile

☐ < 4.52
▨ 4.52 - 21.12
▨ 21.13 - 88.06
■ > 88.06
☐ No estimated use

Estimated Agricultural Use for Glyphosate, 2012

EPest-Low

Estimated use on
agricultural land, in
pounds per square mile

☐ < 4.52
▨ 4.52 - 21.12
▨ 21.13 - 88.06
■ > 88.06
☐ No estimated use

INCREASED USE. These two maps show the increase of glyphosate use over the last two decades. Ohio Northern University chemist Christopher Spiese has found that the increase in glyphosate use is responsible for 20-25% of dissolved reactive phosphorus loads in Lake Erie. He notes that while the map on the right is from 2012, the amount of glyphosate being used since then is virtually unchanged.

FIGURE 8.18   Chris Spies.

different environments, they are all fed the same feed. Soils fertilized/spread with the corn stalk bedding pack manure are maintaining or gaining organic matter and phosphorus in contrast to those soils to which only liquid manure is applied. This observation makes sense to a cattle veterinarian. Fiber is essential when feeding the organisms in a cow's rumen, and it helps beneficial soil organisms as well. When we started practice in 1974, an older landlord renting his ground had "No straw will leave this farm" written into the rental agreement. Now, what little straw we do see is mainly hauled in. Most clients want to avoid labor and do not want to spend the time to make bedding, only to have to haul it out to the field again in the spring. We are now building manure-pitted buildings for cattle. I wonder if we will see a fall-off in value of liquid manure here as well. This might happen even when cattle diets have considerably more fiber in them than hog diets.

Phosphate application can (and does) cause desorption of glyphosate bound to cations in the soil; thus, freeing the glyphosate to be an active chelator–herbicide–antimicrobial agent again. Our relatively new local Amish community is discovering this when their members buy acreages that have seen nothing but Roundup Ready corn and beans for years. If they have applied phosphorus, either commercially or via manure application, before seeding oats and hay, they must often start over. Stands start out looking great. They then succumb to the action of the reactivated herbicide. In this case, the phosphate anions bind to calcium, magnesium, and trace mineral cations in the soil, cations that were previously bound to residual glyphosate. Even with this loss of soluble phosphate through immobilization by binding the cations, chemist Chris Spies has shown that glyphosate's negative effect on soil health can increase leachable phosphorus levels in his Maumee watershed project as referred to in the two previous maps.

Another factor that is probably playing a concomitant role with glyphosate on some hog and poultry farms is the use of the enzyme phytase. Phytate is the part of organic matter in the soil that holds phosphorus and which keeps the phosphorus from leaching without soil loss. Phytate is also what holds the phosphorus in grains, and phytate is hard to digest. Without the addition of the enzyme phytase into the rations of these animals, some of this phytate remains undigested to pass in the feces and carry phosphorus with it. If a farmer adds the enzyme phytase to pig or poultry feed, he does not have to add as much inorganic phosphorus to the mineral premix. The pigs and birds receive more of their needed phosphorus via utilizing that which is naturally found in the grain. This addition of phytase looked like it would always be good environmentally, since less inorganic mineral would need to be added to the rations. Soil phosphorus values initially looked as if they would not get too high, even in the face of heavy application rates of manure. For some reason, it appears that the addition of phytase has sometimes led to crashing soil phosphorus values, and some farmers now add commercial fertilizer phosphorus on top of hog manure for field applications.

BASF was the first company to market a stabilized phytase, Nutriphos® in 2006. This phytase could withstand the heat generated by the pelleting process so that feed manufacturers could put the phytase in their premixes before pelleting and no longer had to spray the phytase onto the pellets after pelleting. The entire industry duplicated this ability shortly after 2006 to lower cost. This stabilized phytase can still continue to work in the manure pit and in the field where it could be breaking down remaining phytate in the pit and in the soil so that the now liberated phosphorus can leach more readily, even in the absence of soil loss. More soluble phosphate means more chance of that phosphate desorbing the competing anion glyphosate. The two anions together could contribute to more algae bloom. Industry claims that phytases are totally metabolized in the intestinal tract. The only research I could find to back this claim was done with older phytases that were not heat stable and that should be used up in the digestive tract [43]. Additional research is needed to understand the nutritional and environmental interactions impacted by this technology. Some of the biological pit additives are probably useful since we are not seeing crashing phosphorus soil tests on all hog farms that are using liquid manure. We need more research like that of Chris Spies. The knowledge gained from such research should provide more than an application chart based solely on nutrient analysis.

8.9 CONCLUSION

Environmental health, soil health, plant health, animal health, and human health are all inevitably linked. Based on my clinical observations of food animals, coupled with the science that is documenting a steady decline in environmental, plant, animal, and human health, it would be wise to discontinue the use of glyphosate, along with the genetically engineered crop systems that depend on it. Critics point out that if the world bans glyphosate, it will probably be replaced by herbicides that are equally, if not more, toxic. I share this fear, and am aware that other herbicides, and herbicide combinations, currently used in GE crop systems are already proving to be as, if not more, dangerous to health.

An example of this is the May 7, 2020, USDA Animal and Plant Health Inspection Service invitation for public comment on a petition by Monsanto-Bayer seeking deregulation of a corn variety intended for hybrid seed production that is genetically engineered for tolerance to dicamba, glufosinate, quizalofop, 2,4-D, and tissue specific glyphosate tolerance. This means that there can be five toxic chemicals added to our foods and feed products made from this corn.

Insecticidal neonicotinoids are routinely applied to soybean and corn seeds that are used throughout the Midwest, USA. Their detrimental effects on pollinators and overall biodiversity are an unfortunate reminder of ignoring such risks. A recent article in Hoard's Dairyman entitled, "There's still a place for granular soil insecticides," mentions a new, novel, integrated pest management approach via a beneficial nematode that feeds on corn rootworm larvae [44]. The article does not mention that the original native populations of naturally occurring beneficial nematodes have been lost because of the crop monoculture, engineered Bt traits, and a continuing dependence on a chemical approach that is failing. Also not mentioned is the fact that you must abandon most of the chemical approach for the beneficial nematode method to have a chance. Therefore, in addition to discontinuing glyphosate use, it is time to move to a smarter, more ecological agriculture that prioritizes better attention to biodiversity in our biosphere. Today's scientific researchers are doing a wonderful job investigating biodiversity's importance in both agricultural landscapes and the animal microbiome; however, current agricultural practices and policies do not reflect these investigations. We already know, for example, that lack of biodiversity on our planet increases the occurrence of nutritional deficiencies and disease; including worrisome human ailments such as Lyme's disease, SARs, malaria, and, most recently, COVID-19 (SARS-CoV-2). The relationship between illnesses and the lack of both nutritional diversity and biodiversity has been well established in both the animal and human medical communities. Recent peer-reviewed honey bee studies by Dr. Motta et al. [45] and studies done by forester Rod Cumberland from New Brunswick are just two examples of wildlife studies that support my food animal observations.

Why, then, does society continue to promote a national agricultural policy that puts extreme pressures on farmers to change once-diverse landscapes into monocultures that depend on widespread chemical use? The current chemical-dependent row crop system has helped the economics of scale but has eliminated many farms along with many small-town jobs. We can, and must, do better than this. I feel for both my animal and crop farming friends and clients. You usually cannot survive in the farming world without an abundant amount of optimism and a willingness to adapt to the sometimes-bizarre political policies that shape modern agricultural decision-making. Unfortunately, these policies steer farmers away from what we have historically done best: the practice of maintaining healthy landscapes and healthy animal populations. If farm policy required farmers to show improvement to and maintenance of soil organic matter to receive payment, I am confident they would comply.

I can see few reasons why we cannot "grow" more than one crop in one field [46]. In the US Midwest, we offer monetary subsidies and crop insurance for field corn and soybeans. Should it surprise anyone that those two crops are what is grown? The current monocrop, chemical system does not align with current research on integrated management and regenerative practices for sustainability, and it also aggravates the conditions described in this chapter. Furthermore, we should find a way to reward farmers who rely on fewer chemicals and who promote diversity, rather than subsidizing farmers who do the opposite. Many farmers appreciate that our present system does not make sense. Farmers are not going to watch their neighbors receive thousands of dollars in subsidies while they ignore a similar insurance. They, too, are bound in a competitive marketplace and have families to feed. Reduced dependence on chemicals like glyphosate, and diversification of mono-cultural agricultural systems, will simultaneously reduce the prevalence of many of the illnesses discussed in this chapter. This will take optimism and a willingness to change at multiple levels, as well as support from consumers and nonagricultural sectors of our economy. Nonetheless, this change is not only imperative to the future of farming, but also to the future of animal and human health as well.

REFERENCES

1. Abraham, W. 2010. Glyphosate formations and their use for inhibitions of 5-enolpyrovylshikimate-3-phosphate synthase. US Patent 7771736 B2.
2. Samsel, A. 2016. Email to author, April 20, 2016.
3. United States Geological Survey (USGS). 2016. Map of estimated agricultural use for glyphosate in 2016. https://water.usgs.gov/nawqa/pnsp/usage/maps/show_map.php?year=2016&map=GLYPHOSATE&hilo=L&disp=Glyphosate.
4. Nutrient Requirements of Dairy Cattle, seventh revised edition, 2001 NRC, page 140 Manganese Requirements.
5. Schefers, J. 2011. Fetal and Perinatal Mortalities associated with manganese deficiency. *Proceedings of the 2011 Minnesota Dairy Health Conference*, pp. 70–73.
6. Reynolds, K.J.M., K. Sheridan, DATE? Preugschas. Posterior paresis and paralysis associated with spinal cord vacuolation in late finisher pigs. *AASV* 339.
7. Aalseth, Dr. Earl, email conversation.
8. Federal Register, Environmental Protection Agency 180.364 Glyphosate Tolerances for residues pp. 512–513.
9. Pacholok, S.M., J.J. Stuart 2011. *Could it Be B12? An Epidemic of Misdiagnoses*. Quill Driver Books, Fresno, CA.
10. Ganson, R.J. 1988. The essential role of cobalt in the inhibition of the cystolic isozyme of 3-deoxy-D-arabino-heptulosonate-7- phosphate synthase from Nicotiana sylvestris by glyphosate. *Arch. Biochem. Biophys.*, 260: 85–93.
11. Streit, B., email confirmation to author, May 28, 2020.
12. Basu, P., J. Stolz, C. Sparacino-Watkins 2014. Nitrate and periplasmic nitrate reductases. HHS Public Access, https://www.ncbi.nlm.nih.gov/pmc/articles/PMC4080430/.
13. Hamish T.D., L. Smith, et al. 1995. The effect of amoxycillin on salivary nitrite concentrations: An important mechanism of adverse reactions? Br. J. Clin. Pharmacol., 39: 460–462.
14. Purdey, M., N. Purdey. 2007. Animal Pharm. Clairview, pp. 109–148.
15. Dr. Don Huber, personal conversation, 2019.
16. Lipinski, B. 2014. Can Selenite be an ultimate inhibitor of Ebola and other viral infections? Br. J. Med. Med. Res., 6: 319–324.
17. Kruger, M., personal communication, 2014, China and 2016, The Netherlands.
18. M. Kruger, A. Rodloff, W. Schrodl, A. Shehata, Glyphosate suppresses the antagonistic effect of Enterococcus spp. on Clostridium botulinum. Sci. Direct, 20:74–78. https://www.sciencedirect.com/science/article/abs/pii/S1075996413000188
19. Boucher, H. et al. Bad Bugs, No Drugs: No Escape! An Update from the Infectious Diseases Society of America. Infectious Disease Society of America. https://www.idsociety.org/globalassets/idsa/policy--advocacy/current_topics_and_issues/antimicrobial_resistance/10x20/statements/010109-bad-bugs-no-drugs-update.pdf.
20. Swerczek, T. 1980. Toxicoinfectious botulism in foals and adult horses. Am. Vet. Med. Assoc., 176(3):217–220. https://pubmed.ncbi.nlm.nih.gov/6988376/.
21. Skarin, H., B. Segerman, 2014. Plasmidome Interchange Between Clostridium Botulinum, Clostridium Novyi and Clostridium Haemolyticum Converts Strains of Independent Lineages Into Distinctly Different Pathogens. Pubmed.org. https://pubmed.ncbi.nlm.nih.gov/25254374/.
22. Jorn, E., email conversation, Sept. 11, 2016.
23. Hoskins, T. 2011. Broad Effect Glyphosate Effect Beyond Weed Control Gaining Attention. Iowa Farmer Today. https://www.sciencedirect.com/science/article/abs/pii/S1161030109000641.
24. Fernandez, M.R., et al., 2005. Crop production factors associated with fusarium head blight in spring wheat in eastern Saskatchewan. Crop Sci., 45, 1908–1916.

25. Huber, D., G. Johal, 2009. Glyphosate effects on diseases of plants. Science Direct. http://www.ask-force.org/web/HerbizideTol/Johal-Glyphosate-Effects-2009.pdf.

26. Smith, T. 2018. Mycotoxins in cattle feed: Detection, consequences and preventive strategies. Driftless Region Beef Conference.

27. Bhattari, P., et al., 2016. Zeolite food supplementation reduces abundance of enterobacteria. Elsevier.

28. Minnesota Dairy Conference, 2014.

29. Niederwerder, M. et al., 2016. Microbiome associations in pigs with the best and worst clinical outcomes following co-infection with porcine reproductive and respiratory syndrome virus (PRRSV) and porcine circovirus type 2 (PCV2). Pubmed.org. https://pubmed.ncbi.nlm.nih.gov/27139023/.

30. Sato, Y., Personal conversation, 2018.

31. Ayyadurai, S., et al., 2015. Integrative modeling of oxidative stress and C1 metabolism reveals upregulation of formaldehyde and downregulation of glutathione. Am. J. Plant Sci., https://www.scirp.org/journal/paperinformation.aspx?paperid=57871 https://www.researchgate.net/publication/281360784_Integrative_Modeling_of_Oxidative_Stress_and_C1_Metabolism_Reveals_Upregulation_of_Formaldehyde_and_Downregulation_of_Glutathione.

32. Dr. Larry Weiss, Presentation, 2019.

33. Dietert, R. 2016. The human super-organism: How the microbiome is revolutionizing the pursuit of a23 healthy life. Dutton.

34. Abraham, W. 2010. Glyphosate formulations and their use for inhibitions of 5-enolpyrovylshikimate-3-phosphate synthase. US Patent 7771736 B2.

35. Kurenbach, B., et al., 2018. Agrichemicals and antibiotics in combination increase antibiotic resistance evolution. https://peerj.com/articles/5801/.

36. A One Health Approach to Antimicrobial Use & Resistance: A dialogue for a common purpose. Symposium in Columbus Ohio. https://www.animalagriculture.org/Resources/Documents/Conf%20-%20Symp/Symposiums/2012%20Antibiotics%20Symposium/NIAA%202012%20AB%20White%20Paper%20-Final.pdf.

37. USDA. 2016. USDA's Response to Antibiotic Resistance. USDA. https://www.usda.gov/oig/web-docs/50601-0004-31.pdf.

38. Alvarez, F.B.J. 2011. Why are so many cows losing pregnancies? Hoards Dairymen. Ishmael, W. 2012. Nowhere Fast. Beef Magazine. 40.

39. Dorton A. and Swerczek, T. 2019. Effects of nitrate and pathogenic nanoparticles on reproductive losses, congenital hypothyroidism and musculoskeletal abnormalities in mares and other livestock: New hypotheses. Anim. Vet. Serv., 7(1): 1–11. http://www.sciencepublishinggroup.com/journal/paperinfo?journalid=212&doi=10.11648/j.avs.20190701.11.

40. Swerczek, T. 2020. An alternative model for fetal loss disorders associated with mare reproductive loss syndrome. Science Direct. https://www.sciencedirect.com/science/article/pii/S2405654520300305?via%3Dihub.

41. A. Samsel, email conversation, June 5, 2020.

42. M. Druille, et al. 2013. Glyphosate reduces spore viability and root colonization of arbuscular mycorrhizal fungi. Appl. Soil Ecol. 64: 99–103. www.elsevier.com/locate/apsoil.

43. AASV-Fact Sheet: Phytase https://www.aasv.org/shap/issues/v18n2/v18n2p90.html.

44. Thomas, E., Wiersma, D. 2020. There's still a place for granular soil insecticides. Hoards Dairyman 281.

45. Motta, E., et.al. Impact of Glyphosate on the Honey Bee Gut Microbiota. https://pubmed.ncbi.nlm.nih.gov/32723788.

46. https://www.youtube.com/watch?v=MDoUDLbg8tg&pbjreload=101.

9 Agricultural Pesticide Threats to Animal Production and Sustainability

Dr. Ted Dupmeier
Dupmeier Veterinary Services

CONTENTS

9.1 INTRODUCTION

Over my 50-year career as a practicing veterinarian, I have had the opportunity to see the deterioration of animal health concurrent with the commercialization of glyphosate-based herbicides (GBH) and especially with the broad acceptance of genetically engineered (GMO) crops tolerant of these herbicides, as well as their use as "harvest aids" for crop desiccation prior to harvest. (For simplicity, glyphosate-based herbicides will be referred to as glyphosate hereafter.) These latter two practices have greatly increased animal exposure to glyphosate well above that when they were used primarily for weed cleanup or preplant in preparation for the new crop. The "New Normal" in veterinary health is dramatically different from the relatively high level of animal health maintained in my early veterinary career. Today, as a food-producing veterinarian, I deal with glyphosate toxicity on a daily basis. If I see a case that does not make diagnostic sense, then I check for glyphosate exposure, and I have rarely been wrong. Glyphosate toxicity is insidious with exposure to low levels for a long period of time (months) because of its indiscriminate use in both agriculture and forestry, and other commercial and domestic uses. We seldom see high levels of glyphosate for a short period of time (days). It bioaccumulates in the animal from feed, bedding, and environmental exposure. Even with laboratory documentation of glyphosate toxicity, owners will almost always deny a possible source for the glyphosate. Although I will not discuss the impact of glyphosate on pets, it is a major concern and must be mentioned because glyphosate in animals and birds is a humane issue of high priority (Hoy et al., 2015). I am always learning and have had many great mentors; however, I must mention Dr. Don Huber who has saved more animal suffering than he will ever realize. Glyphosate toxicity in animals and birds is the most interesting part of my career; however, the ramifications to other toxins and chemicals must always be considered. I hope the information in this chapter will stimulate your thinking because the case studies mentioned are just a few of the real-life situations I personally have dealt with.

9.2 GLYPHOSATE: A MINERAL CHELATOR, ANTIBIOTIC, SYNTHETIC AMINO ACID, AND POWERFUL TOXIN

Glyphosate-based herbicides are the most utilized agricultural chemical in the world. In the 40-year period between glyphosate's commercialization in 1974 and 2014, 1.6 billion kilograms of glyphosate had been applied to soils in the United States, and the global usage was 8.6 billion kilograms. Glyphosate is only part of the formulated product that contains adjuvants, surfactants, and other products to make it even more toxic. Glyphosate itself is a toxin, but the formulated product can be 125–1,000 times more toxic to cells than the glyphosate alone (Clair et al., 2012; Mesnage et al., 2013). It wouldn't kill plants if it was not a powerful chronic toxin! It is an agricultural chemical with far-reaching biological consequences for livestock and all other animals. Glyphosate contamination of livestock bedding and feed for cattle, poultry, sheep, swine, and horses is a major predisposing factor, contributor to, and cause of extremely serious animal health issues. As a patented mineral chelator and antibiotic, glyphosate in livestock feed can cause many unintended biological complications that result in harm, hurt, and failure to thrive (Table 9.1). In addition to dietary intake from feed or water, animal exposure to glyphosate may be from drift and direct contact with residue in the soil.

In the area of Southwest Saskatchewan, Canada, there is a large quantity of glyphosate used. It was first available as 360 g of the active ingredient, glyphosate, per liter; however, the product is now sold as 540 g per liter in the formulated product. Often, the same liters per hectare are utilized because of reduced efficacy against tolerant weed species or lowered cost of the product.

TABLE 9.1

Some Animal Health Findings Associated with Glyphosate Laboratory Analysis That Were Remediated by Removing the Source of the Glyphosate

Dairy cows:
- Botulism syndrome
- Fatty liver
- Hemorrhagic bowel syndrome
- Malformed calves (birth defects)
- Increased mastitis
- Milk production drops
- Reproduction performance decreases (infertility)
- Miscarriage/abortion increases

Beef cows:
- Dead cows, all with symptoms of chronic botulism

Beef calves:
- Aborted/miscarried
- Born weak, birth defects
- Enlarged liver, color variations, friable, fatty
- Compromised immune system
- Joint problems, lame

Feedlot animals:
- Severe acidosis with minor feed changes (rumen bacteria can't adapt)

Young beef calves:
- Bloating
- Nontreatable bloody scours
- Breeding bulls
- Poor semen quality (infertility)

Swine:
- Poor reproductive performance
- Infertility, small litters
- Birth defects
- Tail biting potentiated
- Lameness, especially in fast-growing pigs
- Unusual and difficult to treat scour (especially in weaned pigs)

Horses:
- Agitated and aggressive behavior (even if well trained)
- Untreatable intermittent diarrhea
- Lameness often similar to a laminitis
- Miscarriage
- Pain on palpation of the cecum

Lambs:
- Sudden death resembling acute Clostridial death even though well vaccinated
- Enlarged liver, variable color, friable
- Kidney hemorrhagic and friable
- Poor wool growth
- Stunted growth, weak

Poultry:
- Feed intake and production often within normal limits
- Unexplained consistent number of deaths each day
- Enlarged liver, bronze color, fatty, friable
- For each syndrome seen, there often is variation in symptoms

Although Monsanto, the initial maker of glyphosate herbicides (Roundup®), sometimes included another mineral chelator in the formula so a lower concentration of glyphosate could be used, the Canadian market apparently appeared too small to patent formulations containing the chelators. Canadian producers are reluctant to use chelators in case there should be a chemical problem and they would not be covered by any company guarantees. Overlaps from application equipment and maladjustment often result in even higher amounts being applied in many situations, such as on head-lands or at the end of the field where they turn for the next pass.

9.3.1 GLYPHOSATE AS A CHELATOR OF ESSENTIAL MINERALS

Glyphosate was first patented in 1964 as a strong mineral chelator and originally used to clean steam pipes and boilers. As a mineral chelator (Martel and Smith, 1974; Motekaitis and Martell, 1985), glyphosate binds with and blocks the bioavailability of essential mineral nutrients that function in biological systems as components and cofactors (catalysts) for chemical reactions. When it binds manganese, cobalt, and other essential cationic minerals, physiological functions in the various animal tissues are inhibited. As one of the unintended biological consequences of glyphosate contamination of bedding or animal feed, it not only reduces feed efficiency, but also has a negative effect on vitamin levels in animals, so that affected animals often show, and analysis will prove, a deficiency of vitamin A even in the presence of good supplementation.

In 1974, glyphosate was also patented as a nonselective plant killer or herbicide. Its proposed herbicidal mode of action was inhibition of the Shikimate pathway in plants (Abraham, 2010). Since animals are the only living things that don't directly utilize the Shikimate pathway, it initially was assumed, and promoted, to be nontoxic to animals; however, glyphosate, as a broad-spectrum chelator, can inhibit over 290 enzymes while stimulating others, and animals and humans are dependent on many microorganisms in the GI track and other tissues that depend on the shikimate pathway for function. Animals also are dependent on microorganisms, indirectly in the environment, for balancing the various components of the microbiome; many of these microorganisms are important suppressors of pathogens that have an alternate shikimate pathway that is insensitive to glyphosate in contrast to sensitivity of "good" bacteria and fungi (Krueger et al., 2012, 2013).

9.3.2 GLYPHOSATE AS AN ANTIBIOTIC

Glyphosate is patented as a broad-spectrum antimicrobial in 2010 (US Patent No. 7771736). Just as it inhibits the Shikimate pathway in plants, it also inhibits the Shikimate pathway in many essential beneficial organisms in the environment and animal gut. Animals have millions of good bacteria in their guts that are essential to their health. The gut microbiome isn't just responsible for digestion, but also for vitamin production, mineral absorption, and the strength of the immune system. Glyphosate affects the gut microbiome and compromises the immune system. Some organisms have an alternate shikimate pathway that is not sensitive to glyphosate, such as *Clostridium botulinum*, *C. perfringens*, *C. difficile*, *Salmonella enteritidis*, *S. gallinarum*, *S. typhimurium*, *Escherichia coli*, and *Enterobacter cloacae*. These pathogenic organisms can then overgrow the microbiome and develop gut dysbiosis (Ackermann et al., 2015; Clair et al., 2012; Krueger et al., 2012, 2013; Shehata et al., 2013).

The widespread adoption of Roundup Ready® crops initially patented in 1996, which were genetically engineered to tolerate glyphosate, resulted in as much as a 15-fold increased use of glyphosate. This tolerance meant that glyphosate could be sprayed directly over the crop to kill weeds without killing the engineered (GMO) crop plant. Glyphosate, however, is a highly water-soluble, systemic chemical that bioaccumulates in all plant and animal tissues. Unlike most agricultural pesticides that can be removed from the protected grain surface during harvest operations or processing, glyphosate is systemic, and in every cell of the plant that is used as feed or fodder. Approval for the use of glyphosate as a plant desiccant for non–Roundup Ready crops in 2001 soon became a common agricultural practice. It is applied three to five days prior to harvest to speed the ripening process so the entire field is at the same stage for direct combining without requiring a waiting

TABLE 9.2

Representative Levels of Glyphosate in Some Livestock Feeds Determined by Laboratory Analysis

Product	Glyphosate Level (ppb)
Lentil pellets	3,004.0
Roundup Ready corn (for grazing)	448.0
Calf creep feed (explain what this is)	894.5
Pea stubble (half and half peas and pea straw)	5,186.0
Soybean meal	6,727.0
Yellow hay (desiccated alfalfa grass hay just prior to cutting	19,542.0
Barley grain (desiccated before harvest)	945.9
Layer ration feed	956.1
Commercial horse feed	1,030.0

TABLE 9.3

Physiological Malfunctions in Animals and Birds Resulting from Glyphosate Antibiotic Action

Microbiome is disrupted, gut dysbiosis
Pathogenic bacterial overgrowth (loss of suppression by normal gut microflora)
Digestion in ruminant is hindered
Decreased protozoa, bacteria, etc. lead to starvation of ruminants in the face of plenty
Ketosis and fatty liver develop
Hormones are disrupted
Cytokines are disrupted and hormone relationships are disrupted
Lethal action of chronic botulism
Immunity is compromised and massively jeopardized

period or swathing. Many non-GMO crops that are used for livestock feed, such as wheat, barley, oats, flax, canola, lentils, corn, and sugar beets, are desiccated with glyphosate. Again, glyphosate is a systemic chemical in the plant that swiftly bioaccumulates in all plant tissues. The broad exposure to this broad-spectrum antibiotic throughout the environment is cause for considerable concern, since it also induces pathogen resistance to other antibiotics (Kurenbach et al., 2015, 2017, 2018).

Glyphosate levels in livestock feed are simple to analyze for at an Accredited Laboratory such as A&L Laboratories in London, Ontario, Canada (http://www.alcanada.com/). Representative findings of glyphosate levels in livestock feed are presented in Table 9.2. It should be remembered that the formulated glyphosate herbicides are even more toxic than the glyphosate analyzed by the laboratory. Some of the physiological disruptions caused by glyphosate are presented in Table 9.3.

9.3   REPRESENTATIVE ANIMAL CASE STUDIES OF GLYPHOSATE TOXICITY

In all of these case studies, and numerous more, there is consistency of the presence of glyphosate and the ill-health of the animal. Having two broad modes of action: as a strong mineral chelator and potent antibiotic with both systems working synergistically, the wide variation in symptom expression of glyphosate toxicity can be appreciated. Add to this the synergistic activity of the two modes of action and the fact that the formulated product is 125–1,000 times more toxic, means that glyphosate is a recipe for disaster. Dosage also is always important. Numerous representative cases are presented in summary form below. Variation as to clinical aspects results from environmental factors such as feed, mineral/vitamin program, climate, and other stressors. There is a diverse group of species in the environment, each with its various physiological features. In each case, some factors didn't make clinical sense according to traditional wisdom; therefore, analysis for glyphosate was justified to determine if a significant glyphosate dosage was present. The consistency in the numerous cases presented in summary form is that glyphosate is involved in each case. None of these cases involved the prescription of a traditional antibiotic. Yet, in each case, complete recovery of surviving animals is reported, and each case is a practical, on-farm situation that is often repeated numerous times. Although these summaries are consistent, there is a great deal of additional information that could be added. The following case numbers are organized by species with comments made of each case situation. These are only the tip of the destructive iceberg we call glyphosate. There are numerous similar cases I could cite for each of the animal species presented here as examples of the slaughter and suffering permitted with the use of the glyphosate herbicide.

9.3.1 BEEF CATTLE

9.3.1.1 Purebred Bull Supplier (Case 1)

Yearling horned Herford bulls were split into two groups of 25 each where one group received pea and lentil screenings containing 506.5 ppb glyphosate for 120 days while the other group of 25 did not. Both groups received calculated adequate minerals/vitamins. At semen testing in April, 21 of the 25 bulls that did not get the pea and lentil screenings passed the semen test. In contrast, bulls fed the pea and lentil screenings had poor semen morphology and only four of the 25 passed the semen test. This was economic disaster for the producer since many of the bulls that did not pass the semen test were already sold and buyers depended on the bulls. The problem was diagnosed as glyphosate toxicity and, for bulls not passing the semen test, the feed was changed to non-glyphosate-contaminated feed and retested in 37 days when 19 of the 25 bulls passed the test.

9.3.1.2 Range Herd (Case 2)

This situation was a beef cow herd that was fed lentil straw that had been sprayed with glyphosate for desiccation prior to harvest. Cows received this as their primary feed for 130 days. Their mineral/vitamin supplementation was variable to poor and their "condition" rating of a 2 out of 5 indicated that they were burning some body fat. Twelve of the 21 calves born during early calving had large abdomens and were small in statue (1/2 normal height for Angus calves. One calf died and was examined. It had a large liver (hepatomegaly, 2× normal size) that was friable and would break in your hand. Laboratory diagnosis was of unknown cause; however, they observed lipid-filled hepatocytes (liver cells filled with fat). The cattle manure indicated very poor digestion. Diagnosis was poor rumen bacterial action (microbiome affected). The owner still had 155 cows to calve and was very worried.

The solution was replacement of the glyphosate desiccated lentil straw with good quality hay supplemented with a small amount of grain (1 kg per day per cow) and liquid "Promolas" with appropriate minerals provided to the cows. "Promolas" is a molasses liquid often used as a supplement at ½ kg per day (the level can be controlled by additives to the Promolas) to stimulate rumen bacteria. Although a total of five calves were lost, within 2 weeks, there were no more losses or abnormal calves (Figure 9.1).

9.3.1.3 Small Farm Beef Herd (Case 4)

The herd consisted of 30 cows on poor nutrition (body condition rating of 2/5 indicating very little body fat) that were fed screening pellets containing 2,900 ppb glyphosate for 70 days with variable mineral/vitamin supplementation. Two cows died prior to calving. Their livers were hepatomegaly and friable with lipid-filled hepatocytes. Manganese and vitamin A were deficient, similar to cattle with glyphosate toxicity. Laboratory diagnosis could find no signs of disease. The solution was Promolas, good-quality forage with 1 kg grain per head per day, and removal of all screening pellets. Results were no more death losses and, within 6 weeks, the cows began calving with normal calves.

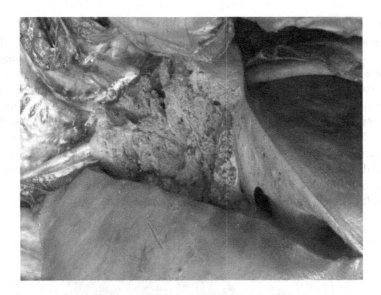

FIGURE 9.1   Cow liver, severe ketosis; variegated, friable, fat-filled sinusoids (Case 9).

9.3.1.4   Large Beef Herd Grazing Corn for Winter (Case 6)

This was a 600-cow beef herd that was grazing Roundup Ready corn, which had received two passes of glyphosate during the growing period. Of the first 30 calves born, ten were late abortions and 20 calves were born very weak that died within 1–2 days. This was a very stressed ranching operation that ran out of grazing corn 3 weeks prior to the first 30 calves being born. A good-quality forage plus grain had to be purchased and a good mineral/vitamin program was started. The remaining 570 calves were born normal and calving went as expected.

9.3.1.5   Cows Grazing Desiccated Lentil/Pea Stubble for Winter (Case 7)

This was a 250-cow beef herd that grazed on glyphosate desiccated lentil and pea stubble containing 5,186 ppb glyphosate for 130 days over the winter. They received an excellent vitamin and mineral supplement and the cows were in excellent body condition (3.5/5) and in an excellent facility for calving in late spring; however, the first 50 calves born were a nightmare as the owner and family were very busy off farm. The first 50 calves had ill navels (infection entering through the navel), were of poor grade, had significant respiratory problems requiring many treatments, had a compromised immune system, and were not nursing well. This resulted in a great deal of labor and stress to keep these animals alive. Three weeks after changing to a good-quality, glyphosate-free roughage feed fed ad lib, along with 1 kg grain per cow per day and Promolas, all calves were healthy with very few calf problems.

9.3.1.6   Cows and Calves Graze RR Corn Stubble in November (Case 8)

Cows and calves (average weight of 650 lbs) grazed on Roundup Ready corn in early November. They had a good water source, but the supplemental vitamin and mineral levels were marginal. By the end of December, the cows were doing poorly and the calves only averaged 560 lbs, for a weight loss of 90 lbs. These animals had poor shelter from the elements and the corn contained 340 ppb glyphosate. Calves were weaned and cows and calves were given shelter and a good nutrition program so that by the end of February, the cows and calves were back to their condition the previous November. This was an economic disaster for this ranch as calves had been contracted for January delivery at 800 lbs. Rather than being able to rely on the grazing corn, outside feed had to be purchased and the contract extended at the farm expense.

9.3.1.7   Very Important, Split Beef Herd and Massive Differences (Case 10)

This farm operation consisted of a 220-head Angus cross beef herd that was split, due to family disruption, where 110 animals were regularly grazing and fed long hay with a vitamin and mineral supplement in the winter following organic standards. The other 110 were grazing Roundup Ready corn containing 400 ppb glyphosate for 127 days that had had a preplant and two in-season glyphosate applications, each of 540 g/acre glyphosate, and Roundup desiccated lentil stubble. The herd had been operated as a closed herd for 20 years and neither group had contact with other cattle after the split. The only problem reported for the organically maintained herd was one weak calf. In contrast, the herd grazing the corn and lentil stubble had nine abortions and ten weak calves born. One weak calf died and showed severe hepatomegaly and lipid-filled hepatocytes with a suspected infectious bovine respiratory disease. Within 2 weeks of changing the nonorganic animals to long hay and Promolas, there were no further problems (Figures 9.2 and 9.3).

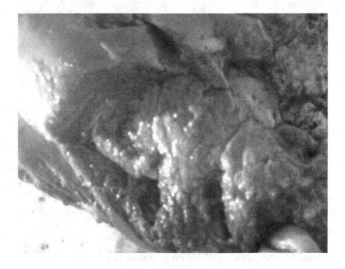

FIGURE 9.2   Friable liver with fat-filled sinusoids (Case 10).

FIGURE 9.3   Enlarged liver for a calf (Case 10).

9.3.1.8 Feedlot Cattle with Flax Straw Bedding (Case 11)

This farm had 125 calves in the feedlot with free choice of rolled grain that was changed to full feed of pea and lentil pellets for 60 days, which had been done for 15-plus years. Flax straw was used as bedding.

Twenty animals appearing normal the evening before went suddenly down with a diagnosis of acute acidosis or grain overload (which did not make much sense). Fifty of the animals that were treated died. The owner reported that the animals had a "pica" or crazy appetite for the flax straw he was using for bedding. The laboratory diagnosis was "unknown cause." Analysis of the flax straw for glyphosate showed that it contained 506.5 ppb glyphosate and the lentil pellets contained 3,004 ppb glyphosate. The glyphosate antibiotic had killed the rumen bacteria and the animals could not handle the change in feed.

It took 3 weeks to get these animals back to normal. Note: Monsanto had a professional representative involved with this case and paid for the laboratory analyses.

9.3.1.9 250 Beef Cows (Case 12)

This farm had 250 crossbreed beef cows grazing 135 days on Roundup Ready corn that had received two passes of Roundup glyphosate during the growing season. These were excellent cows, with excellent body condition (3.5/5), in an excellent calving area receiving an excellent mineral and vitamin supplement. Ten calves were aborted, ten calves were born weak, seven calves had skeletal malformations, and 20+ calves had navel infection related to a depressed immune system. Analysis of the feed showed 448 ppb glyphosate. The mycotoxin panel was negative. Cows were taken off the Roundup Ready corn, and long hay of good quality was provided along with Promolas. The calving area was changed to a new, clean area, and calves were treated with Tender Loving Care as needed. Five of the navel ill calves died. Within 2 weeks of the problem, all problems had ceased. Three weeks after changing the feed, everything appeared normal. This was very important as this ranch was in an expansion phase and had planned to go to 500 cows and more corn the following year except for this change in plans (Figure 9.4).

9.3.1.10 Calves (Case 13)

There was a sudden outbreak of coccidiosis-like disease in calves on cows and creep feeders on this farm. Coccidia is a parasitic disease that is common in calves and causes severe bloody diarrhea. The owner is very experienced and started his treatment for coccidiosis; however, with no success, and four calves out of 30 died. Examination showed that the calves had severe hepatomegaly (lipid-filled sinusoids), friable livers, and blood-filled intestines. The creep feeding program was

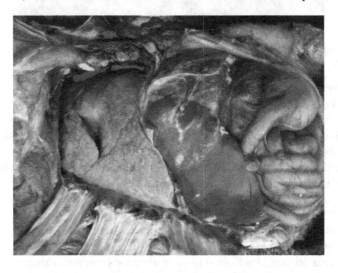

FIGURE 9.4 Umbilical infection – showing infection by the liver (Case 12).

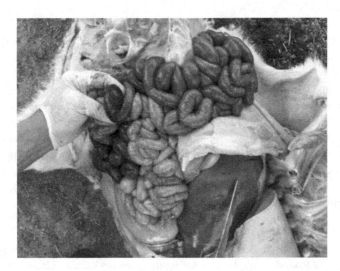

FIGURE 9.5   Blood-filled small intestine in my hand (Case 13). (The liver at pointer hepatomegaly – enlarged and various colors – variegated.)

FIGURE 9.6   Liver friable (Case 13).

good and the calves were eating the creep well. The creep ration contained 894.5 ppb glyphosate as it contained a high percentage of lentils and some canola. For remediation, the creep feed was replaced with an organic product and calves were given an injectable iron and multiple B vitamins to replace their blood loss. Calves also were given electrolytes to replace fluid loss. Three days after the feed change, the calves were normal, and at 5 days, they were sent out to pasture. The rapidity of change appeared to be impossible. This was a very important case as the owner and, of course myself, told everyone not to utilize that creep (Figures 9.5 and 9.6).

9.3.1.11   20 Calves as Orphans (Case 14)

Twenty orphan calves were initially pail fed on milk replacer and then calf starter was added. Calves ate the starter well, but within 2 days all 20 began to bloat and the owner was going hysterical. The owner attempted his usual bloat treatments; however, without success. The starter ration was tested for glyphosate and contained 813 ppb glyphosate. Due to experience with case 13,

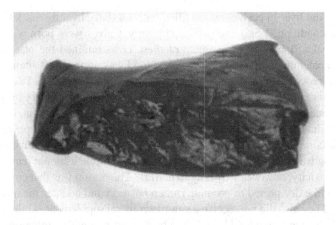

FIGURE 9.7 Normal liver comparison with Case 13.

the owner was advised to change starter rations. Within three days on the glyphosate-free starter ration, only one calf was still bloating and within five days, all calves were normal with no deaths. The owner was very pleased because these calves belonged to his young children as a 4-H youth project (Figure 9.7).

9.3.2 DAIRY

It is not that dairy animals have fewer problems; I have just presented a few examples below. The beef examples give an idea of the massive number of glyphosate-related problems.

9.3.2.1 Dairy (Case 5)

This case is of a well-operated dairy herd of 150 milking cows that were fed for 44 days with Roundup Ready corn silage, which had received two passes of 540 g per acre glyphosate during the growing season for weed control. Milk production dropped 4 kg per cow and the somatic cell count (SCC) tripled due to the immune system being jeopardized. SCC is indicative of udder infection. Out of desperation, the dairyman stopped feeding the glyphosate-laced silage and had to purchase glyphosate-free silage plus long hay and change his vitamin and mineral program. Three weeks after changing the feed to non-GMO feed free of glyphosate, milk production was back up and the SCC were down to normal. Cows late in lactation did not recover, which would be common for such cows. This was a major economic setback for this producer; however, it was in good timing because he was just planning an expansion, which would have been an even greater disaster.

9.3.2.2 Excellent Example, 2 Separate Herds But the Same Management and Feed (Case 9)

This was the most important case that I have been involved with. Dr. Don Huber helped me immensely in this case. There were two identical 250 cow dairy herds, herd A and herd B, with an average of 34 kg milk production and reasonable SCCs. Barns were the same age and equipped the same. They shared the same silage pit and dry feed source, plus all purchased feeds and products were the same. Genetics also were the same as cows were moved between barns as dictated by the quota. The water source was the same as were health protocols. The difference between the two herds was that herd A had very few problems. In contrast, herd B added a product, TMR, to their dry cow program that was supposed to boost their immune system but caused massive health problems. The TMR product caused the cows in herd B to reject their mineral/vitamin intake. The cows that were milking in both herds initially had good vitamin and mineral intake. Dry cows in herd B were very ketotic (burning body fat because of glyphosate toxicity to rumen bacteria) and 13 cows died at freshening. The laboratory could find no diagnosis; however, the liver in each cow was filled with fat (fatty liver disease). Cows that survived were in poor body condition (1.5/5) and produced poorly. Calves were born with limb deformities and took a great deal of labor. Ninety percent of these cows retained the placenta. Downer cows went on to die. The calcium and phosphorus ratio in blood was normal, but vitamin A was deficient. Temperature of the cows was down, pulse was up, and respiration was up. Twenty-two cows died. Necropsy findings were fatty liver, uterine infection, and abomasum lining very hyperemic with some ulcers. Analysis of one of the cows at the Vet lab found severe uterine infection, fatty liver, and manganese, cobalt, and vitamin A below normal; but no diagnosis as to cause except very rapid degeneration due to Clostridial organisms. Analysis of the TMR product showed that it contained 6,717 ppb glyphosate. Drinking water had 3 ppb glyphosate. Mycotoxin levels were not remarkable. This was a typical glyphosate toxicity; however, why only in herd B? The reason was that cows in herd B did not receive Promolas in the dry period to promote rumen bacteria and counter the mineral immobilization by glyphosate. They also did not receive a vitamin/mineral supplement, while dry cows in herd A did. Feed changes were made and herd B was soon back to normal within three weeks! This was very important since, in the face of severe glyphosate toxicity, if all is normal, and there are no associated stresses, animals can perform reasonably well. It is always a concern if they could do better, and any small problem with even one cow can end up as a major problem.

9.3.2.3 Dairy Herd (Case 26)

This excellently managed dairy farm had more than 300 cows milking robotically. They essentially had two serious health problems. The first was that cows would develop "environmental" mastitis at about day 127 in milk and then go on to die. This condition was impossible to treat because of a lowered immune system. The second challenge was that they had to build concrete walls by the waster troughs because the cows would lap so much water that the cow alleys were dangerous for cows or employees slipping. This was due to "Chronic botulism" affecting the cows swallowing so that they would drink like a dog lapping rather than a cow sucking water. The dairy was losing one cow per week. At the time of my visiting the herd, it was evident that 90% of the cows were not passing their placentas after calving, and SCCs were very high.

On further examination, the mastitis was an environmental type caused by *Klebsiella,* even though they had an exceptionally clean barn and did not bed with wood shavings that may sometimes increase the problem. This case didn't make sense until the feed was examined and analyzed. The barley feed grain supplement being fed had been desiccated (burnt) with glyphosate and contained 945.9 ppb glyphosate, ten times higher than that Dr. Monika Krueger, Leipzig University, had shown caused gut dysbiosis and chronic botulism in European cattle (Krueger et al., 2013). The mycotoxin tests were negative. The compromised immune system resulted from the glyphosate in the feed about 127 days after starting on the burnt barley feed and was coincidental with the chronic botulism that had also developed to this point. Both of these problems were solved by immediately replacing the barley feed with quality long hay and adding Promolas to all rations. Within 1 month, they took down the concrete walls by the water troughs. It is very important to use Promolas as both a source for minerals to compensate for those immobilized (chelated) by residual glyphosate from the feed and a nutrient source to stimulate rumen bacteria. The source of the Promolas is a concern at times now that the sugar beets are sometimes desiccated with glyphosate.

9.3.3 SHEEP

9.3.3.1 Sheep Fed Glyphosate Desiccated (Yellow) Feed (Case 3)

Yellow feed is feed that is harvested within 1 week of desiccation with glyphosate. The usual rate of glyphosate applied is 1080 g per acre to be certain of plant death. Desiccation is used in preparation for breaking and preparing the field for a subsequent crop. After harvest, desiccation leaves less stem and root mass to deal with. This is a very common practice that is encouraged by government agencies. Typical glyphosate level in the burnt, harvested material fed to animals is 19,542 ppb.

In this case, after sheep were offered the burnt feed, feed intake dropped to ½ normal intake. Within 1 week, sheep were passing poorly digested feed. When the mineral/vitamin supplementation is variable, the animals show illness more severely, coincidental with a decrease in the immune system. These sheep were lethargic. To resolve the issue, good long hay was provided and yellow feed was limited to approximately ¼ the previous intake. Promolas was also offered and within 1 week, the sheep were back to normal.

9.3.3.2 Sheep on Creep (Case 15)

This was a well-managed sheep flock with lambs on a free choice creep ration. Lambs were well vaccinated and dewormed. The wool coats were exceptional and growth was great until the creep ration was changed. Within 2 weeks of the change in creep, the lambs had a poor wool coat and poor growth. Two lambs died. The dead lambs showed severe hepatomegaly, with very friable livers. Kidneys were enlarged and hemorrhagic. The creep feed contained lentil and pea screenings containing 645.5 ppb glyphosate. Consumption of the creep had declined by 50%. To remediate, the creep was removed immediately and the lambs were turned to pasture since good-quality creep free of glyphosate was not available. This is a common problem, especially for sheep, since the creep for sheep is manufactured as economically as possible. This is important for the sheep industry as a whole since most lambs require a creep ration to grow fast enough to make a profit for their owners (Figure 9.8 and 9.9).

9.3.3.3 110 Dead (Case 22)

This case was the sudden death of a flock of sheep where 110 died in one evening and were all lined up at the edge of a watering hole. Several days earlier, there had been a massive rain with eight inches (20 cm) in two hours' time. The pond drained five sections of continuously and conventionally cropped land. No laboratory diagnosis was available, but symptoms resembled botulism; however, there were no dead birds or other animals in the water course or at the pond typical of acute botulism. Numerous sheep were sent to the Veterinary Pathology Laboratory, and the Veterinary diagnostic team from the Veterinary College visited the case; however, no diagnosis was rendered. Chronic botulism from glyphosate toxicity is a common symptom in ruminant animals; however, it is nearly impossible to send tissues from Canada to the United States for confirmatory analysis, and there aren't laboratories in Canada that can test for botulism since it is considered a biological warfare agent.

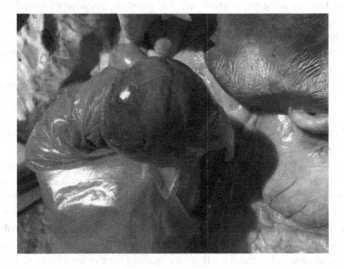

FIGURE 9.8 Kidney with hemorrhage (Case 15).

FIGURE 9.9 Liver (Case 15) hepatomegaly (enlarged), friable (breaks in hand) fat-filled (lipid in sinusoids), variegated (various colors).

9.3.4 SWINE

9.3.4.1 Lactation Variability, Representative of Many Cases (Case 18)

Swine problems related to glyphosate toxicity are very common. In this case, the problems that were seen included:

a. Lactation variability within a breed genetically consistent for milk production.
b. Scours in pigs weaned 3–5 days previously that is always called a nutritional effect because of change in protein source. It may very well be nutritional, however, not from protein; but from glyphosate toxicity from the changed feed. Although various antibiotics are administered, they are not effective and seldom see a clinical improvement associated with the antibiotic administration.
c. Lameness in fast-growing pigs, which are fast growing because of genetics; however, the mineral/vitamin balance must be correct, especially for manganese and vitamins A, D, and E. It is this balance that is disrupted by the chelating and antibiotic activity of glyphosate,
d. Tail biting and other aggressive behaviors from glyphosate's antibiotic activity on gut bacteria producing neural compounds.
e. Leaky gut syndrome from glyphosate's antibiotic activity on gut bacteria causing gut dysbiosis.

In this case, soybean meal is the usual protein supplement and contains approximately 6,700 ppb glyphosate (Table 9.2).

Conventional laboratory diagnosis was inconclusive. The producer wanted to change to a stronger antibiotic or higher antibiotic rate in the feed; however, changing to a feed that was not contaminated with glyphosate solved his problem. There is a critical need for much research regarding glyphosate toxicity to swine. Improvements are always evident if a binder such as humates or clinoptilolite (a zeolite) that remove glyphosate (Krueger et al., 2013), or excellent probiotics to counter the microbiome dysbiosis are added to the feed. Rather than hoping to solve the problem with more antibiotics, we need access to laboratories that can analyze for the antibiotic glyphosate so it can be removed from the feed source.

9.3.5 EQUINE

Equine is an area that requires extensive research as can be said for dogs, cats, and other pets (Fox, 2012), wildlife, and for humans. The following case is a common finding in horses, especially if they have limited grazing. Commercial supplements provided horses often contain 1,000 ppb or higher glyphosate. Consistent health findings with equine ingesting glyphosate include:

a. Cecum painful upon palpation. Glyphosate is a powerful antibiotic to beneficial microorganisms in the cecum, so the toxicity of glyphosate is more intense in the cecum of a horse than the rumen of a cow or sheep.
b. Intermittent diarrhea containing undigested or partially digested material.
c. Laminitis type of syndrome.
d. Decreased performance.
e. Behavioral issues are very important in handling.
f. The solution is removal of feed that contains glyphosate. Within 1 week there will be a noticeable difference in all of the above and much improved within a month.

9.3.5.1 Horses-Related Cases (Case 19)

Horses, with limited grazing opportunity, fed a commercial complete feed supplement containing 55% alfalfa on a continuous basis, developed intermittent diarrhea containing undigested or poorly digested material, laminitis syndrome, painful cecum on palpation, decreased performance, and serious aggressive behavioral issues. Analysis of the feed supplement showed that it contained over 2,000 ppb glyphosate. There was a clinical difference in the horses 1 week after changing to a feed that was not contaminated with glyphosate and, within one month after the change to noncontaminated feed, the cecum could be palpated without signs of discomfort, there were no indications of lameness or digestive issues, and the horse's behavior was significantly altered and described as mellow.

9.3.6 POULTRY

Birds are very susceptible to glyphosate toxicity. Although the case presented here is for laying hens, it is an important concern in all aspects of the poultry industry.

9.3.6.1 Poultry Layers (Case 27)

These birds were housed in a new barn and had 96% egg production on 105 g feed intake (normal for their stage of production and weight). Twenty days after a change of feed, 20–30 birds died per day with no evidence of paralysis. The laboratory analysis cited pleuritis, peritonitis, septic colibacillosis, and two birds with lymphoid leucosis. This diagnosis didn't make much sense and medication would be very difficult. The *E. coli* sensitivity to antibiotics was terrible. Mycotoxins were not of importance. Analysis of the new feed for glyphosate showed that it contained 756.1 ppb glyphosate. The feed was changed from a soybean base to one free of glyphosate, more energy was added to the ration, and a more efficient mineral/vitamin supplement was used. From day 1 to 5, the death losses dropped from the 20–30 per day to 5 per day and was maintained at a low level. The egg production level was maintained for the growth stage without the use of antibiotics, which is paramount in a layer operation. It is important that the egg industry test feed for glyphosate and also measure the glyphosate level in tissues and eggs since it is a systemic, water-soluble toxin.

9.3.7 CROP RESIDUES

9.3.7.1 Pea Stubble (Case 20)

A random sample of pea stubble and residual peas considered for animal consumption, taken 6 weeks after harvest, had glyphosate equivalent levels of 5,286 ppb (78 as AMPA and 5,264 as glyphosate, indicating no significant degradation of the glyphosate in storage). The field had received 540 g per acre glyphosate applied preplant to the soil to burn down weed control, and the peas were desiccated with 540 g glyphosate per acre 1 week prior to harvest.

9.3.7.2 Irrigated Alfalfa Grass Mixture (Case 21)

A field of mixed alfalfa and grass was sprayed with 1080 g/acre RoundUp to facilitate tillage in preparation for a subsequent crop. The hay was cut for feed 3 days after being sprayed with glyphosate (yellow feed). A random sample of bales was taken 1 month after cutting for glyphosate analysis. The yellow hay contained 19,542.5 ppb glyphosate equivalent (19,160 ppb glyphosate and 382.5 ppb AMPA to indicate no significant degradation of the glyphosate applied after storage for one month). It must be remembered that wildlife – birds and animals – are constantly feeding on glyphosate-contaminated feed. A map of chronic wasting disease of deer and elk and glyphosate usage superimpose very closely.

9.4 REMEDIATION, REVERSAL, AND STOPPAGE OF GLYPHOSATE DAMAGE

Certain considerations need to be recognized for remediation to be successful: (i) Glyphosate toxicity is insidious. (ii) In all cases where the clinical syndrome does not respond to treatment, plus the usual management changes that lead to improvement, glyphosate-based toxicity must be seriously considered. (iii) Clinical syndromes are seen with lower levels of glyphosate than are generally considered toxic because there are two factors that determine toxicity: (a) although lower levels may be in the feed, it is generally being fed over an extended period of time where chronic toxicity is not considered, and (b) all "safe" levels of glyphosate are based on acute toxicity of glyphosate and not the much more toxic formulated product. (iv) There is a "bell-shaped" curve in biological systems; therefore, some animals are more susceptible than others, even within the same group. (v) Young animals and stressed or hardworking animals are more susceptible than more relaxed animals. Consider that the toxicity is more evident in the high-producing dairy cow than a beef cow. Ultimately, there are two approaches to remediate glyphosate toxicity: (i) removal of the source of the glyphosate and/or (ii) compensation or countering the pathological action of glyphosate.

9.4.1 REMOVE THE SOURCE OF THE GLYPHOSATE

Changing to glyphosate-free feed, bedding, and water will give dramatic, almost unbelievable, results that are illustrated in the case studies. There is a remarkable change within 5 days even without doing anything else. The dramatic improvement in such a short time seems almost biologically impossible. Yet, within 2 weeks, animals are generally restored to a normal condition and production is greatly improved. For most situations, there is no other intervention required – only removal of the toxic glyphosate feed or water and a short recovery period, even though no antibiotics are administered.

9.4.2 COUNTERING THE PATHOLOGICAL ACTION OF GLYPHOSATE

Since minerals are not only components of biological systems, but also the enzyme cofactors and regulators of physiological processes, it is important to have good-quality, nutrient-dense feed and supplements available to compensate for mineral deficiencies induced by glyphosate chelation. This will also help in restoring biological function in the digestive processes and in restoring the balance needed in the gut microbiome that is trashed by the antibiotic activity of glyphosate. Removal of glyphosate from the animal with humic acid products and/or zeolite (clinoptilolite) in feed also removes pathogenic *Clostridia*, *Salmonella*, and other potentially pathogenic organisms (Krueger et al., 2013; Prasai et al., 2016).

9.5 CONCLUSIONS

1. Glyphosate exposure under current agricultural management is insidious and not sustainable.

2. Glyphosate is a patented antibiotic (Abraham, 2010), and all health professionals are now practicing "Antimicrobial Stewardship" worldwide in order to preserve the efficacy of antibiotics. The massive use of a glyphosate-potent, broad-spectrum antibiotic, indiscriminately applied throughout the environment and to fields at a rate of 2 kg/ha or higher, does not constitute Antimicrobial Stewardship. Glyphosate is known to induce resistance to other known antibiotics (Kurenbach et al., 2015, 2017, 2018). It must be very judiciously regulated.

3. Humane care of animals cannot knowingly condone the feeding of glyphosate-based herbicide toxic feed to animals because the affected animals suffer severely and for a long duration if the problem is not remediated.

4. The drastic loss of profit by feeding glyphosate-contaminated feed or using glyphosate contaminated bedding must stop immediately if food animal agriculture is to be sustained.

5. Consideration for the crop producer must include development and promotion of alternatives to glyphosate management. Recording, storage, and documentation of the numerous case studies available, along with evaluation with documented humane trials with statistical analysis to complement empirical studies, are also needed.

6. Immediate needs are for research that is focused on "one health," the overall consideration of all components of glyphosate toxicity across species, and the complete sharing and storing of information on glyphosate toxicity. This is very important since every animal with glyphosate toxicity has fatty liver disease, and nonalcoholic fatty liver disease (NAFLD) is a major unmet health need in humans worldwide (Callaghan, 2018; Sargious and Swain, 2020; Swain, 2020). Instances of liver disease are rising dramatically – and it's not because of excessive drinking. Since symptoms are not expressed until late in the disease, it can go undetected for a number of years until it reaches a major health crisis. It is often referred to as a silent killer. Prior to 2016, less than one in ten Canadians was diagnosed with NAFLD and, by 2018, it had reached one in four. Maybe we should focus more attention on the "one health" concept and learn from the animals.

REFERENCES

Abraham, W. 2010. Glyphosate formulations and their use for the inhibition of 5-enolpyruvalshikimate-3-phosphate synthase. US 7,771736B2. United States Patent Office.

Ackermann, W., M. Coenen, W. Schrodl, A.A. Shehata, and M. Krueger. 2015. The influence of glyphosate on the microbiota and production of botulinum neurotoxin. During ruminal fermentation. *Curr. Microbiol.* 70: 374–382. doi: 10.1007/s00284-014-0732-3.

Callaghan, C. 2018. Instances of liver disease rising dramatically – and it's not because of excessive drinking. https://globalnews.ca/news/4376238/instances of liver disease rising dramatically.

Clair, E., L. Linn, C. Travert, C. Amiel, G.-E. Seralini, and J.M. Panoff. 2012. Effects of RoundUp® and glyphosate on three food microorganisms: Geotrichum candidum, Lactococcus lactis subsp. cremoris and Lactobacillus delbrueckii subsp. bulgaricus. *Curr. Microbiol.* 64: 486–491.

Fox, M.W. 2012. Pet foods with genetically modified (GMO) ingredients put dogs and cats at risk. Manuscript ipan@erols.com.

Hoy, J., N. Swanson, and S. Seneff. 2015. The high cost of pesticides: Human and animal diseases. *Poult. Fish, Wildl. Sci.* 3: 1–19. doi:10.4172/2375-446X.1000132.

Krueger, M., A. Shehata, J. Neuhaus, T. Muller, M. Kotsch, and W. Schrodl. 2012. Chronic botulism of cattle, a multifactorial disease. What is the role of the herbicide glyphosate? Inst. Fur Bacteriol. Mykol., U. Leipzig, Germany.

Krueger, M., A.A. Shehata, W. Schrodl, and A. Rodloff. 2013. Glyphosate suppresses the antagonistic effect of Enterococcus spp. on Clostridium botulinum. *Anaerobe* 20: 74–78.

Kurenbach, B., D. Marjoshi, C.F. Amabile-Cuevas, G.C. Ferguson, W. Godsoe, P. Gibson, and J.A. Heinemann. 2015. Sub-lethal exposure to commercial formulations of the herbicides dicamba, 2,4-D and glyphosate cause changes in antibiotic susceptibility in *Escherichia coli* and *Salmonella enterica* serovar Typhimurium. *Microbiology* 6: e00009–e00015.

Kurenbach, B., P.S. Gibson, A.M. Hill, A.S. Bitzer, M.W. Silby, W. Godsoe, and J.A. Heinemann. 2017. Herbicide ingredients change *Salmonella enterica* sv. Typhimurium and *Escherichia coli* antibiotic responses. *Microbiology* 163: 1791–1801.

Kurenbach, B., A.M. Hill, W. Godsoe, S. van Hamelsveld, and J.A. Heinemann. 2018. Agrichemicals and antibiotics in combination increase antibiotic resistance evolution. *Microbiol. PeerJ* 6: e5801. doi: 10.7717/peerj.5801.

Mesnage, R., N. Defarge, J. de Vendimois, and G.-E. Seralini. 2013. Major pesticides are more toxic to human cells than their declared active principles. *BioMed Res. Int.* 214: 179691. Hindawi Publishing.

Martell A.E. and R.M. Smith. 1974. *Critical Stability Constants.* Vol. 1,5-6. Polenum Press, New York.

Motekaitis, R.J., and A.E. Martell. 1985. Metal chelate formation by N-phosphonomethyl glycine and related ligands. *J. Coord. Chem.* 14: 139–149.

Prasai, T.P., K.B. Walsh, S.P. Bhattarai, D.J. Midmore, T.T.H. Van, R.J. Moore, and D. Stanley, 2016. Zeolite food supplementation reduces abundance of enterobacteria. *Microbiol. Res.* 195: 24–30.

Sargious, M. and M.G. Swain. 2020. Doctors call fatty liver disease a growing public health concern in Canada. https://globalnews.ca.

Shaheen, A.A., K. Riazi, A. Medellin, D. Bhayana, G.G. Kaplan, J. Jiang, R. Park, W. Schaufert, K.W. Burak, M. Sargious, and M.G. Swain. 2020. Risk stratification of patients with nonalcoholic fatty liver disease using a case identification pathway in primary care: A cross-sectional study. *CMAJ Open* 8: 2. doi:10.9778/cmajo.202000009.

Shehata, A.A., W. Schrodl, Q.Q. Aldin, H.M. Hafez and M. Krueger. 2013a. The effect of glyphosate on potential pathogens and beneficial members of poultry microbiota in vitro. *Cur. Microbiol.* 66: 350–358.

Shehata, A.A., M. Kuehnert, S. Haufe, and M. Krueger. 2013b. Neutralization of the antimicrobial effect of glyphosate by humic acids in vitro. *J. Chemosphere* 10: 258–261.

Sirinathsinghji, E. 2015. A roundup of Roundup reveals converging pattern of toxicity from farm to clinic to laboratory studies. Institute of Science in Society, January 19, 2015.

Subliah, M., S.M. Mitchell and D.R. Cal. 2016. Not all antibiotic use practices in food animal agriculture afford the same risk. *J. Environ. Qual. Special Sec.* 45: 618–629.

Swain, M.G. 2020. Fatty liver disease: A growing public health concern in Canada. https://globalnews.ca/news/4376238/instances-of-fatty-liver-disease-rising.

Thixton, S. 2017. Pet food ingredients. More concerning news linked to glyphosate. Truth About Pet Food, January 25, 2017.

Van Bruggen, A.H.C., M. M. He, K. Shin, V. Mai, K.C. Jeong, M.R. Finckh, and J.G. Morris Jr. 2017. Environmental and health effects of the herbicide glyphosate. *Elsevier Sci. Total Environ.* 616–617: 255–268.

Von Soosten, D., U. Meyer, L. Huether, S. Danicke, M. Lahrssen-Wiederholt, H. Schafft, M. Spolders, and G. Breves. 2016. Excretion pathways and ruminal disappearance of glyphosate and degradation product aminomethylphosphonic acid in dairy cows. *J. Dairy Sci.* 99: 5318–5324.

10 Disruption of the Soil Microbiota by Agricultural Pesticides

Robert J. Kremer

CONTENTS

10.1　INTRODUCTION

Modern conventional cropping systems depend on synthetic agricultural chemicals (fertilizers and pesticides) to produce food, feed, fiber, and bioenergy crops (Arriaga et al. 2017). Recent literature on the impact of pesticides on soil microbiota under conventional crop production provides a framework to base decisions on whether pesticide deployment should continue following routine schedules or if use should be altered or suspended to maintain biological abundance and functional diversity and protect soil and environmental health (Locke and Zablotowicz 2004; Rose et al. 2016). This type of information would be extremely important in pesticide-use decisions concerning food production that ultimately affect food security, land-use policy, and biodiversity conservation if agricultural systems are altered in developing countries of Africa (Travis et al. 2014). The agricultural intensification strategy of these systems is to optimize management to obtain maximum output (yields) with little regard to environmental factors and long-term soil productivity or health. Soil microbial biomass and metabolism decline in fields under long-term continuous annual crops because of the combined effects of synthetic chemical inputs, tillage, and loss of organic matter on soil physicochemical and biological properties and altered composition and quantity of crop root exudates.

Currently most agroecosystems under agricultural intensification include cultivation of genetically engineered (GE) crops, first introduced in 1996. Traits genetically engineered into these crops primarily include herbicide tolerance (HT) and insect resistance as well as gene stacking in maize (*Zea mays*), cotton (*Gossypium hirsutum*), and soybean (*Glycine max*) for tolerance to one or more herbicides, resistance to multiple insect pests with several incorporated *Bacillus thuringiensis* (Bt) toxins (cry proteins), and combinations of HT and cry genes. Use of GE crops contributes to a dependence on continuous inputs of herbicides to the environment. A critical outcome of the continued use of various herbicides and other pesticides in modern agriculture is the buildup of these synthetic chemicals in our soil resource that inevitably impact soil biology and environmental quality.

Smallholder farms in developing countries produce 80% of the global food supply, therefore the sustainability and productivity of these production systems are of vital importance (Clark 2019). On smallholder farms, including those of sub-Saharan Africa, about 80% of dryland crop production is dedicated to multiple- or inter-cropping where two or more crops are grown simultaneously. The use of pesticides in these systems is prohibitive due to the difficulty and expense in preparation of mixes of various compounds to manage an array of weeds, insect pests, and pathogens that could potentially affect the different crops. These situations likely contribute to very low pesticide use in Africa, which is only 2.1% of the global total pesticide usage (Meena et al. 2020). However, traditional land-use practices have degraded the physical, chemical, and biological properties of most soils, decreasing fertility and productivity across sub-Saharan Africa. Consequently, most cropland is characterized by poor soil health, threatening the productivity and resilience of agricultural systems that depend on them (Vanlauwe et al. 2014). Soil health is related to production of abundant and high-quality food, thus maintaining and improving soil biological processes, and conservation within smallholder agricultural systems is critically important. Good soil health practices include organic inputs to soil, crop variety selection, crop rotation, reduced tillage, residue utilization and continuous soil cover, and integrated management of weeds and pests. Unfortunately, adoption of soil health practices in sub-Saharan Africa is very low (Giller et al. 2009). This is partially attributed to competing uses for crop residues and organic matter (for feed and fuel), which deplete soil organic matter, lack of markets for rotation crops, and lack of immediate financial benefits because productivity yield gains only materialize over years as soil health improves gradually.

However, threats due to cultural shifts in ownership, perceived needs to produce more food or a switch to industrial crops for international trade and economic incentives, or land grabs by outside investors that force out smallholder farmers would likely establish the commercial agriculture model with reliance on monocropping on large tracts of land and inorganic and synthetic chemical inputs, leading to further ecological degradation (Clark 2019). Several African nations are under pressure to adopt GE crops that could further change smallholder farm numbers and/or alter traditional production practices including mandatory use of pesticides. However, current adoption of GE crops in Africa is practiced on only about 3 million hectares, compared with 300,000 million hectares currently claimed to be planted globally to GE crops (ISAAA 2018). Modern industrial agricultural production systems in the developed countries have degraded soil health due to monocropping, pesticide inputs, lack of cover cropping, and intensive tillage (Stirling et al. 2016). The adoption of such practices in Africa on a majority of smallholder farms would exacerbate the low microbial structural and functional diversity associated with poor soil health and create an agricultural disaster where expensive inputs of chemicals would be required to balance the inadequacies of the biological processes to assure acceptable crop yields. Such agricultural intensification would essentially compound soil and ecosystem degradation and biodiversity loss. Thus, it is important to provide an overview of soil microbiota, biological process, and the impacts that pesticides have on the biological component of soil health to serve as a guide for management decisions for food production in current and future agricultural systems in Africa. The objective of this chapter is to provide background on mechanisms by which pesticides disrupt soil microbiota comprising the microbiome and their biological processes.

10.2 THE SOIL MICROBIOME

From an ecological perspective, soil microbiota may be considered as the soil microbiome, defined by Whipps et al. (1988) as the characteristic microbial community occupying specified microhabitats with distinct physiochemical properties. More recently, the microbiome has been described as only the genomes or the collective genes of the microbiota individuals; however, here we broaden the soil microbiome to represent the array of soil-inhabiting organisms including viruses, bacteria, archaea, fungi, protists, nematodes, micro-, meso-, and macroinvertebrates, earthworms, and burrowing mammals and encompass their activities and inter-organismal relationships, contributing to an environment that mediates one the most complex biological communities on earth. A realistic view of microorganisms in soils is their existence within communities consisting of millions to billions of individual species representing millions of different taxonomic groups. Interactions among groups within communities mediate numerous important functions including decomposition of natural and xenobiotic organic substances (including pesticides) and carbon cycling; nutrient mineralization and cycling; soil structure formation; plant growth promotion; and soil organic matter formation (Whalen and Sampedro 2010; Torsvik et al. 1990). When evaluating the impacts of environment and management on soils, microbial diversity in these ecosystems must be considered from standpoints of structural diversity: the species abundance and their distribution within a microbial community based on taxonomic characteristics; and functional diversity: the distribution and abundance of microbial groups based on metabolic functions (Whalen and Sampedro 2010). Modern developments in molecular methods have advanced studies of soil microbial communities by expanding the concept of microbial diversity through detection of nonculturable microorganisms and functional genes that can be matched with soil biological processes and the microbial source of origin (Fierer et al. 2013) within the soil microbiome. The microbiome can be further characterized based on the total proteins of the community (proteomics) responsible for microbial activity and on the metabolites of the community (metabolomics) that identify the products of microbial activity (Bouchez et al. 2016). Soil biology encompassing the composition of soil microbiomes, or biodiversity, and biological functions is associated with soil health (Pankhurst 1997).

Soil biological activities are influenced by interactions with soil chemical and physical properties influenced by plant roots, environment, and land management. For example, soil microbiomes and their activities are highly dependent on quantity and quality of above- and below-ground inputs of plant-derived organic matter (Meyers et al. 2001; Zak et al. 2003). Soil enzymes produced by the soil microbiota rapidly respond to changes in soil management (Bandick and Dick 1999). Soil enzymes mediate key biochemical processes including organic matter decomposition, nutrient mineralization and cycling, biodegradation of synthetic compounds, and synthesis of plant-growth regulating substances and play central roles in most biochemical and ecological processes in the soil ecosystem (Sinsabaugh et al. 1991; Bardgett and van der Putten 2014). Soil enzyme activities are sensitive indicators of microbial activity and are used to differentiate various soil and crop management regimes and to quantify specific soil biological processes (Bandick and Dick 1999). Other soil biological processes and organismal groups sensitive to soil management include soil C mineralization, active C (AC or permanganate-oxidizable C), water-extractable (soluble) C, soil enzymes, soil microbial community structure and biodiversity components, soil fauna (i.e., earthworms), and plant disease criteria (Morrow et al. 2016; Stott et al. 2013; van Bruggen and Semenov 2000). A functioning microbiome in a healthy soil depends on interactions with properly functioning soil chemical and physical factors (Stirling et al. 2016); thus, these bioindicators will be considered in describing impacts of pesticides on microbiota and their activities.

10.3 SOIL HEALTH

Soil biology embodies a composite of soil-inhabiting organisms ranging from viruses and microorganisms to macroinvertebrates and vertebrates, including their activities and inter-organismal relationships, resulting in an environment sustaining a complex microbiome that is essential for soil health.

Soil health is the capacity of a living soil to function, within natural or managed ecosystem boundaries, sustain plant and animal productivity, maintain or enhance water and air quality, and promote plant and animal health (Doran 2002; van Bruggen and Semenov 2000). Optimal soil health depends upon the balance between soil productivity, environmental quality, and plant and animal health, all of which interact with soil's chemical, physical, and biological properties and with management and land-use decisions. Good management practices that consider soil health must consider all ecosystem functions, rather than focusing on single functions, such as crop productivity (Doran 2002).

Current molecular microbiology methods allow detailed characterization of the biodiversity of soil microbiomes; however, microbial diversity and microbial functional groups, which provide a foundation for soil function, are not yet used as general standard soil health indicators due to limited databases and difficulty in devising in-field sampling methods that maintain in situ conditions (Lehman et al. 2015). Soil health assessment models for evaluating the effects of land management on soil ecosystems (Morrow et al. 2016; Stott et al. 2013) are deficient in variables of important soil biological properties that contribute to an overall soil health rating (Stott et al. 2013). Bioindicators based on key microbial groups and processes are sensitive to environmental changes and should be used for the overall soil health assessment. In this chapter, the impacts of pesticides used in the management of agricultural ecosystems, both worldwide and in Africa, will be described relative to bioindicators of soil microbiome abundance, diversity, and function. The majority of pesticides target plant growth and yield improvement, thus impacts on rhizosphere microbial function will be considered based on the influence of roots on soil biology including microbiome characteristics (Garbeva et al. 2004; Marschner 2012).

Understanding the impacts of specific practices, such as pesticide use, on soil microbiome biodiversity using a soil health assessment model is very difficult because such comparisons are practically nonexistent. Comparative assessments of organic and conventional farms in California, where pesticide application rates were highly and inversely related to soil health indicator scores, suggested that pesticide use on conventional farms was a major factor in lower soil health ratings (Andrews et al. 2002a, b). Also, assessment of various production systems within landscapes of similar soils reveals that pastures and grasslands and long rotations under no-till with no or minimal chemical inputs yielded higher soil health ratings than conventional, tilled, and short rotation systems receiving high pesticide inputs (Veum et al. 2015). Highly rated biological indicators found in nonconventional or alternative systems suggest that microbiome biodiversity was also high because functional activity was not detrimentally affected by pesticide inputs, even though abundance or biodiversity indices were not measured (Pankhurst 1997).

10.4   EARLY PESTICIDE SOIL MICROBIOTA RESEARCH INVESTIGATIONS

Shortly after World War II and through the Green Revolution era, synthetic chemicals were developed and introduced as fertilizers and pest management practices; evaluation of the potential for nontarget impacts of pesticides were limited to effects on phytopathogens and diseases of plants (Altman and Campbell 1977; Katan and Eshel 1973). Environmental biodegradation was assessed mainly for potential phytotoxicity toward subsequent crops by residual pesticide concentrations (Audus 1964; Kaufman and Kearney 1976). Effects of pesticides on specific beneficial biological processes such as nitrogen transformations were addressed in limited research studies (Simon-Sylvestre and Fournier 1979). Generally, however, pesticide fates and nontarget effects were investigated in the context of the evolving modern crop production systems in which synthetic pesticide inputs were readily adopted with little concern for soil biology or soil health. Early research was aimed to assure the effectiveness and reliability of pesticides, as expressed by Skipper et al. (1986), "modern agriculture and industry depend upon the wide-scale use of complex, synthetic compounds that may ultimately contaminate the soil environment. The soil microflora, if they are to survive, must adapt to the introduction of these chemicals."

Early studies on pesticide–soil microbiome interactions thus focused on the effects of pesticides on populations of broad microbial components and on specific groups of soil microorganisms.

A wide variety of compounds, target microorganisms, soils, and management practices were examined under various experimental conditions, making it virtually impossible to find general trends regarding pesticide–microorganism interactions (Johnsen et al. 2001; Simon-Sylvestre and Fournier 1979). Typically, the effects on populations were studied by culture-dependent methods for selective examination of bacterial groups with known, specific functional roles in soil. This approach overlooked a vast segment of the soil microbiome that contributes to overall functional diversity because methods for easily detecting nonculturable microorganisms were not available until the late 1990s. This early research provided evidence for direct effects of pesticides on populations of a wide range of soil microorganisms, but offered no information on short-term or long-term changes in microbial diversity on a community scale (Johnsen et al. 2001). Bollen (1961) first recognized the potential detrimental effects of introducing numerous active ingredients in various formulations and their widespread use on soil microorganisms and their activities. He also noted the necessity to avoid "serious injury to the great variety of microbes" that mediated vital functions in soil, an early indication of the importance of biodiversity for optimal biological function and soil health.

10.5 PESTICIDES AND SOIL BIOLOGY: GENERAL PESTICIDE USAGE CONSIDERATIONS, MICROBIAL ABUNDANCE, AND SYMBIOTIC ASSOCIATIONS

Of the millions of tons of pesticides applied to production fields annually, it is estimated that less than 5% of these products reach the target organisms with the remainder deposited on the soil, some of which moves into the atmosphere and water (Van Eerd et al. 2003). Also, many pesticides applied directly to crop foliage are systemic with a major proportion of applied compounds released through roots into soil; thus, pesticides readily contact the nontarget soil and rhizosphere microbiota. Pesticide effects on the soil microbiome are ultimately influenced by complex properties of the soil environment (Kaufman and Kearney 1976). Thus, effects on the microbiome directly due to pesticides must be distinguished from those facilitated by biological, chemical, or physical soil processes. Furthermore, relationships between soil microbial diversity and management practices in the field are often overlooked. Rose et al. (2016) suggest that variability in pesticide dosages applied to the environment combined with soil microbiome diversity renders a full assessment of effects on the soil biota and their functions almost impossible.

10.5.1 EFFECTS ON SOIL MICROBIAL ABUNDANCE AND DIVERSITY

Pesticide effects on microbial biomass are often negligible, although effects on specific microbiome components are more frequent (Table 10.1) (Rose et al. 2016). Barriuso and coworkers adapted next-generation sequencing, a highly sensitive molecular technique for detecting both culturable and nonculturable microbial taxa, to show a considerable reduction in diversity of the soil microbial community by the herbicide glyphosate alone or combined with acetochlor and terbuthylazine (Barriuso et al. 2011; Barriuso and Mellado 2012). Glyphosate treatment of sensitive plants, as in burndown (or "knockdown") practices for weeds or cover crops prior to no-tillage or direct seeding, boosted the abundance of soil bacteria and protists due to increased availability of carbon substrates from glyphosate-killed roots (Imparato et al. 2016). Roots of glyphosate-treated plants are heavily colonized by soilborne fungi, affecting the diversity of the soil and rhizosphere microbiomes (Johal and Huber 2009; Levesque and Rahe 1992). Glufosinate, which inhibits glutamine synthetase in plants and microorganisms, may restrict the growth and activity of beneficial soil fungi such as *Trichoderma* spp. and inhibit the growth and propagation of several pathogenic fungal species (Kortekamp 2011). *Trichoderma* growth suppression negates biological control of soilborne phytopathogenic fungi *Fusarium oxysporum* and *Nectria ochroleuca*, which are not affected by glufosinate, and shifts fungal diversity toward a higher proportion of phytopathogens (Ahmed et al. 1995).

TABLE 10.1

Impacts of Some Pesticides on Soil Microbiome Abundance and Diversity

Pesticide	Impact	Comments	Reference
	Fungicides		
Ametoctradin	Decrease soil microbiome diversity	Soil mesocosm	Whittington et al. (2020)
Benomyl	Increased detrimental rhizobacteria, opportunistic pathogens on various ornamental crops	Field, nursery, greenhouse studies	Mills et al. (1996), Kloepper et al. (2013)
	Herbicides		
Alachlor	AMF spore density, sporulation, hyphal infection, reduced	Field studies	Pasaribu et al. (2013)
Atrazine	Shift soil microbiome diversity toward bacterial degraders	Field studies	Krutz et al. (2010)
	AMF-enhanced root uptake, degradation of atrazine in maize	Greenhouse studies	Huang et al. (2007)
Chlorotoluron	Shift in soil microbial structure (diversity)	Field studies	Carpio et al. (2020)
Flufenacet + Diflufenican	Shift in soil microbial structure (diversity)	Field studies	Carpio et al. (2020)
Glyphosate	Decreased growth of *Bradyrhizobium* sp. (*Lupinus*)	Laboratory culture	de Maria et al. (2005)
	AMF spore density reduced	Field studies	Pasaribu et al. (2013)
	Reduce AMF colonization of grasses	Grassland field studies	Helander et al. (2018)
	Shift diversity of rhizobacteria on *Avena sativa*	Field studies	Allegrini et al. (2015)
	Decrease ammonium-oxidizing bacteria and archaea	Field studies	Allegrini et al. (2015)
	Decrease Mn-reducing bacteria and fungi; decrease rhizosphere fluorescent pseudomonads	Greenhouse, field studies	Kremer and Means (2009), Zobiole et al. (2010)
	Reduced soil pseudomonad growth	Laboratory studies	Aristilde et al. (2017)
	Soil bacterial components reduced in maize rhizosphere	Growth chamber study	Banks et al. (2014)
	Increase soil bacteria and protists	Field, burndown	Imparato et al. (2016)
	Decreased soil and rhizosphere microbial diversity	Field studies	Newman et al. (2016)
	Decreased microbial biomass – soybean rhizosphere	Field studies	Lane et al. (2012)
	Rhizobacteria community diversity decreased	Field studies	Barriuso et al. (2011)
	Increase GE soybean and maize root-colonizing *Fusarium* spp.	Greenhouse, field studies	Kremer and Means (2009), Zobiole et al. (2010)
	Increase fungal root phytopathogens on various plant hosts	Greenhouse; field studies	Levesque and Rahe (1992), Johal and Huber (2009)
	Suppress saprophytic fungus (*Aspergillus nidulans*) growth	Laboratory studies	Nicolas et al. (2016)
	Minor reductions in soil faunal members of soil food web	Field, burndown	Hagner et al. (2019)

(Continued)

TABLE 10.1 (*Continued*)
Impacts of Some Pesticides on Soil Microbiome Abundance and Diversity

Pesticide	Impact	Comments	Reference
	Juvenile earthworm weight loss on exposure to AMPA	Soil mesocosms	Dominguez et al. (2016)
	Decrease litter decomposition activity by *Collembola*	Soil mesocosms	Maderthaner et al. (2020)
Glufosinate	Restrict growth of fungal phytopathogens and beneficial *Trichoderma viride* fungus	Field and greenhouse studies	Ahmed et al. (1995), Kortekamp (2011)
Trifluralin	Reduced bacteri- and fungivorous nematodes	Field studies	Yardim and Edwards (1998)
Insecticides			
Imidacloprid	Earthworm body mass loss (*Lumbricus terrestris*)	Seed treatments	Van Hoesel et al. (2017)
	Reduced soil arthropod community by 54%–62%	Field studies	Peck (2009)
Carbaryl	Reduced bacteri- and fungivorous nematodes	Field studies	Yardim and Edwards (1998)
Chlorpyrifos	Damage growth and development of earthworm	Lab bioassays and mesocosm	Muangphra et al. (2016)
Nematicides			
Metam sodium; 1,3 dichloropropane; fluensulfone	Reduce beneficial nematode (soil food web members) abundance; suppress antagonists of parasitic nematodes	Field studies	Stirling et al. (2016)

AMF, arbuscular mycorrhizal fungi; AMPA, aminomethylphosphonic acid.

Some pesticides applied to soil either annually or multiple times within a season may radically shift abundance and composition of the soil microbiome, which adapts to rapidly degrade the pesticide using it as a source of carbon, nitrogen, and energy. This "enhanced or accelerated pesticide degradation" develops in soil previously treated with the pesticide or a compound of similar structure and is considerably faster relative to its rate of degradation in a comparable untreated soil (Racke 1990). Enhanced degradation is associated with repeated use of the same pesticide at application dosages ranging from 2.2 to 5.6 kg ha^{-1}, providing sufficient pesticide chemical to stimulate certain segments of the soil microbiome to adapt metabolism for degradative pathways. This led to pest control failures in fields treated with soil applications of early carbamate and thiocarbamate pesticides that were readily degraded via "adapted microbial catabolism" (Racke 1990). Enhanced degradation is currently suspected for soils repeatedly treated with s-triazine herbicides for weed control in maize and rotation crops when application frequency in maize is greater than once every 4 years (Krutz et al. 2010). Gram-negative and actinobacterial species are implicated in mediating the enhanced degradation based on detection of genes coding for triazine metabolism within the bacterial genomes. It is speculated that the genes are shared through horizontal gene transfer or plasmid conjugation within the soil bacterial community (Krutz et al. 2010). More recent evidence for enhanced degradation in a Brazilian Red Latosol found that atrazine-degrading genes were stimulated after atrazine applications causing a soil microbial consortium to establish the complete degradation pathway suggesting that the increased abundance of atzA and trzN degradative genes was due to increases in certain bacterial phyla due to atrazine exposure (Fernandes et al. 2020). More recently, Whittington et al. (2020) demonstrated that a soil prokaryotic component could develop to rapidly degrade the triazolopyrimidine fungicide ametoctradin within 14 days post-soil incorporation resulting in only 24% of the parent compound remaining.

10.5.2 Effects on Microbial–Plant Symbioses

Mycorrhizal fungi are critical for productivity of at least 90% of food crops due to their multiple effects on plant nutrient acquisition, suppression of soilborne phytopathogens, access to water, and soil aggregation (Smith and Read 2008). Mycorrhizal spore numbers in soil are affected by alachlor and glyphosate application, with spore densities decreasing with increased herbicide application rates (Pasaribu et al. 2013). External hyphal development of *Glomus mosseae* was not affected by either soil or foliar-applied herbicides (alachlor and glyphosate, respectively) due to tolerance of these structures to the herbicides. Mycorrhizal sporulation, hyphal infectivity, and P inflow to plant roots were significantly affected by alachlor, but unaffected by glyphosate. In contrast, *Glomus caledonium* in symbiosis with maize enhanced accumulation and facilitated degradation of atrazine in roots (Huang et al. 2007) resulting in decreased transport of intact atrazine into maize shoots and reduced soil accumulation of atrazine. In this case, the mycorrhizal component of the soil microbiome was beneficial for reducing herbicide carryover in agricultural soils and could potentially phytoremediate contaminated soils. Effects of herbicide application on symbioses are seemingly variable and depend on herbicide chemistry, formulation compounds application rates, and the various transformations the host-associated mycorrhizal species are capable of carrying out.

Pesticides affect bacterial–plant symbioses including the association of nitrogen-fixing rhizobia with leguminous plants. For example, methyl-parathion and DDT affect *Sinorhizobium meliloti* associated with alfalfa by disrupting rhizobial–host plant signaling and subsequent nodulation, reducing nitrogen fixation, and decreasing overall plant yield (Fox et al. 2007). Glyphosate disrupts nodule ultrastructure in lupin (*Lupinus albus*) and alters nodule cytosol and bacteroid proteins, possibly due to direct herbicide effects on the rhizobia and indirect effects on plant photosynthetic mechanisms (de Maria et al. 2005). A reduction in rhizobia community diversity and activity in soils is detrimental not only to host plant nodulation but also from reduced benefits of plant growth promotion and stress tolerance when rhizobia occupy rhizosphere environments of nonhost plants (Antoun et al. 1998).

10.5.3 Pesticide Effects on Biodiversity

Biodiversity relative to soil health and ecosystem services generally refers to the total of all species in any habitat. Studies on biodiversity of soils have primarily focused on the microbial components of the food web with less attention to other soil ecosystem members including meso- and macrofauna represented by micro-arthropods, nematodes, earthworms, and larger organisms such as arthropods, crustaceans, and small mammals. Biodiversity is considered a critical functional component of agroecosystems because it has a fundamental input role to ecosystem services (Kremen and Miles 2012).

Continuous use of the systemic fungicide benomyl drastically altered the balance of diversity in the rhizosphere of a variety of ornamental nursery crops toward a predominance of deleterious rhizobacteria consisting mainly of fluorescent pseudomonads that eventually colonized endophytically resulting in foliar distortion symptoms (Mills et al. 1996; Kloepper et al. 2013). Similar increases in detrimental microorganisms concomitant with reductions in beneficial microbial components have been noted for production systems relying on continuous annual applications of glyphosate either as burndown in preparation for no-till direct seeding or associated with cropping systems with GE varieties of maize and soybean (Johal and Huber 2009; Kremer and Means 2009; Lane et al. 2012; Newman et al. 2016).

10.5.4 Pesticide Effects on Microbiota-Mediated Biological Processes

Effects of pesticides on biological processes have been traditionally limited to functions in soil of agronomic importance, primarily carbon and nitrogen transformations. Therefore, much literature provides evidence for direct effects of pesticides on a wide range of soil microorganisms, but offers little information on short- or long-term changes of microbial activity or function on a community scale (Johnsen et al. 2001). Numerous previous reviews offer in-depth coverage of the effects of pesticide on activities of soil and rhizosphere microorganisms (Audus 1964; Bollen 1961; Curl and Truelove 1986; Greaves et al. 1976; Simon-Sylvestre and Fournier 1979).

No consistent standardized approach has been used in the majority of studies investigating effects of pesticides on soil biological processes. Only a few studies have used realistic field application dosages under various conditions ranging from artificial laboratory culture assays, growth chamber, and greenhouse trials with processed soils, to actual in-field assessments. Thus only selected studies covering a variety of experimental designs are illustrated in Table 10.2. Some basic studies (i.e., culture-based) are useful in ascertaining the mode of action of pesticide compounds on microbial activity that potentially occur in soil microsites, such as in-vitro assays examining effects on plant growth regulator synthesis by rhizobacteria (Shahid et al. 2019). Basic research showing mechanisms of some pesticides that disrupt protein syntheses in beneficial soil bacteria at the molecular level is critically important for monitoring activity in soils to avoid depletion of microbiome components that protect crops from stress conditions (Shahid and Khan 2020).

TABLE 10.2

Impacts of Some Pesticides on Soil Biological Processes

Pesticide	Impact	Comments	Reference
Fungicides			
Kitazin	Disrupt protein synthesis in PGPR	Laboratory bioassay	Shahid and Khan (2020)
	Decreased phytohormone synthesis by PGPR	Laboratory bioassay	Shahid and Khan (2020)
	Decrease IAA production in *Azotobacter vinelandii*	Laboratory bioassay	Shahid et al. (2019)
Metalaxyl	Decrease IAA production in *Azotobacter vinelandii*	Laboratory bioassay	Shahid et al. (2019)
Herbicides			
2,4-D	Suppressed carbonic anhydrase, C sequestering enzyme	Soil microcosms	Nathan et al. (2020)
Atrazine	Decrease IAA production in *Azotobacter vinelandii*	Laboratory bioassay	Shahid et al. (2019)
	Suppress nitrification/denitrification	Field studies	Zhang et al. (2018)
Chlorotoluron	Reduced soil dehydrogenase	Field studies	Carpio et al. (2020)
Flufenacet + Diflufenican	Reduced soil dehydrogenase	Field studies	Carpio et al. (2020)
Glyphosate	Reduce IAA synthesis by PGPR	Laboratory bioassay	Shahid et al. (2019)
	Decrease lupine nodule leghemoglobin content; alter nodule ultrastructure	Growth chamber study	de Maria et al. (2005)
	Disruption of fungal metabolism	Laboratory bioassay	Poirier et al. (2017)
	Decreased phytohormone synthesis by PGPR	Laboratory bioassay	Shahid et al. (2019)
	Decrease IAA production in *Azotobacter vinelandii*	Laboratory bioassay	Shahid et al. (2019)
	Suppress nitrification/denitrification	Field studies	Zhang et al. (2018)
	Reduce nitrogenase activity, nodule mass, N fixation in soybean	Field studies	Zablotowicz and Reddy (2007)
	Reduced nodulation in soybean	Field studies	Kremer and Means (2009), Zobiole et al. (2010)
Paraquat	DNA mutation in earthworms	Laboratory bioassay; mesocosm	Muangphra et al. (2012)
Quizalofop	Decrease IAA production in *Azotobacter vinelandii*	Laboratory bioassay	Shahid et al. (2019)

(Continued)

TABLE 10.2 (*Continued*)
Impacts of Some Pesticides on Soil Biological Processes

Pesticide	Impact	Comments	Reference
		Insecticides	
Imidacloprid	Reduce litter decomposition by earthworms	Seed treatments; mesocosm studies	Van Hoesel et al. (2017)
Phorate	Reduce siderophore production in selected soil bacteria	Laboratory culture studies	Kumar et al. (2019)
Acephate	Reduce siderophore production in selected soil bacteria	Laboratory culture studies	Kumar et al. (2019)
Monocrotophos	Reduce siderophore production in selected soil bacteria	Laboratory culture studies	Kumar et al. (2019)

IAA, indole-3-acetic acid; PGPR, plant growth-promoting rhizobacteria.

Soil enzymes primarily originating from the soil microbiome represent metabolic processes driven by the microbiome and are indicators of overall microbial activity or of specific nutrient transformations. Soil enzyme activity is also considered as an important indicator of the biological component of soil health (Stott et al. 2010). Pesticides likely influence various metabolic processes by interacting with soil enzymes and other soil components. For example, atrazine and glyphosate decrease nitrification and denitrification rates in Australian sugarcane soils by inhibiting microbial functional genes involved in each process (Zhang et al. 2018). However, in measuring the impact of pesticides on soil enzyme activity, pesticide concentrations used in many studies may range from 10 to 1,000 times the recommended field dose, a questionable means of simulating repeated and long-term use of pesticides in the field (Riah et al. 2014). Even if pesticides applied at recommended rates lead to slight and transient changes to soil microbial abundance or activities (Johnsen et al. 2001), it is obvious that long-term recurrent applications of pesticides are known to interfere with the biochemical balance, which can reduce soil fertility and productivity by affecting local metabolism and enzymatic activities. Information provided by analyses of soil enzyme–pesticide interactions to date may yield insights to potential effects on soil biological processes and soil health, however standardized methods are still needed to assure adequate representation of field-based activity.

Although genotoxicity studies on commercial glyphosate products and chlorpyrifos revealed effects on DNA due to pesticide exposure, they were conflicting because toxicity depends not only on the absolute concentrations and purity of the active agent, but also on the nature of the carrier components (Muangphra et al. 2012, 2016). Similar concerns have been expressed regarding testing of formulated and pure active compounds of glyphosate on activity of springtail species (*Collembola*) whereby each exhibited different effects (Maderthaner et al. 2020) leading to suggestions that pesticide assessments should be performed with the actual pesticide product used in the field. Similarly, Banks et al. (2014) reported that glyphosate combined with the surfactant alkylphenol ethoxylate plus alcohol ethoxylate decreased soil bacterial components 7 weeks after application at labeled dosages to silt loam soil (pH 5.6). However, no effect on any microbiome component was detected in silty clay loam (pH 4.5). Changes in the microbiome due to herbicides or surfactants were minimal in this study of a single application of these chemicals, but could be indicators of potential long-term effects. Aristilde et al. (2017) strongly recommended the evaluation of synergistic or counteractive effects of glyphosate, its breakdown products, and formulation components for impacts on growth and activity of soil microorganisms.

10.6 PESTICIDES AND THE SOIL MICROBIOME IN AGRO-ECOSYSTEMS WITH GENETICALLY ENGINEERED (GE) CROPS

Current soil health assessments primarily focus on broad comparisons of management systems that vary in general practices including crop rotation, tillage, fertilizer or manure amendment, and soil conservation (Karlen et al. 2008, 2014; Stott et al. 2013). Understanding how more specific practices, such as inclusion of GE crops, affect soil biodiversity is very difficult because these types of comparisons are practically nonexistent. Development of GE crops has emphasized selection of plant phenotype characteristics, while important beneficial plant–microbe interactions that impact soil health and environmental services have been apparently overlooked (Berg 2009; Morrissey et al. 2004). Subsequent increased use of pesticides in GE cropping as well as conventional agricultural production systems in continuous short rotation practices with crops modified for resistance to the same herbicide (i.e., maize-soybean) has hastened drastic decreases in soil microbiome diversity (Table 10.1) (Barriuso et al. 2011; Newman et al. 2016). With the potential incursion of GE crops into Africa as national government agencies are pressed to include these crops as part of agricultural intensification programs, it is important to briefly consider the impacts this system could have on soil microbiota and biological function.

10.6.1 HERBICIDE-TOLERANT CROPS AND EFFECTS ON SOIL MICROBIOTA

Potential threats to farmland and natural habitats are apparent with the cultivation of herbicide-tolerant GE crops. Approximately 80% of GE crops under cultivation have modified genes expressing tolerance to glyphosate or glufosinate herbicides and/or stacked with insect resistance. Apart from toxicity to plants themselves, the possibility of toxicity to other life forms also exists (Tsatsakis et al. 2017). Several studies report that glyphosate weakens plant defense and increases root pathogen virulence in both glyphosate-resistant and susceptible plants (Johal and Huber 2009; Kremer and Means 2009). Glyphosate inhibits the plant's defense and structural barriers and immobilizes micronutrients such as manganese (Mn), which play vital roles in disease resistance and plant nitrogen metabolism. Other indirect effects of herbicide tolerance include changes in the biodiversity of weeds, weed inhabiting arthropods, pollinators, parasitoids, predators, and decomposers, which may lead to imbalanced symbiotic relationships, decreased beneficial insect populations, and rapid changes in ecosystem food chains (Schutte et al. 2017). Exposure of soils to frequent applications of glyphosate and the subsequent deployment of dicamba and 2,4-dichlorophenoxyacetic acid (2,4-D) to control glyphosate-resistant weeds has induced development of multiple antibiotic-resistance phenotypes in bacteria, which is of considerable concern as posing a major threat to human and livestock health (Ramakrishnan et al. 2019).

Widespread herbicide-resistant (HR) weed infestations due to overuse of glyphosate on GE HT crops may alter the soil microbiome and soil biological processes that differ from weed-free GE crop monocultures. Depending on the HR weed biotype, infestations may change the amount and type of labile carbon released from roots into soil, affect mycorrhizal viability and abundance, and alter plant nutrient uptake in production fields (Kremer 2014). Both soil health and ecosystem services may be negatively impacted. Understanding all complex interactions of HR weeds with soil microorganisms exposed to extensive infestations that develop due to failed weed management with GE crops is important to consider whether such a weed management system should be adopted in regions currently not under GE crop production.

10.6.2 INSECT-RESISTANT CROPS AND THE SOIL MICROBIOME

Varieties of GE crops modified to produce insecticidal compounds (i.e., the cry toxins encoded by Bt genes originating from the bacterium *Bacillus thuringiensis*) affect soil and rhizosphere microbiome diversity. Root exudates of Bt cotton promoted spore germination and mycelial growth of the

F. oxysporum wilt pathogen compared with root exudates from the non-GE isolines (Li et al. 2009). Resistance to *Verticillium* wilt was lower in transgenic cotton and their root exudates also promoted growth of the *Verticillium* wilt pathogen. Spore germination and differentiation of presymbiotic hyphae of the mycorrhizal fungus *Rhizophagus irregularis* decreased in the transgenic Bt cotton rhizospheres compared with corresponding parental nontransgenic isolines; subsequent appressorium number, colonization intensity, and arbuscule abundance were lower in Bt plant roots (Chen et al. 2016). It was concluded that the Bt-trait significantly contributed to the inhibition of presymbiotic development and mycorrhizal colonization, which may be attributed to either Bt toxicity or interference of signaling between mycorrhizae and host roots during initiation of the symbiosis.

Residues of Bt maize and cotton often reach streams or water bodies where the Cry toxins may leach from plant tissues and contact aquatic organisms. A survey of 217 streams in Indiana found that 85% of the streams contained maize plant tissues and the Cry1Ab toxin was detected in water from 23% of the sites (Tank et al. 2010). The toxins were detected 6 months after maize harvest in nearby fields, illustrating the prolonged persistence in aquatic environments. Exposure of the freshwater crustacean *Daphnia magna* to Cry toxins caused high mortality, small body size, and low reproductive rates (Bohn et al. 2016). Because *Daphnia magna* is very sensitive to changes in aquatic quality, the adverse response to Cry toxins suggests that these compounds may interrupt the aquatic food web with further potential disruption of ecosystem functioning.

10.7   RECENT INSECTICIDES INTRODUCED FOR MODERN CROP PRODUCTION

New insecticides include the recently commercialized neonicotinoid compounds for controlling a number of pest insects and for use as soil, seed, or foliar applications. Many (>90%) of the neonicotinoid-active ingredients enter the soil environment and remain with half-lives exceeding 6,900 d (Goulson 2013). Little is known regarding the impacts of these chemicals on the composition and activity of the soil microbiome. Current deployment of neonicotinoids likely detrimentally affects a broad range of nontarget soil and aquatic invertebrates (Goulson 2013), potentially altering the diversity and activity of certain trophic groups within the soil food web and posing ultimate consequences on overall ecosystem functioning (Wagg et al. 2014). Limited degradation of selected neonicotinoids occurs in both rice paddy (Mulligan et al. 2016) and dryland soils (Liu et al. 2010) and within specific soil bacterial groups (Pandy et al. 2009). Residual components of several neonicotinoid active ingredients remain in the soil after the cropping season, and impacts of their soil concentrations on the soil community remain to be determined. Imidacloprid, a systemic neonicotinoid applied early in the life cycle to control neonate white grubs, had a discernible impact on nontarget soil microbiota by suppressing abundance of broad taxonomic soil arthropod groups by 54%–62% (Peck 2009). Seed treatment of winter wheat with imidacloprid also detrimentally affects micro- and mesofauna after a single seed dressing application and macrofauna after a second seed dressing application (Van Hoesel et al. 2017).

10.8   CONCLUSION

Information on impacts of pesticides on structural and functional diversity of the soil microbiome and on biodiversity in general and specifically on soil health indicators is limited to mostly nonstandardized procedures. As a consequence, a broad range of conclusions regarding pesticide impacts is found in the literature, which is not unexpected considering the multiple modes of action associated with the more than 500 pesticides available in over 5,000 commercial formulations (Skipper et al. 1986). General statements suggesting negligible impacts of pesticides on the soil microbiome and beneficial soil processes at recommended field doses are not substantiated (Rose et al. 2016), especially when specific characteristics such as diversity, distinct functions, as well as spatial heterogeneity of microbial activity are considered (Dechesne et al. 2014). Insightful recommendations in consideration of pesticide impacts on environmental biodiversity and soil health based on Rose et al. (2016) are: consider results from studies of certain microbial groups sensitive to specific pesticides; consider evidence on effects of specific pesticides on certain biological functions; use and/or develop standardized response variables of community structure and function measured with consistent methodology, including advanced molecular techniques (Kumar et al. 2017), for interpretation of results; and conduct long-term, field plot or on-farm studies for better assessment of potential impacts of pesticides on soil health and the microbiota in realistic agricultural production systems. Regarding GE cropping systems, many studies suggest that these practices do not promote soil health and biodiversity unless sustainable practices including no-tillage, cover cropping, or organic soil amendments and/or mixed crop rotation are included in the management plan (Bedano and Dominguez 2016; Bedano et al. 2016). Consistent records of negative effects of continuous planting of GE crops and use of glyphosate on soil fungi and mycorrhizae (see Table 10.1) are of great concern due to the critical functions these groups of microorganisms perform relative to soil structure, soil organic matter formation, and nutrient cycling.

What is the relevance of the current status of pesticide use and the impacts on soil microbiota to African agriculture? If pesticide use were to be promoted as a means for improving food production or as an incentive for producing commodity grains for international trade, most soils under agricultural production are likely poorly prepared for conversion to intensified agriculture, especially for smallholder farms based on traditional management practices that have depleted soils of structure, fertility, and biology (Vanlauwe et al. 2014; Giller et al. 2009; Clark 2019). The African agroecosystems are already compromised and introduction of pesticides will quickly further deteriorate an already degraded microbiome, which is reflected in poor soil health and inadequate ecosystem services. Smallholder farmers should be encouraged, and perhaps incentivized, to improve soil health on their production fields whether they intend to continue with their current food production or if pursuing other production considerations. A regenerative agricultural or sustainable model (LaCanne and Lundgren 2018) could be used as a guide in implementing the restoration of soil productivity, soil health, and biological function. Key practices necessary for restoring soil health include organic soil amendments, continuous cover on soil, crop rotation, reduced tillage, and integration of livestock. A management integrating these practices will build up a diverse and functional soil microbiome that is necessary for improving food production or moderating detrimental effects of inputs such as pesticides if management plans also include this approach in future production systems.

REFERENCES

Ahmed, I., Bissett, J., and Malloch, D. "Influence of bioherbicide phosphinothricin on interaction between phytopathogens and their antagonists." *Canadian Journal of Botany* 73 (1995): 1750–1760.

Allegrini, M., Zabaloy, M.C., and Gómez, E.V. "Ecotoxicological assessment of soil microbial community tolerance to glyphosate." *Science of the Total Environment* 533 (2015):60–68.

Altman, J., and Campbell, C.L. "Effect of herbicides on plant diseases." *Annual Reviews of Phytopathology* 15 (1977): 361–385.

Andrews, S.S., Mitchell, J.P., Mancinelli, R., Karlen, D.L., Hartz, T.K., Horwath, W.R., Pettygrove, G.S., Scow, K.M., and Munk, D.S. "On-farm assessment of soil quality in California's Central Valley." *Agronomy Journal* 94 (2002a): 12–23.

Andrews, S.S., Karlen, D.L., and Mitchell, J.P. "A comparison of soil quality indexing methods for vegetable production systems in Northern California." *Agriculture Ecosysems and Environment* 90 (2002b): 25–45.

Antoun, H., Beauchamp, C.J., Goussard, N., Chabot, R. and Lalande, R. "Potential of *Rhizobium* and *Bradyrhizobium* species as plant growth promoting rhizobacteria on non-legumes: Effect on radishes (*Raphanus sativus* L.)." *Plant and Soil* 204 (1998): 57–67.

Aristilde, L., Reed, M.L., Youngster, T., Kukurugya, M.A., Katz, V., and Sasaki, C.R. "Glyphosate- induced specific and widespread perturbations in the metabolome of soil *Pseudomonas* species." *Frontiers in Environmental Science* 5 (2017): 34.

Arriaga, F.J., Guzman, J., and Lowery, B. "Conventional agricultural production systems and soil functions". In M.M. Al-Kaisi and B. Lowery (eds.) *Soil Health and Intensification of Agroecosystems*, pp. 109–125. San Diego: Elsevier Academic Press, 2017.

Audus, L.J. "Herbicide behavior in the soil II. Interactions with soil microorganisms". In L. Audus (ed.) *The Physiology and Biochemistry of Herbicides*, pp. 163–206. New York: Academic Press, 1st edition, 1964.

Bandick, A.K., and Dick, R.P. "Field management effects on soil enzyme activities." *Soil Biology and Biochemistry* 31 (1999):1471–1479.

Banks, M.L., Kennedy, A.C., Kremer, R.J., Eivazi, F. "Soil microbial community response to soil texture, surfactants and herbicides." *Applied Soil Ecology* 74 (2014): 12–20.

Bardgett, R.D., and van der Putten, W.H. "Belowground biodiversity and ecosystem functions." *Nature* 515 (2014):505–511.

Barriuso, J., Marin, S., and Mellado, R.P. "Potential accumulative effect of the herbicide glyphosate on glyphosate-tolerant maize rhizobacterial communities over a three-year cultivation period." *PLoS One*, 6 (2011): e27558.

Barriuso, J., and Mellado, R.P. "Relative effect of glyphosate on glyphosate-tolerant maize rhizobacterial communities is not altered by soil properties." *Journal of Microbiology and Biotechnology* 22 (2012): 159–165.

Bedano, J.C., and Dominguez, A. "Large-scale agricultural management and soil meso- and macrofauna conservation in the Argentine Pampas." *Sustainability*, 8 (2016): 653.

Bedano, J.C., Dominguez, A., Arolfo, R., and Wall, L.G. "Effect of Good Agricultural Practices under no-till on litter and soil invertebrates in areas with different soil types." *Soil Tillage Research* 158 (2016): 100–109.

Berg, G. "Plant-microbe interactions promoting plant growth and health: perspectives for controlled use of microorganisms in agriculture." *Applied Microbiology and Biotechnology* 84 (2009): 11–18

Bohn, T., Rover, C.M., and Semenchuk, P.R. "*Daphnia magna* negatively affected by chronic exposure to purified Cry-toxins." *Food Chemistry and Toxicology* 91 (2016): 130–140.

Bollen, W.B. "Interactions between pesticides and soil microorganisms." *Annual Reviews of Microbiology* 15 (1961): 69–92.

Bouchez, T., Blieus, A.L., Dequiedt, S., Domaizon, I., Dufresne, A., Ferreira, S., Godon, J.J., Hellal, J., Jouliam, C., Quaizer, A., Martin Laurent, F., Mauffret, A., Monier, J.M., Peyret, P., Schmitt- Koplin, R., Sibourg, O., D'oiron, E., Bispo, A., Deprotes, I., Grand, C., Cuny, P., Maron, P.A., and Ranjard, L. "Molecular microbiology methods for environmental diagnosis." *Environmental Chemistry Letters* 14 (2016): 423–441.

Carpio, M.J., García-Delgado, C., Marín-Benito, J.M. Sánchez-Martín, M.J., and Rodríguez-Cruz, M.S. "Soil microbial community changes in a field treatment with chlorotoluron, flufenacet and diflufenican and two organic amendments." *Agronomy* 10 (2020): 1166.

Chen, X.-H., Wang, F.-L., Zhang, R., Jib, L.-L., Yang, Z.L., Lin, H., and Zhao, B. "Evidences of inhibited arbuscular mycorrhizal fungal development and colonization in multiple lines of Bt cotton." *Agriculture Ecosystems and Environment* 230 (2016): 169–176.

Clark, K.M. "The solar corridor crop system: A pivotal production system for African smallholder farmers". In R.J. Kremer, and C.L. Deichman (eds.) *The Solar Corridor Crop System: Implementation and Impacts*, pp. 145–162. San Diego: Elsevier Academic Press, 2019.

Curl, E.A., and Truelove, B. *The Rhizosphere*. New York: Springer-Verlag, 1986.

Dechesne, A., Badawi, N., Aamand, J., and Smets, B.F. "Fine scale spatial variability of microbial pesticide degradation in soil: Scales, controlling factors, and implications." *Frontiers in Microbiology* 5 (2014): 667.

Dominguez, A., Brown, G.G., Sautter, K.D., de Oliveira, C.M., de Vasconcelos, E.C., Niva, C.C., Bartz, M.L.C., and Bedano, J.C. "Toxicity of AMPA to the earthworm Eisenia andrei Bouché, 1972 in tropical artificial soil." *Scientific Reports* 6 (2016): 19731.

Doran, J.W. "Soil health and global sustainability: translating science into practice." *Agriculture, Ecosystems and Environment* 88 (2002): 119–127.

Fernandes, A.F.T., Wang, P., Staley, C., Moretto, J.A.S., Altarugio, L.M., Campanharo, S.C., Stehling, E.G., and Sadowsky, M.J. "impact of atrazine exposure on the microbial community structure in a Brazilian tropical latosol soil." *Microbes and Environments* 35 (2020). doi: 10.1264/jsme2.ME19143

Fierer, N., Ladau, J., Clemente, J.C., Leff, J.W., Owens, S.M., Pollard, K.S., Knight, R., Gilbert, J.A., and McCulley, R.L. "Reconstructing the microbial diversity and function of pre- agricultural tallgrass prairie soils in the United States." *Science*, 342(6158) (2013): 621–624.

Fox, J.E., Gulledge, J., Engelhaupt, E., Burow, M.E., and McLachlan, J.A. "Pesticides reduce symbiotic efficiency of nitrogen-fixing rhizobia and host plants." *Proceedings of the National Academy of Science USA* 104 (2007): 10282–10287.

Garbeva, P., van Veen, J.A., and van Elsas, J.D. "Microbial diversity in soil: Selection of microbial populations by plant and soil type and implications for diseased suppressiveness." *Annual Review of Phytopathology* 42 (2004): 243–270.

Giller, K.E., Witter, E., Corbeels, M., and Tittonell, P. "Conservation agriculture and smallholder farming in Africa: The heretics' view." *Field Crops Research* 114 (2009): 23–34.

Goulson, D. "An overview of the environmental risks posed by neonicotinoid insecticides." *Journal of Applied Ecology* 50 (2013): 977–987. doi: 10.1111/1365-2664.12111.

Greaves, M.P., Davies, H.A., Marsh, J.A.P., Wingfield, G.I., and Wright, J.L. "Herbicides and soil microorganisms." *CRC Critical Reviews in Microbiology* 5 (1976): 1–38.

Hagner, M., Mikola, J., Saloniemi, I., Saikkonen, K., and Helander, M. "Effects of a glyphosate-based herbicide on soil animal trophic groups and associated ecosystem functioning in a northern agricultural field." *Scientific Reports* 9 (2019): 8540. doi:10.1038/s41598-019-44988-5.

Helander, M., Saloniemi, I., Omancini, M., Druille, M., Salminen, J.-P., and Saikkonen, K. "Glyphosate decreases mycorrhizal colonization and affects plant-soil feedback." *Science of the Total Environment* 642 (2018): 285–291.

Huang, H., Zhang, S., Shan, X., Chen, B.-D., Zhu, Y.-G., and Bell, J.N.B. "Effect of arbuscular mycorrhizal fungus (*Glomus caledonium*) on the accumulation and metabolism of atrazine in maize (*Zea mays* L.) and atrazine dissipation in soil" *Environmental Pollution* 146 (2007):452–457.

Imparato, V., Santos, S.S., Johansen, A., and Geisen, S. "Stimulation of bacteria and protists in rhizosphere of glyphosate-treated barley." *Applied Soil Ecology* 98 (2016): 47–55.

ISAAA. "Global Status of Commercialized Biotech/GM Crops in 2018: Biotech Crops Continue to Help Meet the Challenges of Increased Population and Climate Change." ISAAA Brief No. 54. ISAAA: Ithaca, NY. (2018).

Johal, G.S., and Huber, D.M. "Glyphosate effects on diseases of plants." *European Journal of Agronomy* 31(2009): 144–152.

Johnsen, K., Jacobsen, C.S., Torsvik, V., and Sorensen, J. "Pesticide effects on bacterial diversity in agricultural soils—a review." *Biology and Fertility of Soils* 33 (2001): 443–453.

Karlen, D.L., Andrews, S.S., Wienhold, B.J., and Zobeck, T.M. "Soil quality assessment: Past, present and future." *Electronic Journal of Integrated Biosciences* 6 (2008): 3–14.

Karlen, D.L., Stott, D.E., Cambardella, C.A., Kremer, R.J., King, K.W., and McCarty, G.W. Surface soil quality in five Midwestern cropland conservation effects assessment project watersheds. *Journal of Soil and Water Conservation* 69 (2014): 393–401

Katan, J., and Eshel, Y. "Interactions between herbicides and plant pathogens." *Residue Reviews* 45 (1973): 145–177.

Kaufman, D.D., and Kearney, P.C. "Microbial transformations in the soil". In L. Audus (ed.) *Herbicides: Physiology, Biochemistry, Ecology*, vol. 2, 2nd ed., 29–64. New York: Academic Press, 1976.

Kloepper, J.W., McInroy, J.A., Liu, K., and Hu, C.-H. "Symptoms of fern distortion syndrome resulting from inoculation with opportunistic endophytic fluorescent pseudomonas spp." *PLoS ONE*. 8 (2013): e58531. doi: 10.1371/journal.pone.0058531.

Kortekamp, A. *Unexpected Side Effects of Herbicides: Modulation of Plant-pathogen Interactions*. Rijeka, Croatia: InTech, 2011.

Kremen, C., and Miles, A. "Ecosystem services in biologically diversified versus conventional farming systems: Benefits, externalities, and trade-offs." *Ecology and Society* 17 (2012):40. 10.5751/ES-05035-170440

Kremer, R.J. "Environmental implications of herbicide-resistant weeds: Soil biology and ecology." *Weed Science* 62 (2014):415–426.

Kremer, R.J., and Means, N.E. "Glyphosate and glyphosate-resistant crop interactions with rhizosphere microorganisms" *European Journal of Agronomy* 31(2009):153–161.

Krutz, L.J., Shaner, D.L., Weaver, M.W., Webb, R.M.T., Zablotowicz, R.M., Reddy, K.N., Huang, Y., and Thomson, S.J. "Agronomic and environmental implications of enhanced s-triazine degradation." *Pest Management Science* 66 (2010): 461–481.

Kumar, M., Prasad, R., Goyal, P., Teotia, P., Tuteja, N., Varma, A., and Kumar, V. "Environmental biodegradation of xenobiotics: Role of potential microflora," In M. Hashmi, and V. Kumar (eds.) *Xenobiotics in the Soil Environment, Soil biology* vol. 49, 319–334. Cham, Switzerland: Springer International, 2017.

Kumar, V., Simranjeet, S., and Upadhyay, N. "Effects of organophosphate pesticides on siderophore producing soil microorganisms." *Biocatalysis and Agricultural Biotechnology* 21 (2019): 101359.

Lane, M., Lorenz, N., Saxena, J., Ramsier, C., and Dick, R.P. "Microbial activity, community structure and potassium dynamics in rhizosphere soil of soybean plants treated with glyphosate." *Pedobiologia* 55 (2012): 153–159.

Levesque, C.A., and Rahe, J.E. "Herbicide interactions with fungal root pathogens with special reference to glyphosate." *Annual Review of Phytopathology* 30 (1992):597–602.

LaCanne, C.E., and Lundgren, J.G. "Regenerative agriculture: Merging farming and natural resource conservation profitably." *Peer Journal* (2018): e4428.

Lehman, R.M., Acosta-Martinez, V., Buyer, J.S., Cambardella, C.A., Collins, H.P., Ducey, T.F., Halvorson, J.J., Jin, V.L., Johnson, J.M.F., Kremer, R.J., Lundgren, J.G., Manter, D.K., Maul, J.E., Smith, J.L., and Stott, D.E. "Soil biology for resilient, healthy soil." *Journal of Soil& Water Conservation* 70 (2015): 12A–18A.

Li, X., Liu, B., Heia, S., Liu, D., Han, Z., Zhou, K., Cui, J., Luo, J., and Zheng, Y. "The effect of root exudates from two transgenic insect-resistant cotton lines on the growth of *Fusarium oxysporum*." *Transgenic Research* 18 (2009): 757–767.

Liu, Z., Dai, Y., Huang, G., Gu, Y., Ni, J., Wei, H., and Yuan, S. "Soil microbial degradation of neonicotinoid insecticides imidacloprid, acetamidprid, thiacloprid and imidaclothiz and its effect on the persistence of bioefficacy against horsebean aphid *Aphis craccivora* Koch after soil application." *Pest Management Science* 67 (2010): 1245–1252.

Locke, M.A., and Zablotowicz, R.M. "Pesticides in soil – benefits and limitations to soil health", In B.T. Christensen, S. Elmholt, and P. Schjønning (eds.) *Managing Soil Quality: Challenges in Modern Agriculture*, pp. 239–260. Oxford: CAB International, 2004.

Maderthaner, M., Weber, M., Takáca, E., Mörtl, M., Leisch, F., Römbke, J., Querner, P., Walcher, R., Gruber, E., Székács, A., and Zaller, J.G. "Commercial glyphosate-based herbicides effects on springtails (Collembola) differ from those of their respective active ingredients and vary with soil organic matter content." *Environmental Science and Pollution Research* 27 (2020): 17280–17289.

de Maria, N., de Felipe, M.R., and Fernandez-Pascual, M. "Alterations induced by glyphosate on lupin photosynthetic apparatus and nodule ultrastructure and some oxygen diffusion related proteins." *Plant Physiology and Biochemistry* 43(2005): 985–996.

Marschner, P. *Mineral Nutrition of Higher Plants*, 3rd edition, Amsterdam: Academic Press, 2012.

Meena, R.S., Kumar, S., Datta, R., Lal, R., Vigayakumar, V., Brtnicky, M., Sharma, M.P., Yadav, G.S., Jhariya, M.K., Jangir, C.K., Pathan, S.I., Doculilova, T., Pecina, V., and Marfo, T.D. "Impact of agrochemicals on soil microbiota and management: A review." *Land* 9 (2020): 34.

Meyers, R.T., Zak, D.R., White, D.C., and Peacock, A. Landscape-level patterns of microbial community composition and substrate use in upland forest ecosystems. *Soil Science Society of America Journal* 65 (2001): 359–367.

Morrissey, J.P., Dow, J.M., Mark, G.L., and O'Gara, F. "Are microbes at the root of a solution to world food production?" *EMBO Reports* 5 (2004): 922–926.

Morrow, J.G., Huggins, D.R., Carpenter-Boggs, L.A., and Reganold, J.P. Evaluating measures to assess soil health in long-term agroecosystem trials. *Soil Science Society of America Journal* 80 (2016): 450–462.

Mills, H.A., Sasseville, D.N., and Kremer, R.J. "Effects of Benlate on leatherleaf fern growth, root morphology and rhizosphere bacteria." *Journal of Plant Nutrition* 19 (1996): 917–937.

Muangphra, P., Kwankua, W., and Gooneratne, R. "Genotoxic effects of glyphosate or paraquat on earthworm coelomocytes." *Environmental Toxicology* 29 (2012): 612–620.

Muangphra, P., Tharapoom, K., Euawong, N., Namchote, S., and Gooneratne, R. "Chronic toxicity of commercial chlorpyrifos to earthworm *Pheretima peguana*." *Environmental Toxicology* 31 (2016): 1450–1459.

Mulligan, R.A., Tomco, P.L., Howard, M.W., Schempp, T.T., Stewart, D.J., Stacey, P.M., Ball, D.B., and Tjeerdema, R.S. "Aerobic versus anaerobic microbial degradation of clothianidin under simulated California rice field conditions." *Journal of Agricultural and Food Chemistry* 64 (2016): 7059–7067.

Nathan, V.K., Jasna, V., and Parvathi, A. "Pesticide application inhibits the microbial carbonic anhydrase–mediated carbon sequestration in a soil microcosm." *Environmental Science and Pollution Research* 27 (2020): 4468–4477.

Newman, M.M., Hoilett, N.O., Lorenz, N., Dick, R.P., Liles, M.R., Ramsier, C., and Kloepper, J.W. "Glyphosate effects on soil rhizosphere-associated bacterial communities." *Science of the Total Environment* 54 (2016): 155–160.

Nicolas, V., Oestreicher, N., and Vélot, C. "Multiple effects of a commercial Roundup® formulation on the soil filamentous fungus *Aspergillus nidulans* at low doses: Evidence of an unexpected impact on energetic metabolism." *Environmental and Pollution Research* 23 (2016): 14393–14404.

Pandy, G., Dorrian, S.J., Russell, R.J., and Oakeshott, J.G. "Biotransformation of the neonicotinoid insecticides imidacloprid and thiamethoxam by Pseudomonas sp. 1G" *Biochemistry and Biophysics Research Communications* 380 (2009): 710–714.

Pankhurst, C.E. "Biodiversity of soil organisms as an indicator of soil health". In V.V.S.R. Gupta, C.E. Pankhurst, and B. Doube (eds.) *Biological Indicators of Soil Health*, pp. 297–324. Oxford: CAB International, 1997.

Pasaribu, A., Mohamad, R.B., Hashim, A., Rahman, Z.A., Omar, D., and Morshed, M.M. "Effect of herbicide on sporulation and infectivity of vesicular arbuscular mycorrhizal (*Glomus mosseae*) symbiosis with peanut plant." *Journal of Animal and Plant Sciences* 23 (2013): 1671–1678.

Peck, D.C. "Comparative impacts of white grub (Coleoptera: Scarabaeidae) control products on the abundance of non-target soil-active arthropods in turfgrass." *Pedobiologia* 52 (2009): 287–299.

Poirier, F., Boursier, C., Mesgage, R., Oestreicher, N., Nicolas, V., and Vélot, C. "Proteomic analysis of the soil filamentous fungus *Aspergillus nidulans* exposed to a Roundup formulation at a dose causing no macroscopic effect: A functional study." *Environmental Science and Pollution Research* 24 (2017): 25933–25946.

Racke, K.D. "Pesticides in the soil-microbial ecosystem". In J.R. Coats, and K.D. Racke *Enhanced Biodegradation of Pesticides in the Environment*, pp. 1–12. Washington: American Chemical Society, Symposium Series 426, 1990.

Ramakrishnan, B., Venkateswarlu, K., Sethunathan, N., and Megharaj, M. "Local applications but global implications: Can pesticides drive microorganisms to develop antimicrobial resistance?" *Science of the Total Environment* 654 (2019): 177–189.

Riah, W., Laval, K., Laroche-Ajzenberg, E., Mougin, C., Latour, X., and Trinsoutrot-Gattin, I. "Effects of pesticides on soil enzymes: A review." *Environmental Chemistry Letters.* 12 (2014):257–273. 10.1007/s10311-014-0458-2.

Rose, M.T., Cavagnaro, T.R., Scanlan, C.A., Rose, T.J., Vancov, T., Kimber, S., Kennedy, I.R., Kookana, F.S., and van Zwieten, L. "Impact of herbicides on soil biology and function." *Advances in Agronomy* 136 (2016): 133–220.

Schutte, G., Eckerstorfer, M., Rastelli, V., Reichenbecher, W., Restrepo-Vassalli, S., Ruohonen-Lehto, M., Wuest Saucy, A.G., and Mertens, M. "Herbicide resistance and biodiversity: agronomic and environmental aspects of genetically-modified herbicide-resistant plants." *Environment Science Europe* 29 (2017): 5.

Shahid, M., and Khan, M.S. "Pesticide-induced alteration in proteins of characterized soil microbiota revealed by sodium dodecyl sulphate–polyacrylamide gel electrophoresis (SDS-PAGE)." *Journal of Proteins and Proteomics* 11 (2020): 1–9.

Shahid, M., Zaidi, A., Ehtram, A., and Khan M.S. "In vitro investigation to explore the toxicity of different groups of pesticides for an agronomically important rhizosphere isolate *Azotobacter vinelandii*." *Pesticide Biochemistry and Physiology* 157 (2019): 33–44.

Simon-Sylvestre, G., and Fournier, J.C. "Effects of pesticides on the soil microflora." *Advances in Agronomy* 31 (1979): 1–92.

Sinsabaugh, R.L., Antibus, R.K., and Linkins, A.E. An enzymic approach to the analysis of microbial activity during plant litter decomposition. *Agriculture Ecosystems and Environment* 34 (1991):43–54

Skipper, H.D., Mueller, J.G., Ward, V.L., and Wagner, S.C. "Microbial degradation of herbicides." In N.D. Camper (ed.) *Research Methods in Weed Science*, 3rd edition, pp. 430–455. Champaign, IL: Southern Weed Science Society, 1986.

Smith, S.E., and D.J. Read. *Mycorrhizal Symbiosis*, 3rd edition, New York: Academic Press, 2008.

Stirling, G., Hayden, H., Pattison, T., and Stirling, M. *Soil health, Soil Biology, Soilborne Diseases and Sustainable Agriculture*. Clayton South: CSIRO Publishing, 2016.

Stott, D.E., Andrews, S.S., Liebig, M.A., Wienhold, B.J., and Karlen, D.L. "Evaluation of β-glucosidase activity as a soil quality indicator for the soil management assessment framework." *Soil Science Society of America Journal* 74 (2010): 107–119.

Stott, D.E., Cambardella, C.A., Karlen, D.L., and Harmel, R.D. "A soil quality and metabolic activity assessment after fifty-seven years of agricultural management." *Soil Science Society of America Journal* 77 (2013): 903–913.

Tank, J.L., Rosi-Marshall, E.J., Royer, T.V., Whiles, M.R., Griffiths, N.A., Frauendorf, T.C., and Treering, D.J. "Occurrence of maize detritus and a transgenic insecticidal protein (Cry1Ab) within the stream network of an agricultural landscape." *PNAS* (2010): 17645–17650. https://www.pnas.org/cgi/doi/10.1073/pnas.1006925107.

Torsvik, V., Goksoyr, J., and Daae, F.L. "High diversity in DNA of soil bacteria." *Applied and Environmental Microbiology* 56 (1990): 782–787.

Travis, D.A., Chapman, D.W., Craft, M.E., et al. "One health: Lessons learned from East Africa." *Microbiology Spectrum* 2 (2014). doi:10.1128/microbiolspec.OH-OO17-2012.

Tsatsakis, A.M., Nawaz, M.A., Kouretas, D., Balias, G., Savolainen, K., Tutelyan, V.A., Golokhvast, K.S., Lee, J.D., Yang, S.H., and Chung, G. "Environmental impacts of genetically modified plants: A review." *Environmental Research* 156 (2017): 818–833.

Van Bruggen, A.H.C., and Semenov, A.M. "In search of biological indicators of soil health and disease suppression." *Applied Soil Ecology* 15 (2000): 13–24.

Van Eerd, L.L., Hoagland, R.E., Zablotowicz, R.M., and Hall, J.C. 'Pesticide metabolism in plants and microorganisms." *Weed Science* 51 (2003): 472–495.

Van Hoesel, W., Tiefenbacher, A., König, N., Dorn, V.M., Hagenguth, J.F., Prah, U., Widhalm, T., Wiklicky, V., Koller, R., Bonkowski, M., Lagerlöf, J., Ratzenböck, and Zaller, J.G. "Single and combined effects of pesticide seed dressings and herbicides on earthworms, soil microorganisms, and litter decomposition." *Frontiers in Plant Science* 8 (2017): 215. doi: 10.3389/fpls.2017.00215.

Vanlauwe, B., Coyne, D., Gockowski, J., Hauser, S., and Huising, J., "Sustainable intensification and the African smallholder farmer." *Current Opinion in Environmental Sustainability* 8 (2014): 15–22.

Veum, K.S., Kremer, R.J., Stott, D.E., Sudduth, K.A., Kitchen, N.R., Lerch, R.N., Baffaut, C., Karlen, D.L., and Sadler, E.J. "Conservation effects on soil quality indicators in the Missouri Salt River Basin." *Journal of Soil and Water Conservation* 70 (2015): 232–246.

Wagg, C., Bender, S.F., Widmer, F., and van der Heijden, G.A. "Soil biodiversity and soil community composition determine ecosystem multifunctionality." *PNAS* 111 (2014): 5266–5270.

Whalen, J.K., and Sampedro, L. *Soil Ecology & Management*, Oxfordshire: CAB International, 2010.

Whipps, J.M., Lewis, K., and Cooke, R.C. "Mycoparasitism and plant disease control". In M.N. Burge (ed.) *Fungi in Biological Control Systems*, pp. 161–187. Manchester: Manchester University Press, 1988.

Whittington, H.D., Singh, M., Ta, C., Azcárate-Peril, and Bruno-Bárcena. "Accelerated biodegradation of the agrochemical ametoctradin by soil-derived microbial consortia." *Frontiers in Microbiology* 11 (2020): 1898.

Yardim, E.N., and Edwards, C.A. "The effects of chemical pest, disease and weed management practices on the trophic structure of nematode populations in tomato agroecosystems." *Applied Soil Ecology* 7 (1998): 137–147.

Zablotowicz, R.M., and Reddy, K.N. "Nitrogen activity, nitrogen content, and yield responses to glyphosate in glyphosate-resistant soybean." *Crop Protection* 26 (2007): 370–376.

Zak, D.R., Holmes, W.E., White, D.C., Peacock, A.D., and Tilman, D. Plant diversity, soil microbial communities, and ecosystem function: Are there any links? *Ecology* 84 (2003): 2042–2050.

Zhang, M., Wang, W., Tang, L., Heenan, M., and Xu, Z. "Effects of nitrification inhibitor and herbicides on nitrification, nitrite and nitrate consumptions and nitrous oxide emission in an Australian sugarcane soil." *Biology and Fertility of Soils* 54 (2018): 697–706.

Zobiole, L.H., Kremer, R.J., Oliveira, R.S., and Constantin, J. "Glyphosate affects microorganisms in rhizospheres of glyphosate-resistant soybeans." *Journal of Applied Microbiology* 110 (2010): 118–127.

11 Bio-decontamination of Mycotoxin Patulin

Hongyin Zhang and Qiya Yang
Jiangsu University

Gustav Komla Mahunu
University for Development Studies

CONTENTS

11.1 INTRODUCTION

In recent years, global discussions on mycotoxins have been very high since there is growing pressure on the limited food resources to cater for the increasing human population. Approximately 25% of global crops produced are contaminated with mycotoxins and result in significant losses of about 1 billion metric tons of raw produce and other food products (Lukwago et al., 2019). They occur naturally, while the type of fungi, extent of growth, and consequent mycotoxin contamination depend on internal and external environmental factors (Fernández-Cruz et al., 2010). The major groupings of mycotoxins most relevant to the food industries are aflatoxins (AFs), ochratoxins-A (OTA), zearalenone (ZEA), moniliformin, deoxynivalenol (DON), fumonisins, and patulin (PAT). Although specific mycotoxin is known to contaminate particular food items, other previous reports acknowledge the possible occurrence of more than one mycotoxin in food at the same time (Speijers and Speijers, 2004). Such occurring phenomenon though less reported may require further studies to establish clear mechanisms of action and respective control strategy. According to Ianiri et al. (2016), food and feed contaminated with mycotoxins are harmful to human and animal health, together with significant economic losses to especially exporting countries, where contaminated produce are damaged and rejected. The presence of mycotoxins in these food and feed constituents is attributed to fungal infection of host plants at preharvest or postharvest stages.

Mycotoxins are known to be low-molecular-weight fungal metabolites capable of causing harm after an extended period of consuming the contaminated food (Marroquín-Cardona et al., 2014). Due to their toxicological properties, maximum levels (MLs) for mycotoxins have been established for some food and feed to safeguard animal and human health and assure consumer safety. In 2003, minimum of 99 countries around the world had established regulations on mycotoxin for food and/ or feed (FAO and FOODS, 2004).

The purpose of this chapter is to present the available information on PAT in apple fruits and their processed products and the implications on human health. The effect of fungicides on postharvest fungi control and PAT level will be discussed. Additionally, various biological control methods using yeast antagonistic microbes with promising applications in postharvest diseases control and PAT reduction as well as their corresponding modes of action will be highlighted.

11.2   DISTRIBUTION AND ECONOMIC IMPLICATIONS OF PAT

Most of the fruits, PAT contaminates are seasonal, suitable for consumption for short time, therefore they are usually processed or treated in order to be commercially available throughout the year. Since PAT can also be generated in food during storage and maintain its stability through food processing, the constant monitoring of different fruit-based products must to be conducted to provide proper appraisal of human exposure to this toxin. Table 11.1 provides a list of countries in which PAT was detected in food and related products.

Consumers do not only derive satisfaction from eating fresh apple fruits, but they also provide important nutrients to the human body. Several literature sources have reported on the health benefits of apples (Matsuoka, 2019; Yahia et al., 2019). Processing of fresh apple fruits into various products contributes to value addition and improves all-year-round availability. However, PAT is one of the important mycotoxins produced by certain number of fungi associated with fruits and vegetable-based products of which apple is highly susceptible to PAT. However, apart from pome fruits such as apples, loquat, and pear, PAT has also been found toxic to other higher plants including cucumber, wheat, peas, corn and flax. *Penicillium expansum* alone was projected to cause postharvest decay between 70% and 80% with the possibility of PAT contamination (Barkai-Golan and Paster, 2011). The possible PAT contamination intensifies the need for food-processing methods that can control and remove the toxic substance from the fruit products, to reduce cost to the food industry (Moake et al., 2005). At refrigeration temperatures, a number of penicillia can produce PAT because it is stable in the contaminated products (Stott and Bullerman, 1975), and this occurrence is of particular concern to producers. Common rotting molds found on refrigerated apple products for instance are frequent PAT producers during cold storage. Refrigerated storage of foods will not necessarily avoid PAT as many molds are capable of producing PAT at low temperatures. However, the absence of PAT is a direct and very good quality index of apple products.

TABLE 11.1

Depicts the Countries Around the World Where PAT Was Detected in Food and Other Products

Continent	Country
Asia	India (Saxena et al., 2008), Malaysia (Lee et al., 2014), Japan (Watanabe and Shimizu, 2005), China (Guo et al., 2013)
Africa	South Africa (Shephard et al., 2010), Tunisia (Zaied et al., 2013; Zaied et al., 2012)
America	USA (Reddy et al.), Argentina (Funes and Resnik, 2009), and Brazil (Sylos, 1999)
Europe	UK (Atkins, 1994), Turkey (Gökmen and Acar, 1998), Spain (Marín et al., 2011), Sweden (Josefsson and Andersson, 1977), Italy (Spadaro et al., 2007)
Australia	Australia (Cressey, 2009)

11.3 TOXICOLOGY OF PATULIN

Investigators including Birkinshaw in the early 1943 undertook the first PAT isolation from *Penicillium expansum* searching for new fungal molecules containing antibiotic properties after discovery of penicillin by Fleming. Later the interest in PAT as a potential antibiotic declined after it was found to be toxic to both humans and animals. In recent decades, PAT has become an important chemical contaminant among the various mycotoxins in food (Puel et al., 2010) due to its potential to cause harm and pollution (Spring and Fegan, 2005). PAT is a secondary metabolite with a molar mass of 154.12 g mol^{-1} and its molecular formula is 4-hydroxy-4H-furo [3, 2-c] pyran-2(6H)-one (Figure 11.1). During PAT synthesis, one acetyl-CoA molecule in addition to three malonyl-CoA molecules undergoes reduction, decarboxylation, and oxidation. The 6-MS molecule is converted to m-OH-benzaldehyde, which then undergoes a rearrangement leading to one molecule of PAT (Figure 11.2).

However, several countries across the world have set regulations in order to decrease PAT in products to the lowest level to protect the health and safety of consumers. For the European countries, the regulation 1425/3003 was set in 2003, and they were one of the first countries to set up such a regulation (Table 11.2). PAT regulation was fixed at 50 µg L^{-1} as the maximum level for fruit juice and derived products, solid apple products (25 µg L^{-1}) and juices, and foods (10 µg L^{-1}) as raw materials for babies and young infants. Similarly, the US Food and Drug Administration (FDA) adopted 50 µg L^{-1} for PAT concentration limit in fruits and their products (Van Egmond et al., 2008). In the case of China, the Gulf Cooperation Council (GCC) adopted same limit for PAT regulations in apple fruits and derived products (Table 11.3) (FAO and FOODS, 2004). The highest PAT intake of 1.04 ng kg^{-1} body weight (bw)/day for apple juice consumption was reported by the World Health Organization's (WHO's) Global Environment Monitoring System/Food Contamination

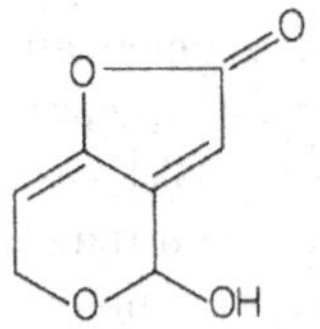

FIGURE 11.1 Structure of patulin (Stott and Bullerman, 1975).

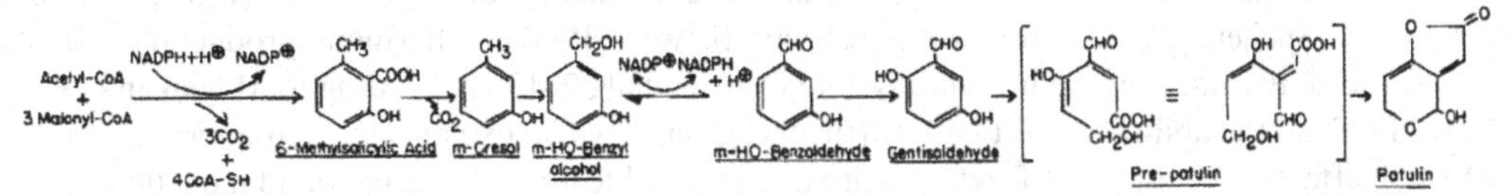

FIGURE 11.2 The main pathway for patulin biosynthesis in *Penicillium* spp. (Forrester and Gaucher, 1972).

TABLE 11.2
European Union Maximum Levels for PAT in Fruits and Their Processed Products

Patulin/Product	Maximum Level
Fruit juice, concentrate fruit juice as reconstituted, and fruit nectars	50.0
Spirit drinks, cider, and other fermented drinks derived from apples or containing apple juice	50.0
Solid apple products, including apple compote, apple puree intended for direct consumption	25.5
Apple juice and solid apple products, including apple compote and apple puree for infants and young children and labeled and solid as such	10.0
Baby foods other than processed cereal-based foods for infants and young children	10.0

TABLE 11.3
Recommended Maximum Levels for PAT in Apple Fruits and Their Processed Products According to Countries/Agencies Worldwide (FAO and FOODS, 2004)

Country	Products	PAT µg kg^{-1}
EU	Fruit juices, concentrated fruit juice as reconstituted, and fruit nectars.	50
China	Fruit products containing apple or hawthorn (excluding Guo Dan Pi, a Chinese-style fruit snack)	50
	Fruit or vegetable juice containing apple or hawthorn juice	50
	Alcoholic beverages containing apple or hawthorn	50
Codex, GCC, Kenya, Nigeria	Apple juice	50
South Africa	Apple juice, apple juice ingredients in other juices	50
USA	Apple juice, apple juice concentrates and products	50
India	Apple juice, apple juice ingredients in other beverages	50
Japan	Apple juice, food made using only apple juice as raw material	50

Monitoring and Assessment Programme (GEMS/Food) (Organization, 2010). The estimated PAT intakes through apple juice among these groups of the population were adults (28.1 ng kg^{-1} bw/day), children (67.5 ng kg^{-1} bw/day), and babies (110 ng kg^{-1} bw/day). The Joint Expert Committee on Food Additives (JECFA) operating under the WHO and the Food and Agriculture Organization (FAO) (2005) of the United Nations (UN) presented reports that offered in-depth data on PAT toxicity and other mycotoxins (Van Egmond et al., 2008). In addition, there has been adequate documentation of the adverse effects of PAT on health of consumers after long exposure to contaminated foods.

According to the International Agency for Research on Cancer (IARC), PAT is classified as category 3 of carcinogens (Zoghi et al., 2017). PAT has been found to cause different acute effects (gastrointestinal tract distension, intestinal hemorrhage, and epithelial cell degeneration), chronic effects (genotoxic, neurotoxic, teratogenic, immunotoxic, cytotoxic), inhibition of protein synthesis, and inhibition of DNA and RNA synthesis (Abrunhosa et al., 2016; Glaser and Stopper, 2012; Moake et al., 2005).

As a food contaminant, PAT has the potential to cause oxidative damage to cells (Liu et al., 2006; Speijers and Speijers, 2004), while the interaction between PAT and hormone-production systems can be destructive and alter the immune system (Marin et al., 2013). According to Magan and Olsen (2004), PAT has the ability to react with sulfhydryl groups, which explains the cytotoxic and certain genotoxic effects. Generally, PAT cytotoxicity has a typical feature of causing rapid induction of calcium influx and total lactic dehydrogenase release causing the damage of structural integrity of the plasma membrane in mammalian cells (Riley and Showker, 1991) and a reduction on gap junction mediated intercellular communications (Burghardt et al., 1992). It was also reported that after penetrating the gastric wall, PAT degraded rapidly and its toxicity was not systemic. However, the reaction with glutathione and maybe with proteins could contribute partially to PAT degradation, whereas a major diminution of glutathione in gastric tissue is likely to influence PAT to produce localized toxic effects (Rychlik et al., 2004). It was reported that foods high in protein and low in carbohydrates seem unfavorable for excessive PAT production. The combination of this phenomenon with PAT reactivity with sulfhydryl groups appears to minimize the risk of PAT incidence in these foods.

Early investigations indicated that PAT can be removed with active fermentation through yeasts or by aqueous solution with cysteine or glutathione and PAT can disappear in 4–5 days (Halász et al., 2009; Stinson et al., 1979; Valletrisco et al., 1988).

The main PAT-producing organism in apple is known to be ubiquitous (Coulombe Jr., 1993), and this characteristic leads to the development of varied symptoms in mycotoxicoses. PAT is involved in three major mechanisms of action: (i) alteration in the content, absorption, and metabolism of nutrients; (ii) damage to organs and their functions; and (iii) suppression of the immune system. However, the effect of PAT on the immune system predisposes the organism to disease infection and consequently production loss occurs. The exposure of organisms to the toxin suppresses the immune system, which is complicated by opportunistic diseases.

11.4 OCCURRENCE OF PATULIN IN APPLE FRUIT AND ITS DERIVED PRODUCTS

Different species of fungi produce mold (*Penicillium, Aspergillus, Paecilomyces,* and *Byssochlamys*) (Varga et al., 2010). Three Aspergillus species of the Clavati group produce PAT: A. clavatus, A. giganteus, and A. longivesica. PAT is produced by 13 species of *Penicillium*: *P. carneum, P. clavigerum, P. concentricum, P. coprobium, P. dipodomyicola, P. expansum, P. gladioli, P. glandicola, P. griseofulvum, P. marinum, P. paneum, P. sclerotigenum,* and *P. vulpinum* (Frisvad et al., 2004). Out of the 13 *Penicillium* species, *P. expansum* commonly known as blue mold is the main cause of rot in pome fruits, characterized by rapid soft rot and formation of the blue mold (Figure 11.3).

Various parameters influence PAT, they include the amount of free water ($_{aw}$), temperature, oxygen, the condition of the substrate, and pH (Yiannikouris et al., 2007). The interrelation of these factors affects PAT production. To be able to develop effective techniques for controlling PAT accumulation in foods, the mechanisms by which the various parameters influence PAT production need to be well understood (Barkai-Golan and Paster, 2011). Most *Penicillium* species are saprophytic based on their ecological classification, thus surviving on dead and decaying matter or soil (Pitt, 2000). PAT is a colorless, crystalline compound with melting point of 110°C. It is stable in water solution up to 105°C–125°C and pH 3.5–5.5 (González-Osnaya et al., 2007). Similarly, a pH range of 3.2–3.8 supports active production of PAT by the fungus (Moss, 1991).

In order to establish the link between pH of the flesh of apple fruit and *P. expansum* colonization, sections of apple fruit were studied. From these studies, it was observed that pH was between 4 and 6 in nondecayed tissues, whereas in the decayed tissues the pH declined from 3.6 to 4.1 within the 4–6 days after inoculation (Prusky et al., 2014, 2004). As a response to modification of pH (upward or downward) in the host, the pH activity could decline, which could also enhance the pathogen virulence. In response to host signals, alkalization by ammonification of the host tissue in *Colletotrichum* and *Alternaria* or acidification by secretion of organic acids in *Penicillium, Botrytis,* and *Sclerotinia* were observed (Prusky et al., 2014). The rise in pH of a solution can increase the breakdown of PAT content; thus, when pH was 6 at 100°C, PAT declined by 50% (Collin et al., 2008).

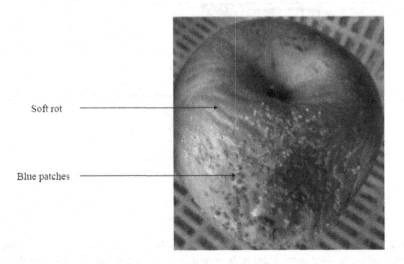

Soft rot

Blue patches

FIGURE 11.3   Blue patches and soft rot apple fruit.

PAT is not easily affected by heat (it is stable and survives the pasteurization process); for this reason, PAT will require higher temperature activity to break down (Salas et al., 2012). PAT production is temperature-dependent, and during incubation high heat applied alone or combination with other factors can kill or reduce the fungi spores and mycelia, even though it may not be able to destroy the toxin totally. According to an earlier report, *Penicillium expansum* is a psychrophilic fungus that is capable of growing in low temperatures (Deming, 2002). The storage of fruit in cold temperatures may not prevent decay caused by blue mold (Morales et al., 2007a) and subsequent buildup of PAT. Within the range of 20°C–25°C is considered the optimum temperature for PAT production, but as low as 5°C can also elicit its production. Some varieties of apples (such as Golden delicious and Fuji) with higher amounts of organic acids reported more PAT accumulation in storage at room temperature. Similarly, Golden delicious and Red delicious apple varieties exhibited significantly higher PAT accumulation as a result of their lower acidity (Konstantinou et al., 2011). This implies that cultivars of apples must be considered a critical factor influencing the biocontrol of *P. expansum* and subsequent accumulation of PAT in the fruit (Spadaro et al., 2013). It was reported that pH value of the apple varieties regulated the accumulation of PAT during storage; as such at 0°C–1°C storage, fresh fruits containing lower pH were reported to be more predisposed to the PAT accumulation (Morales et al., 2008b). Clearly, fruit cultivars have differences in their vulnerability to blue mold decay and possible PAT accumulation (Neri et al., 2010).

Controlled atmosphere (CA) storage functions effectively in mold control and PAT production. Low O_2 (1.5%–2.5%) and high CO_2 (up to 3%) contents provide favorable conditions for extended storage of apple fruit (Juhneviča et al., 2011). Irrespective of the positive impact of CA storage on the prolonged shelf life, it has also the potential to compromise fruit quality since mold growth and PAT pollution increase with extended storage time (de Souza Sant'Ana et al., 2008). CA stored apples appeared to intensify PAT concentration in juice as compared to juice extracted directly from freshly harvested/picked apples and more so, the extended floor storage time prior to processing causes a significant increase in PAT accumulation in apple juice (Baert et al., 2012).

The distribution of PAT in rotten apple fruit showed PAT content present in juice extracted from the different sections of decay (Bandoh et al., 2009). In Figure 11.4, PAT concentration varied from the outer layer toward the inner part of apple pulp where "section a" represents complete decay with PAT content of 40 µg kg^{-1}, "section b" (after the completely decayed area and extending up to 5 mm) with PAT 0.14 µg kg^{-1}, and "section c" (the section within 5–10 mm) with only PAT 0.003 µg kg^{-1}.

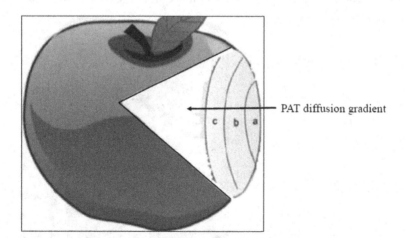

FIGURE 11.4 Distribution of patulin through a rotten apple fruit. The concentration of PAT decreases from "a" (the visibly decayed outer area) to b toward "c" (the inner cortex). Modified illustration from Mahunu et al. (2015).

Welke et al. (2009) reported that PAT can penetrate up to 10 mm of the surrounding healthy-looking pulp. There is evidence that PAT pollution is not restricted to the decayed tissues alone. This is supported by other reports (Laidou et al., 2001), where diffusion of PAT concentration through the pulp declines with increasing distance from the completely rotten region to the central part of the fruit that seems healthy. Similarly, small infected sections on apples were found to contain PAT, and this did not correlate with lesion diameter (Beretta et al., 2000).

Although the rational is that PAT is commonly detected in P. expansum-infected apples, PAT could be absent in infected apple fruits (Neri et al., 2010). The absence of PAT in P. expansum-infected apples could be attributed to the incompetence of the fungus to produce PAT (Barad et al., 2014; Sanzani et al., 2012). Such incidences may be controlled by a number of factors such as fruit variety and degree of ripening, the fungal strain, the presence of other microbes, or conditions of postharvest storage (Ballester et al., 2015).

Large proportion of sugar in apple fruits is sucrose; changes of this component during ripening of fruit may be linked to fungal metabolism and PAT synthesis. The presence of other sugars such as glucose or maltose also lead to high production of PAT by *P. expansum*. Sugars are carbon sources for growth of pathogen in apples and subsequent production of PAT and induced expression of pH modulators (Barad et al., 2016a). The decline in the growth of *P. expansum* may correspond with the decrease in the available sugars or other nutrients, indicating that PAT production may be connected to nutritional stress (McCallum et al., 2002).

Pathogenicity demonstrated by *P. expansum* has a direct effect on PAT content by means of either gene-disruption or RNAi mutants (Barad et al., 2016b; Sanzani et al., 2012). For instance, the mutants still produce some PAT to explain the lack of correlation between the degree of pathogenicity expressed by the *Penicillium* strain and its ability to produce PAT (Ballester et al., 2015).

Estimations of PAT in both raw and processed apple fruits in different countries worldwide have been well described. Murillo-Arbizu et al. (2009) reported on the test results from 100 apple juice samples collected from separate markets in Spain, where PAT concentrations were between 0.7 and 118.7 $\mu g \, L^{-1}$ (average 19.4 $\mu g \, l^{-1}$) compared to the level across Europe. However, 11% of the samples were beyond the highest permissible limits in the EU (50 $\mu g \, L^{-1}$). Spadaro et al. (2007) also indicated that 34.8% of 135 apple products collected in Italy were contaminated with PAT at concentrations of 1.58–55.41 $\mu g \, kg^{-1}$. Further, out of 45 apple product samples obtained in Argentina, ten were contaminated with PAT concentration between 17 and 221 $\mu g \, kg^{-1}$, with mean concentration of 61.7 $\mu g \, kg^{-1}$ (Funes and Resnik, 2009). They found that 50% of the apple fruit jam samples studied were PAT contaminated with an average concentration of 123 $\mu g \, kg^{-1}$. The following PAT levels were detected in samples collected from various countries such as Iran (50–285 $\mu g \, L^{-1}$) (Cheraghali et al., 2005), Saudi Arabia (5–152.5 $\mu g \, L^{-1}$) (Al-Hazmi, 2010), and Romania (0.7–101.9 $\mu g \, L^{-1}$) (Oroian et al., 2014).

Also, varied PAT concentrations have been reported in China. The Chinese Academy of Preventive Medicine noticed that 76.9% of fruit juice, fruit jam, and other semimanufactured products were found contaminated with PAT with concentrations between 18 and 953 ppm. In this survey, rate of PAT occurrence in fresh fruit samples was lower (19.6%) with concentrations of 4–262 $\mu g \, kg^{-1}$ (Zhang et al., 2009). In another study, 1987 apple juice concentrates were collected from Shanxi Province in China on four sequential processing seasons (2006–2010) to study the dietary exposure to PAT content. In processing season of 2007 and 2008, only four product samples out of the total mentioned above were PAT (78.0 $\mu g \, kg^{-1}$) contaminated. Samples collected in 2006–2007, 2008–2009, and 2009–2010 processing seasons recorded PAT concentrations of 29.0, 22.7, and 20.0 $\mu g \, kg^{-1}$, respectively. From the results, PAT was detected in more than 90% of the samples collected with lower than 30 $\mu g \, kg^{-1}$ concentration, which is important for the apple juice industry.

Over the years, apples have been treated with fungicides to prevent *P. expansum* infection during storage. It was reported that neither selection of fruits that appear healthy will prevent PAT in the final products (Chen et al., 2012). It implies that the initial occurrence of postharvest fungal pathogens on harvested fruits must be checked in order to guarantee PAT-free products.

Similarly, growth studies of *P. expansum* on the surface of apple jam for the period of 28 d at 15°C were conducted to determine the production and distribution of fungal metabolites in the sample (approximately 6 cm high divided into three equal layers). First, the study indicated that the growth rate of the pathogen increased with storage time, which corresponded with decrease in temperature. Secondly, PAT was detected in the entire 2 cm layers of the apple jam. The concentrations of PAT in the upper two layers of the jam matched the exposures above the health-based guidance value (HBGV) for a normal serving size. Although reduction of PAT content is often based on the removal of the moldy fragments of the food, the study finally confirmed that this practice is inadequate to avert unhealthy contact (Olsen et al., 2019).

11.5 EFFECT OF POSTHARVEST FUNGICIDE APPLICATION ON DECAY AND PATULIN LEVELS IN APPLES

Traditional methods including sorting, washing, density segregation, and thermal treatments have been used for control of PAT-producing pathogens (Hojnik et al., 2017). These methods can produce substantial prevention and decontamination of PAT, whereas thermal treatment, for instance, can cause thermal degradation of toxins. Usually, processing period is very long with high energy demand and eventually increases the cost, and the treated food products lose quality significantly after the heat treatment (Jouany, 2007). As research in the food industry advances in the search for effective decontamination efficiency in the mycotoxin, novel nonthermal methods (UV- and gamma-irradiation, pulsed-light treatments, and control atmospheric pressure) are being developed. The change in chemical structure of PAT resulting in its degradation is influenced by nonthermal methods. The presence of water in the treated products, the extent of PAT contamination, and intensity of exposure determine the decontamination efficiency (Smith and Girish, 2012). Some of the nonthermal methods such as gamma-irradiation (dosage between 1 and 20 kGy) have achieved mycotoxin removal rate of about 90% in solution. In effect, the degradation of the mycotoxin in the solution was attributed to perhaps the formation of free radicals produced by the radiolysis of water in the product (Hojnik et al., 2017).

In the case of contaminated solid and dry food products (containing low moisture content), the use of gamma-irradiation for decontamination of the mycotoxin has proven less effective (Calado et al., 2014). Nevertheless, the attempt to introduce high doses of gamma-irradiation could affect food product quality negatively (Kottapalli et al., 2003). UV light irradiation has been used directly for decontaminating PAT in apple juice. Application of UV light irradiation at 222 nm produced 99% PAT content decrease, although some photosensitive substances in the juice (healthy ascorbic acid) were lowered by the treatment (Zhu et al., 2014).

For many years, chemical agents including many synthetic compounds have produced significant decontamination effects, probably as cost-effective treatment method. Notwithstanding the outcomes of synthetic fungicides, synthetic fungicides are very expensive with high environmental and human health risks, and prolonged time of treatment is not appropriate for high-quality food preservation.

Globally, each year about 23 million kg fungicides are used to reduce the incidence of diseases (Karabulut et al., 2005). Among the schemes employed, the application of fungicides such as thiabendazole have been relied upon for many years, since it is difficult to cull rotten apples unless the decayed areas are visible on the surface. Synthetic fungicides, therefore, are conventionally used for the control of postharvest diseases and mycotoxin decontamination of food. Irrespective of the role fungicides have played in postharvest treatment, chemical residues remain a potential risk coupled with the development of fungicide-resistant species. Given the fact that continuous mishandling of fungicides is also characterized as an environmental hazard, most countries have gradually restricted or banned completely their applications. Fungal pathogens such as *P. italicum* and *P. digitatum* in citrus fruits have developed resistance to fungicides (Fogliata et al., 2000), whereas 77% of the *P. expansum* isolates were identified as resistant to thiabendazole (Morales et al., 2008a). Also, 70% of *P. expansum* isolates from apples, pears, and grapes demonstrated resistance to thiabendazole (Cabañas et al., 2009). According to Baraldi et al. (2003), the magnitude of *P. expansum* and increased severity of blue mold on fruits could be attributed to the presence of thiabendazole. Following this, it was also revealed that captan was accountable for enhancing PAT production (Paterson, 2007). Fungicides have been reported as stress drivers for mycotoxin-producing fungi, for which reason mycotoxin content can increase even though *P. expansum* growth is suppressed (Kazi et al., 1997).

In order to determine the ability to decrease the resistance of fungal strains to fungicides, several new "low-risk fungicides" (LRFs) have been studied. Some of the LRFs include fludioxonil (Calvert et al., 2008), iprodione (Ochiai et al., 2002), azoxystrobin and pyrimethanil (Kanetis et al., 2007). The filtrates of the LRFs have shown the capacity to reduce fungal growth (Kuiper-Goodman, 2004). The inhibitory properties of iprodione and fludioxonil against *Candida albicans* growth were investigated, and the fungicides' toxicities were capable of suppressing hyphal formation (Ochiai et al., 2002). In as much as fruit producers are likely to continue to use these LRFs in excess, it may eventually render their potency weak (Morales et al., 2010). The next section focuses on the application and manipulation of biocontrol microbes as antagonistic agents and their associated potential to decontaminate PAT in fruit.

11.6 BIOCONTROL MICROBES AS ANTAGONISTIC AGENTS AND POTENTIAL PATULIN DECONTAMINANTS

Biological methods fall in the category of mycotoxin decontamination processes, which are based on the ability of representative microorganisms including yeast, bacteria, molds, actinomycetes, and algae to either eliminate or reduce mycotoxins present in foods and other products. The obvious benefit of biocontrol decontamination is the involvement of zero-chemicals and quality of treated foods not being compromised. Biological control agents (BCAs) represent less than 1% of the annual market share of the chemicals used for crop protection, estimated to cost US$ 16,000 million. More commercially viable microbial antagonists (such as bacteria and yeast) have the tendency to biodegrade or detoxify PAT in order to protect commodities in storage without compromising produce quality. Generally, yeasts respond favorably to different extreme stressful environmental conditions (low and high temperatures, desiccation, wide range of relative humidity, law oxygen levels, pH changes, UV radiation) that exist before and after harvest (Droby et al., 2016). These authors also described yeasts as possessing special qualities to adapt to the microenvironment such as high sugar concentration, high osmotic pressure, and low pH existing in the wounded tissues of fruit. Yeasts are easy to multiply in large quantities, and most of the species are able to grow speedily on inexpensive substrates in fermenters (Spadaro et al., 2010). Compared to filamentous fungi, yeasts exhibit simple nutritional needs that permit them to colonize dry surfaces for extended durations (Droby et al., 2016).

In apples, blue mold has been controlled using *Cryptococcus albidus* (Fan and Tian, 2001), *Candida sake* and *Pantoea agglomerans* (Morales et al., 2008b), *Rhodotorula mucilaginosa* (Li et al., 2011, 2019), *Yarrowia lipolytica* (Yang et al., 2017; Zhang et al., 2017a), *Sporidiobolus pararoseus* (Abdelhai et al., 2019; Li et al., 2017a), *Candida guilliermondii* (Chen et al., 2017; Coelho et al., 2009), *Cryptococcus podzolicus* (Wang et al., 2018), *Pichia caribbica* (Mahunu et al., 2018), and *Hanseniaspora uvarum* (Apaliya et al., 2018).

According to Castoria et al. (2005), *Rhodotorula glutinis* LS11 cells demonstrated ability to resist PAT concentration and further degraded it in vitro. Thus, *Rhodotorula glutinis* LS11 cells were able to metabolize PAT and/or reduce its accumulation or synthesis. A recent study conducted by Li et al. (2019) also validated *R. mucilaginosa* JM19 stain as a promising candidate for control of PAT contamination in food and raw materials. This stain (*R. mucilaginosa* JM19 1×10^8 cells L^{-1}) was able to significantly reduce PAT by 90% after 21 h in MES buffer at 35°C. With at initial PAT concentration (100 μg mL^{-1}), the *R. mucilaginosa* JM19 was able to detoxify above 50%. By this process of detoxification, *R. mucilaginosa* JM19 degraded PAT into desoxypatulinic acid, a lesser nontoxic compound.

Similarly, Coelho et al. (2007) indicated that initial PAT concentration (223 µg) declined by >83% after coincubation with *Pichia ohmeri* 158 for 2 d at 25°C, followed by 99% reduction after 5 days, and eventually untraceable after 15 d. Also, PAT incubated with *Pichia caribbica* yeast for 15 d at 20°C decreased contamination in apples than in the control and same response by *P. caribbica* against PAT was noticed in the in vitro tests (Cao et al., 2013). Despite the performance of antagonistic yeasts, recent studies have reported a significant interest in substances capable of augmenting their biocontrol efficacy against postharvest storage pathogens (Droby et al., 2016; Wisniewski et al., 2016). Tolaini et al. (2010) in their study revealed the application of *Lentinula edodes* (LS28) against *P. expansum* in apples. This study also showed that *Lentinula edodes* (LS28) reduced approximately 50% of apple decay caused by *P. expansum*. It also enhanced the activity of antioxidant enzymes catalase (CAT) and glutathione peroxidase as host defense response to the presence of disease infection in the biocontrol system. Other studies on the efficacy of antagonistic yeast indicated that *C. laurentii* is correlated with its level of superoxide dismutase (SOD) and CAT production, responsible for the yeast resistance to oxidative stress (Castoria et al., 2005). The application of antioxidant quercetin significantly reduced the incidence of apple blue mold. It is evident that quercetin may induce resistance to *P. expansum* in apples, through acting on gene transcription levels involved in several distinct metabolic processes (Romanazzi et al., 2016; Sanzani et al., 2012).

11.7 IMPROVING THE PERFORMANCE AND STABILITY OF ANTAGONISTIC YEAST FOR PATULIN REDUCTION

The satisfactory and stable aproaches for PAT detoxification are extremely important in order to achieve success with any biocontrol representative. Several investigations have reported on various techniques to boost the consistency and effectiveness of the postharvest biocontrol agents. The agents are categorized as salts and organic acids, food additives, glucose analogs, and various physical (thermal and nonthermal) treatments. Each antagonistic yeast within the biocontrol system has its specific unique characteristic response to the amending agent and therefore needs precise protocol to undertake commercial assessment. Indeed, addition of substances at nontoxic but acceptable concentrations has enriched the bio-efficiency of yeasts, where the outcome of the interaction between the two or more agents essentially yielded higher yeast population or improved the use of yeast biomass at lower concentration (Apaliya et al., 2018). Some of the substances include salicylic acid (Babalar et al., 2007; Zhang et al., 2010, 2008), brassinosteroids (Zhu et al., 2010), sodium bicarbonate (Cerioni et al., 2012; Youssef et al., 2014; Zhu et al., 2012), chitosan (Meng et al., 2010; Romanazzi and Feliziani, 2016), phytic acid (Mahunu et al., 2016; Yang et al., 2015), indole-3-acetic acid (Yu et al., 2008), boron (Cao et al., 2012; Qin et al., 2010), harpin (Tang et al., 2015; Zhu and Zhang, 2016), phosphatidylcholine (Li et al., 2016), glycine betaine (Zhang et al., 2017b), jasmonic acid (Li et al., 2017b), potassium phosphite (Lai et al., 2017), trehalose (Apaliya et al., 2017, 2018), bamboo (Mahunu et al., 2018), β-glucan (Wang et al., 2018), methyl thujate (Ji et al., 2018), $CaCl_2$ (Tournas and Katsoudas, 2019), and baobab (Abdelhai et al., 2019).

Given the possibility of enhancing yeast performance, it is pivotal to search for substances that exhibit the ability to counter the mycotoxin activity as well as be relevant in PAT detoxification. Recently, a biocompatibility study by Mahunu et al. (2018) showed that *Pichia caribbica* supplemented with bamboo leaf flavonoid (0.01% w/v) reduced or inhibited PAT accumulation concentration both *in vitro* and *in vivo* after 20 d incubation period at 20°C. Also, an examination showed that *R. mucilaginosa* augmented with phytic acid (4 µmol mL⁻¹) was able to decrease the initial PAT concentration of 20 µg mL⁻¹–1.561 µg mL⁻¹, and the degradation rate was 92.2% after 24 h *in vitro* incubation at 28°C (Yang et al., 2015).

Furthermore, the distinct and multifaceted modes of action exhibited by biocontrol yeasts did reduce PAT contamination on the basis of their competition for space, nutrients, and their wounding competence in apples (Spadaro et al., 2013). Likewise, the improvement of biocontrol efficacy is influenced by yeast resistance to the oxidative stress that is caused by the reactive oxygen species (ROS) generated in fruit wounds (Castoria et al., 2003) and the lytic enzymes production (Castoria et al., 2001). Findings of Lima et al. (2011) indicated that the fungal pathogen cannot readily develop resistance against biocontrol methods due to the multifaceted mode of action of BCAs.

Similarly, the direct application of physical treatment such as ultraviolet-C (UV-C) to control fungi has been well investigated beside other methods for postharvest fruit diseases prevention. UV-C was found to stimulate responses in fruits (Stevens et al., 2005), which confirmed earlier studies on the relationship between UV-C dosage, the number of fungal-spore germination, and the number of infected wound lesions on the fruits' surface (Stevens et al., 1998). Indeed, the assimilation of low doses (200 and 400 Gy) of γ-irradiation together with a BCA (*P. fluorescens*) was described as suitable for extension of the physicochemical qualities and reduce the postharvest losses of apples (Mostafavi et al., 2013). Application of UV-C at a low dose, which reduced pathogen inoculum and induced host resistance, was considered as doubly effective on fungi but did not show evidence of PAT degradation in apples. As a result, UV-C inhibited *P. expansum* infection and growth, but it was unable to prevent PAT distribution in the apple once inoculation had occurred. After γ-irradiation treatment, phenolic compounds rapidly accumulated in the fruit tissues commonly attributed to the ability of UV light to stimulate the activity of PAL, which is the key enzyme in the biosynthesis of phenols in plant tissues (Charles et al., 2008). Generally, there is a connection between phenolic content and antioxidant activity, and some phenolic compounds are known for their antioxidant activities (Zhang et al., 2013). Similarly, antioxidant enzymes and related enzymes were shown to play a key role in eradicating ROS (Wang et al., 2011). It was also reported that the oxidant activities can elicit and regulate the biosynthesis of mycotoxins produced by various *Aspergillus* species (*Aspergillus flavus, A. parasiticus*, and *A. ochraceus*) (Reverberi et al., 2008). Previous studies also showed that natural antioxidants from compounds have been recognized for postharvest fungal control and inhibition of other mycotoxins including aflatoxin, ochratoxin, and PAT (Reverberi et al., 2005; Ricelli et al., 2007).

On the other hand, the exhibition of microbial activity by PAT appears to play a direct role in microbial competition and survival, thus protects the fungal pathogen, and functions partly as a defense mechanism against yeast antipathy (Castoria et al., 2008). From investigations, PAT inhibited the growth of 75 species of gram-positive and gram-negative bacteria and also expressed antiviral and antiprotozoal actions (Moake et al., 2005). This also implies that PAT activity has the capacity to enhance fungal growth and improve their competing ability in an environment with other microorganism (Tikekar, 2009).

In general, the function of mycotoxin-producing fungi is to activate pathogenesis. However, fungal mutants unable to produce toxins might demonstrate substantially less virulence than their toxin-producing counterparts. As initially described by Desjardins and Hohn (1997), *Penicillium* produces PAT mycotoxicity, which plays a role in the pathogenesis. This observation was confirmed by Harris et al. (1999), who indicated that trichothecene mycotoxin produced by *Fusarium graminearum* plays a crucial role in pathogenicity in maize; thus, the activity undoubtedly weakens the defense mechanism of the host cell, causing the host to become vulnerable to fungal infection. The presence of trichothecene might modify the penetrability of the host cell membrane, whereas PAT generates excessive oxidative stress resulting in cell death (Speijers and Speijers, 2004). Tikekar (2009) indicated that fungal toxins are able to increase the competitiveness and pathogenicity of fungi and subsequently support the fungi to survive proficiently in the environment they occupy. Since characteristically PAT is unstable, the BCAs screened and selected to control PAT-producing fungi must have the capacity to defend the antimicrobial activity of the PAT. The assumption is that the antimicrobial activity produced by PAT can persist on fruit surfaces all through preharvest and postharvest stages; therefore when the bioefficacy of agents is enhanced, it will be able to destroy the toxicity and compete with the pathogen (Paster and Barkai-Golan, 2008). In other words, the yeasts, for instance, are able to colonize, survive, and persist on the dry fruit surfaces for longer periods of time during storage. Interestingly, the majority of reports on postharvest BCAs coincubated with yeast cells showed a superior efficacy for eradicating PAT or inhibited fungal infection. This approach based on the use of mediating technologies at critical points of the disease control process also offers possible energy sources for microbes or adds to oxidative stress against the fungi in the biocontrol system.

11.8 MECHANISMS FOR ANTAGONISTIC YEASTS ACTION IN PATULIN REDUCTION

So far studies have shown advances in the mechanisms of PAT degradation through the use of BCAs. Expounding the mechanisms of degradation as well as identifying key functional factors in yeasts will promote the development of systems that will assure efficient removal of PAT in foods. Biocontrol agents have convincingly proven to carry the efficacy to overcome PAT, yet the mechanisms of action of degradation are repeatedly poorly understood. The most important mechanisms of biological removal or degradation of PAT suggested include the microbial cells biosorption of PAT (Guo et al., 2012; Wang et al., 2015; Yue et al., 2013), PAT degradation by microbial enzymes (Zhu et al., 2015), PAT-glutathione adducts, or reaction of PAT with thiol group present in protein extracts and for that reason terminating its functional properties (Luz et al., 2017). It has also been observed that the presence of PAT in growth media induces the synthesis of PAT degrading enzymes in PAT-resistant/degrading yeasts (Zheng et al., 2017). Also earlier report by Park and Troxell (2002) indicated that the capacity of BCAs to reduce mycotoxin levels is governed by the following factors: (i) the stability of the mycotoxin, (ii) the nature of the degradation process, (iii) interactions between mycotoxin and food or host, and (iv) mycotoxin–mycotoxin interactions. Similarly, the control of PAT contamination in fruits is achieved by any of these following three categories: (i) through prevention of initial PAT pollution during harvesting, processing, and storage; (ii) PAT removal during processing; and (iii) use of postproduction treatment agents for PAT removal or detoxification.

The use of supplementary agents has shown the potency to enrich the antifungal activity of yeast as an important mechanism of action against PAT, although some yeasts including *Pichia ohmeri* strain 158 yeast alone in broth cultures decreased PAT production (Coelho et al., 2007). *Lentinula edodes* yeast develops poorly in a highly oxidative environment (Tolaini et al., 2010), whereas *Rhodotorula glutinis* LS11 strain reinforces its enzymatic antioxidant potential under similar conditions in order to prevent PAT biosynthesis (Castoria et al., 2008). From literature, *R. glutinis* LS11 strain is capable of resisting extremely high concentrations of PAT (500 µg mL^{-1}) and metabolizes it under aerobic conditions. Using thin-layer chromatography (TLC)-based detection, two products were formed and their retention factors (Rf) were 0.46 and 0.38 (Castoria et al., 2005). From the authors' report, the formation of the compound(s) with an Rf of 0.46 seemed to be more stable than that of the other compound. The compound with Rf of 0.46 was the main product detected at the end of the degradation process, whereas that with Rf of 0.38 could not be detected anymore. The study by Ricelli et al. (2007) showed that in apple juice, approximately 96% of PAT was degraded after incubation with *Gluconobacter oxydans*. The use of *G. oxydans* converted PAT to ascladiol, a less-toxic compound. *Sporobolomyces sp.* strain IAM 13481 was also found to transform PAT into two different end products (desoxypatulinic acid and (Z)-ascladiol) (Ianiri et al., 2013). Again, *Candida guilliermondii* was found to be a major yeast that degraded PAT to E-ascladiol less toxic compound (Figure 11.5) (Chen et al., 2017). Essentially, rapid biodegradation of PAT depends on the viable/living yeast cells with greater efficacy to eliminate PAT than dead cells. According to Zhu et al. (2015) supported by Chen et al. (2017), intercellular protein showed the ability to degrade PAT, which implies that the biodegradation of PAT is an intercellular occurrence in yeast.

According to Coelho et al. (2007), the presence of antagonistic yeast cells *in vitro* decreased PAT contamination through two main mechanisms: (i) by PAT adsorption on the yeast cell wall and (ii) PAT absorption into yeast cells. With respect to the mode of action, the microbes and additives (such as zeolites, bentonites, clays, and activated carbons) function together to bind toxin to their surface through adsorption or transform it into less toxic metabolites known as biotransformation

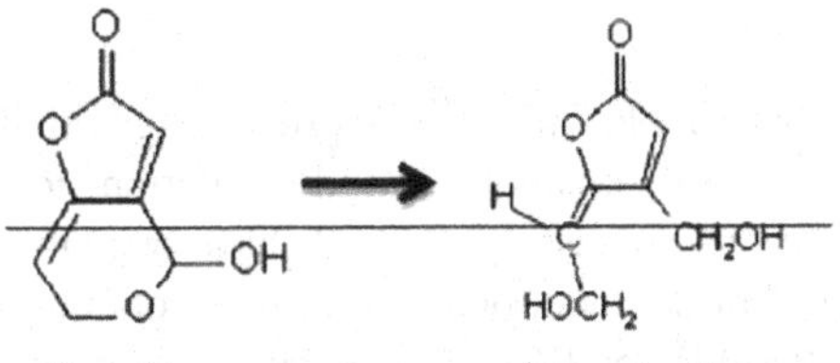

Patulin   E-ascladiol

FIGURE 11.5 Biodegradation by *C. guilliermondii* to convert patulin to E-ascladiol (Chen et al., 2017).

(Kolosova and Stroka, 2011). The studies performed by Yiannikouris et al. (2004) reported that yeast strains with a high cell-wall glucan (HCWG) content are tightly bound to toxins than strains with low cell-wall glucan (LCWG) content. This occurrence was the case, when their total cell wall (TCW) fraction and alkali-insoluble glucan (AIG) fraction were exposed to the toxins. The authors observed that the adsorption efficacy of AIG fraction was the highest when its chitin content was lower than that of the TCW fraction, which increased glucan flexibility and accessibility of the toxin to the glucan network. In similar studies, Moran (2004) described the yeast cell wall as a natural source of two oligosaccharides (β-glucan and mannan), most potent in inducing defense. Mostly β-D-glucans are known to assist in inducing defense against fungal infections. Novak and Vetvicka (2008) and Moran (2004) had previously demonstrated the mechanism of action involved in the mycotoxin–glucan interaction.

Tian et al. (2007) reported on the enzyme β-1, 3-glucanase as one of the well-categorized pathogenesis-related (PR) proteins, acting directly in pathogen cell wall degradation or indirectly releasing oligosaccharides to elicit defense reactions. Also, Janisiewicz and Korsten (2002) stated that the yeast extracellular β-glucanases and chitinases depolymerize fungal cell walls, symbolizing an essential biocontrol mechanism of fungi in postharvest. Following reports of previous authors (Reverberi et al., 2005; Ricelli et al., 2007). Zhang et al. (2013) also confirmed the importance of enzymes in pathogen cell-wall degradation and antioxidant function in toxigenic fungi.

A nonspecific phytotoxic substance, oxalic acid (OA) is also able to elicit plant defense reactions through suppression of the oxidative burst of host plant (Kim et al., 2008). It plays a key role in pathogenesis and fungal growth. The presence of postharvest treatment with BCAs induces host resistance of fruits (apples and citrus) and stimulation of enzymatic activities including phenylala-nine ammonia lyase (PAL) (Droby et al., 2003; El Ghaouth et al., 2003).

Extracellular compounds produced by yeast strains have the potential to prevent and possibly kill mycotoxin-producing strains (Coelho et al., 2007). To achieve this, the yeasts secrete a low-molecular-mass protein or glycoprotein toxin that can kill sensitive cells belonging to the same or related yeast genera, but this does not involve direct cell–cell contact. Furthermore, the biocontrol activities of such yeasts do not depend on the production of harmful antibiotics, but rather are influenced by their ability to compete for space and nutrients on fruit surfaces or in wounded tissues (Paster and Barkai-Golan, 2008). The findings of Castoria et al. (2005) also identified lower PAT accumulation in infected apples that were pretreated with low concentrations of *Rhodotorula gluti-nis* strain LS11, as compared with the level in control (nontreated) infected fruits. These authors also indicated that PAT degradation was hastened after the yeast cells survived and increased population in the infected apples. Similar trend of results was detected in an in vitro test, where surviving yeast cells in decayed apples may possibly metabolize PAT and/or reduce its accumulation. PAT production was estimated to decrease in accordance with the incubation time (20 days at 15°C) and possibly due to organic acids present in the culture that cause PAT to leach from the vacuole and mycelium of *P. expansum* itself (Morales et al., 2007b). In the same study, the growth of *P. expansum* was detected to be slower at the early stages of incubation while sufficient energy source was present in the medium. However, growth of *P. expansum* increased in the late phase after the energy source was almost exhausted in addition to the presence of many intermediaries.

11.9 CONCLUSION

Fungal infections contaminate raw apple fruit prior to processing with the possibility of PAT synthesis. Therefore, there are well-established standards and regulations to prevent PAT crossing over to humans, especially infants. Various processing methods have been used satisfactorily, but still cannot guarantee complete removal or degradation of PAT from contaminated commodities. Synthetic fungicides have been found feasible for PAT control but by themselves are capable of acting as potential stress driver for the mycotoxin-producing fungi. Recent searches point to potential biocontrol agents being used after a significant progress has been made in their commercialization. Other research directions pointing to enhanced biocontrol efficacy with two or more bioactive substances or compounds without compromising consumer safety have been possible. It is expected that biocontrol agents or products will gain increasing recognition with wide acceptability in the coming years, as a component of postharvest disease management. Lastly, molecular studies, when conducted, could further explain the complex mechanisms involved in mycotoxin elimination. From this review, PAT can be degraded in several ways. They are biotransformation or biodegradation or removal of PAT through microbial cells biosorption of PAT, microbial enzymes degrading PAT, and PAT-glutathione adducting, or PAT reacting with thiol group existing in protein extracts. Finally, the presence of PAT in the growth media induces the production of degrading enzymes in PAT-resistant/degrading yeasts.

ACKNOWLEDGMENTS

This work was supported by the National Natural Science Foundation of China (31772369, 31571899), the National key research project (sub-project) of China (2016YFD0400902-04), and the Agricultural Independent Innovation Fund in Jiangsu Province [CX(18)2028]. The authors thank some of the experts in this field for sharing their valuable knowledge over the years.

REFERENCES

Abdelhai, M.H., Tahir, H.E.O.J., Zhang, Q., Yang, Q., Ahima, J., Zhang, X., Zhang, H., 2019. Effects of the combination of Baobab (Adansonia digitata L.) and Sporidiobolus pararoseus Y16 on blue mold of apples caused by Penicillium expansum. *Biological Control* 134, 87–94.

Abrunhosa, L., Morales, H., Soares, C., Calado, T., Vila-Chã, A.S., Pereira, M., Venâncio, A., 2016. A review of mycotoxins in food and feed products in Portugal and estimation of probable daily intakes. *Critical Reviews in Food Science and Nutrition* 56, 249–265.

Al-Hazmi, N.A., 2010. Determination of Patulin and Ochratoxin A using HPLC in apple juice samples in Saudi Arabia. *Saudi Journal of Biological Sciences* 17, 353–359.

Apaliya, M.T., Zhang, H., Yang, Q., Zheng, X., Zhao, L., Kwaw, E., Mahunu, G.K., 2017. Hanseniaspora uvarum enhanced with trehalose induced defense-related enzyme activities and relative genes expression levels against Aspergillus tubingensis in table grapes. *Postharvest Biology and Technology* 132, 162–170.

Apaliya, M.T., Zhang, H., Zheng, X., Yang, Q., Mahunu, G.K., Kwaw, E., 2018. Exogenous trehalose enhanced the biocontrol efficacy of Hanseniaspora uvarum against grape berry rots caused by Aspergillus tubingensis and Penicillium commune. *Journal of the Science of Food and Agriculture* 98, 4665–4672.

Atkins, D., 1994. The UK's food chemical surveillance programme. *British Food Journal* 96, 24–29.

Babalar, M., Asghari, M., Talaei, A., Khosroshahi, A., 2007. Effect of pre- and postharvest salicylic acid treatment on ethylene production, fungal decay and overall quality of Selva strawberry fruit. *Food Chemistry* 105, 449–453.

Baert, K., Devlieghere, F., Amiri, A., De Meulenaer, B., 2012. Evaluation of strategies for reducing patulin contamination of apple juice using a farm to fork risk assessment model. *International Journal of Food Microbiology* 154, 119–129.

Ballester, A.-R., Marcet-Houben, M., Levin, E., Sela, N., Selma-Lázaro, C., Carmona, L., Wisniewski, M., Droby, S., González-Candelas, L., Gabaldón, T., 2015. Genome, transcriptome, and functional analyses of Penicillium expansum provide new insights into secondary metabolism and pathogenicity. *Molecular Plant-Microbe Interactions* 28, 232–248.

Bandoh, S., Takeuchi, M., Ohsawa, K., Higashihara, K., Kawamoto, Y., Goto, T., 2009. Patulin distribution in decayed apple and its reduction. *International Biodeterioration & Biodegradation* 63, 379–382.

Barad, S., Espeso, E.A., Sherman, A., Prusky, D., 2016a. Ammonia activates pacC and patulin accumulation in an acidic environment during apple colonization by Penicillium expansum. *Molecular Plant Pathology* 17, 727–740.

Barad, S., Horowitz, S.B., Kobiler, I., Sherman, A., Prusky, D., 2014. Accumulation of the mycotoxin patulin in the presence of gluconic acid contributes to pathogenicity of Penicillium expansum. *Molecular Plant-Microbe Interactions* 27, 66–77.

Barad, S., Sionov, E., Prusky, D., 2016b. Role of patulin in post-harvest diseases. *Fungal Biology Reviews* 30, 24–32.

Baraldi, E., Mari, M., Chierici, E., Pondrelli, M., Bertolini, P., Pratella, G., 2003. Studies on thiabendazole resistance of Penicillium expansum of pears: Pathogenic fitness and genetic characterization. *Plant Pathology* 52, 362–370.

Barkai-Golan, R., Paster, N., 2011. *Mycotoxins in Fruits and Vegetables*. Academic Press, Cambridge, MA.

Beretta, B., Gaiaschi, A., Galli, C.L., Restani, P., 2000. Patulin in apple-based foods: occurrence and safety evaluation. *Food Additives & Contaminants* 17, 399–406.

Burghardt, R.C., Barhoumi, R., Lewis, E.H., Bailey, R.H., Pyle, K.A., Clement, B.A., Phillips, T.D., 1992. Patulin-induced cellular toxicity: a vital fluorescence study. *Toxicology and Applied Pharmacology* 112, 235–244.

Cabañas, R., Abarca, M., Bragulat, M., Cabañes, F., 2009. Comparison of methods to detect resistance of Penicillium expansum to thiabendazole. *Letters in Applied Microbiology* 48, 241–246.

Calado, T., Venâncio, A., Abrunhosa, L., 2014. Irradiation for mold and mycotoxin control: a review. *Comprehensive Reviews in Food Science and Food Safety* 13, 1049–1061.

Calvert, G.M., Karnik, J., Mehler, L., Beckman, J., Morrissey, B., Sievert, J., Barrett, R., Lackovic, M., Mabee, L., Schwartz, A., 2008. Acute pesticide poisoning among agricultural workers in the United States, 1998–2005. *American Journal of Industrial Medicine* 51, 883–898.

Cao, B., Li, H., Tian, S., Qin, G., 2012. Boron improves the biocontrol activity of Cryptococcus laurentii against Penicillium expansum in jujube fruit. *Postharvest Biology and Technology* 68, 16–21.

Cao, J., Zhang, H., Yang, Q., Ren, R., 2013. Efficacy of Pichia caribbica in controlling blue mold rot and patulin degradation in apples. *International Journal of Food Microbiology* 162, 167–173.

Castoria, R., Caputo, L., De Curtis, F., De Cicco, V., 2003. Resistance of postharvest biocontrol yeasts to oxidative stress: A possible new mechanism of action. *Phytopathology* 93, 564–572.

Castoria, R., De Curtis, F., Lima, G., Caputo, L., Pacifico, S., De Cicco, V., 2001. Aureobasidium pullulans (LS-30) an antagonist of postharvest pathogens of fruits: study on its modes of action. *Postharvest Biology and Technology* 22, 7–17.

Castoria, R., Morena, V., Caputo, L., Panfili, G., De Curtis, F., De Cicco, V., 2005. Effect of the biocontrol yeast Rhodotorula glutinis strain LS11 on patulin accumulation in stored apples. *Phytopathology* 95, 1271–1278.

Castoria, R., Wright, S.A., Droby, S., 2008. Biological control of mycotoxigenic fungi in fruits. In Barkai-Golan, R., and Paster, N. (eds.) *Mycotoxins in Fruits and Vegetables*, pp. 311–333. Elsevier, Amsterdam.

Cerioni, L., Rodríguez-Montelongo, L., Ramallo, J., Prado, F.E., Rapisarda, V.A., Volentini, S.I., 2012. Control of lemon green mold by a sequential oxidative treatment and sodium bicarbonate. *Postharvest Biology and Technology* 63, 33–39.

Charles, M.T., Benhamou, N., Arul, J., 2008. Physiological basis of UV-C induced resistance to Botrytis cinerea in tomato fruit: III. Ultrastructural modifications and their impact on fungal colonization. *Postharvest Biology and Technology* 47, 27–40.

Chen, X., Li, J., Zhang, L., Xu, X., Wang, A., Yang, Y., 2012. Control of postharvest radish decay using a Cryptococcus albidus yeast coating formulation. *Crop Protection* 41, 88–95.

Chen, Y., Peng, H.-M., Wang, X., Li, B.-Q., Long, M.-Y., Tian, S.-P., 2017. Biodegradation Mechanisms of Patulin in Candida guilliermondii: An iTRAQ-Based Proteomic Analysis. *Toxins* 9, 48.

Cheraghali, A.M., Mohammadi, H.R., Amirahmadi, M., Yazdanpanah, H., Abouhossain, G., Zamanian, F., Khansari, M.G., Afshar, M., 2005. Incidence of patulin contamination in apple juice produced in Iran. *Food Control* 16, 165–167.

Coelho, A.R., Celli, M.G., Ono, E.Y.S., Wosiacki, G., Hoffmann, F.L., Pagnocca, F.C., Hirooka, E.Y., 2007. Penicillium expansum versus antagonist yeasts and patulin degradation in vitro. *Brazilian archives of Biology and Technology* 50, 725–733.

Coelho, A.R., Tachi, M., Pagnocca, F.C., Nobrega, G.M.A., Hoffmann, F.L., Harada, K.-I., Hirooka, E.Y., 2009. Purification of Candida guilliermondii and Pichia ohmeri killer toxin as an active agent against Penicillium expansum. *Food Additives and Contaminants* 26, 73–81.

Collin, S., Bodart, E., Badot, C., Bouseta, A., Nizet, S., 2008. Identification of the main degradation products of patulin generated through heat detoxication treatments. *Journal of the Institute of Brewing* 114, 167–171.

Coulombe Jr., R.A., 1993. Biological action of mycotoxins. *Journal of Dairy Science* 76, 880–891.

Cressey, P., 2009. Mycotoxin risk management in New Zealand and Australian food. *World Mycotoxin Journal* 2, 113–118.

de Souza Sant'Ana, A., Rosenthal, A., de Massaguer, P.R., 2008. The fate of patulin in apple juice processing: a review. *Food Research International* 41, 441–453.

Deming, J.W., 2002. Psychrophiles and polar regions. *Current Opinion in Microbiology* 5, 301–309.

Desjardins, A.E., Hohn, T.M., 1997. Mycotoxins in plant pathogenesis. *Molecular Plant-Microbe Interactions* 10, 147–152.

Droby, S., Wisniewski, M., El Ghaouth, A., Wilson, C., 2003. Influence of food additives on the control of post-harvest rots of apple and peach and efficacy of the yeast-based biocontrol product Aspire. *Postharvest Biology and Technology* 27, 127–135.

Droby, S., Wisniewski, M., Teixidó, N., Spadaro, D., Jijakli, M.H., 2016. The science, development, and commercialization of postharvest biocontrol products. *Postharvest Biology and Technology* 122, 22–29.

El Ghaouth, A., Wilson, C.L., Wisniewski, M., 2003. Control of postharvest decay of apple fruit with Candida saitoana and induction of defense responses. *Phytopathology* 93, 344–348.

Fan, Q., Tian, S., 2001. Postharvest biological control of grey mold and blue mold on apple by Cryptococcus albidus (Saito) Skinner. *Postharvest Biology and Technology* 21, 341–350.

FAO, J., Foods, M.H.I., 2004. Food and Agriculture Organization of the United Nations. Rome.

Fernández-Cruz, M.L., Mansilla, M.L., Tadeo, J.L., 2010. Mycotoxins in fruits and their processed products: Analysis, occurrence and health implications. *Journal of Advanced Research* 1, 113–122.

Fogliata, G., Torres, L., Ploper, L., 2000. Detection of imazalil-resistant strains of Penicillium digitatum Sacc. in citrus packinghouses of Tucumán Province (Argentina) and their behavior against currently employed and alternative fungicides. *Revista industrial y agrícola de Tucumán* 77, 71–75.

Forrester, P., Gaucher, G., 1972. m-Hydroxybenzyl alcohol dehydrogenase from Penicillium urticae. *Biochemistry* 11, 1108–1114.

Frisvad, J.C., Frank, J.M., Houbraken, J., Kuijpers, A.F., Samson, R.A., 2004. New ochratoxin a producing species of Aspergillus section Circumdati. *Studies in Mycology* 50, 23–43.

Funes, G.J., Resnik, S.L., 2009. Determination of patulin in solid and semisolid apple and pear products marketed in Argentina. *Food Control* 20, 277–280.

Glaser, N., Stopper, H., 2012. Patulin: mechanism of genotoxicity. *Food and Chemical Toxicology* 50, 1796–1801.

Gökmen, V., Acar, J., 1998. Incidence of patulin in apple juice concentrates produced in Turkey. *Journal of Chromatography A* 815, 99–102.

González-Osnaya, L., Soriano, J.M., Moltó, J.C., Mañes, J., 2007. Exposure to patulin from consumption of apple-based products. *Food Additives and Contaminants* 24, 1268–1274.

Guo, C., Yue, T., Yuan, Y., Wang, Z., Guo, Y., Wang, L., Li, Z., 2012. Biosorption of patulin from apple juice by caustic treated waste cider yeast biomass. *Food Control* 32, 99–104.

Guo, Y., Zhou, Z., Yuan, Y., Yue, T., 2013. Survey of patulin in apple juice concentrates in Shaanxi (China) and its dietary intake. *Food Control* 34, 570–573.

Halász, A., Lásztity, R., Abonyi, T., Bata, Á., 2009. Decontamination of mycotoxin-containing food and feed by biodegradation. *Food Reviews International* 25, 284–298.

Harris, L., Desjardins, A.E., Plattner, R., Nicholson, P., Butler, G., Young, J., Weston, G., Proctor, R., Hohn, T., 1999. Possible role of trichothecene mycotoxins in virulence of Fusarium graminearum on maize. *Plant Disease* 83, 954–960.

Hojnik, N., Cvelbar, U., Tavčar-Kalcher, G., Walsh, J., Križaj, I., 2017. Mycotoxin decontamination of food: Cold atmospheric pressure plasma versus "classic" decontamination. *Toxins* 9, 151.

Ianiri, G., Idnurm, A., Castoria, R., 2016. Transcriptomic responses of the basidiomycete yeast Sporobolomyces sp. to the mycotoxin patulin. *BMC Genomics* 17, 1–15.

Ianiri, G., Idnurm, A., Wright, S.A., Durán-Patrón, R., Mannina, L., Ferracane, R., Ritieni, A., Castoria, R., 2013. Searching for genes responsible for patulin degradation in a biocontrol yeast provides insight into the basis for resistance to this mycotoxin. *Applied and Environmental Microbiology* 79, 3101–3115.

Janisiewicz, W.J., Korsten, L., 2002. Biological control of postharvest diseases of fruits. *Annual Review of Phytopathology* 40, 411–441.

Ji, D., Chen, T., Ma, D., Liu, J., Xu, Y., Tian, S., 2018. Inhibitory effects of methyl thujate on mycelial growth of Botrytis cinerea and possible mechanisms.

Josefsson, E., Andersson, A., 1977. Analysis of patulin in apple beverages sold in Sweden. *Archives de l'Institut Pasteur de Tunis* 54, 189–196.

Jouany, J.P., 2007. Methods for preventing, decontaminating and minimizing the toxicity of mycotoxins in feeds. *Animal Feed Science and Technology* 137, 342–362.

Juhneviča, K., Skudra, G., Skudra, L., 2011. Evaluation of microbiological contamination of apple fruit stored in a modified atmosphere. *Environmental and Experimental Biology* 9, 53–59.

Kanetis, L., Förster, H., Adaskaveg, J.E., 2007. Comparative efficacy of the new postharvest fungicides azoxystrobin, fludioxonil, and pyrimethanil for managing citrus green mold. *Plant Disease* 91, 1502–1511.

Karabulut, O.A., Arslan, U., Ilhan, K., Kuruoglu, G., 2005. Integrated control of postharvest diseases of sweet cherry with yeast antagonists and sodium bicarbonate applications within a hydrocooler. *Postharvest Biology and Technology* 37, 135–141.

Kazi, S., Paterson, R., Abo-Dahab, N., 1997. Effect of 2-deoxy-D-glucose on mycotoxins from apples inoculated with Penicillium expansum. *Mycopathologia* 138, 43–46.

Kim, K.S., Min, J.-Y., Dickman, M.B., 2008. Oxalic acid is an elicitor of plant programmed cell death during Sclerotinia sclerotiorum disease development. *Molecular Plant-Microbe Interactions* 21, 605–612.

Kolosova, A., Stroka, J., 2011. Substances for reduction of the contamination of feed by mycotoxins: A review. *World Mycotoxin Journal* 4, 225–256.

Konstantinou, S., Karaoglanidis, G., Bardas, G., Minas, I., Doukas, E., Markoglou, A., 2011. Postharvest fruit rots of apple in Greece: Pathogen incidence and relationships between fruit quality parameters, cultivar susceptibility, and patulin production. *Plant Disease* 95, 666–672.

Kottapalli, B., Wolf-Hall, C.E., Schwarz, P., Schwarz, J., Gillespie, J., 2003. Evaluation of hot water and electron beam irradiation for reducing Fusarium infection in malting barley. *Journal of Food Protection* 66, 1241–1246.

Kuiper-Goodman, T., 2004. Risk assessment and risk management of mycotoxins in food. *Mycotoxins in Food*, pp. 3–31. Elsevier. doi: 10.1533/9781855739086.1.3.

Lai, T., Wang, Y., Fan, Y., Zhou, Y., Bao, Y., Zhou, T., 2017. The response of growth and patulin production of postharvest pathogen Penicillium expansum to exogenous potassium phosphite treatment. *International Journal of Food Microbiology* 244, 1–10.

Laidou, I., Thanassoulopoulos, C., Liakopoulou-Kyriakides, M., 2001. Diffusion of patulin in the flesh of pears inoculated with four post-harvest pathogens. *Journal of Phytopathology* 149, 457–461.

Lee, T.P., Sakai, R., Manaf, N.A., Rodhi, A.M., Saad, B., 2014. High performance liquid chromatography method for the determination of patulin and 5-hydroxymethylfurfural in fruit juices marketed in Malaysia. *Food Control* 38, 142–149.

Li, Q., Li, C., Li, P., Zhang, H., Zhang, X., Zheng, X., Yang, Q., T. Apaliya, M., Nana Adwoa Serwah, B., Sun, Y., 2017a. The biocontrol effect of Sporidiobolus pararoseus Y16 against postharvest diseases in table grapes caused by Aspergillus niger and the possible mechanisms involved. *Biological Control* 113, 18–25.

Li, R., Zhang, H., Liu, W., Zheng, X., 2011. Biocontrol of postharvest gray and blue mold decay of apples with Rhodotorula mucilaginosa and possible mechanisms of action. *International Journal of Food Microbiology* 146, 151–156.

Li, T., Xu, Y., Zhang, L., Ji, Y., Tan, D., Yuan, H., Wang, A., 2017b. The Jasmonate-activated transcription factor MdMYC2 regulates Ethylene Response Factor and ethylene biosynthetic genes to promote ethylene biosynthesis during apple fruit ripening. *The Plant Cell* 29, 1316–1334.

Li, W., Zhang, H., Li, P., Apaliya, M.T., Yang, Q., Peng, Y., Zhang, X., 2016. Biocontrol of postharvest green mold of oranges by Hanseniaspora uvarum Y3 in combination with phosphatidylcholine. *Biological Control* 103, 30–38.

Li, X., Tang, H., Yang, C., Meng, X., Liu, B., 2019. Detoxification of mycotoxin patulin by the yeast Rhodotorula mucilaginosa. *Food Control* 96, 47–52.

Lima, G., Castoria, R., De Curtis, F., Raiola, A., Ritieni, A., De Cicco, V., 2011. Integrated control of blue mould using new fungicides and biocontrol yeasts lowers levels of fungicide residues and patulin contamination in apples. *Postharvest Biology and Technology* 60, 164–172.

Liu, B.-H., Wu, T.-S., Yu, F.-Y., Su, C.-C., 2006. Induction of oxidative stress response by the mycotoxin patulin in mammalian cells. *Toxicological Sciences* 95, 340–347.

Lukwago, F.B., Mukisa, I.M., Atukwase, A., Kaaya, A.N., Tumwebaze, S., 2019. Mycotoxins contamination in foods consumed in Uganda: A 12-year review (2006–2018). *Scientific African* 3, e00054.

Luz, C., Saladino, F., Luciano, F., Mañes, J., Meca, G., 2017. In vitro antifungal activity of bioactive peptides produced by Lactobacillus plantarum against Aspergillus parasiticus and Penicillium expansum. *LWT-Food Science and Technology* 81, 128–135.

Magan, N., Olsen, M., 2004. *Mycotoxins in Food: Detection and Control*. Woodhead Publishing, Cambridge, UK.

Mahunu, G.K., Zhang, H., Apaliya, M.T., Yang, Q., Zhang, X., Zhao, L., 2018. Bamboo leaf flavonoid enhances the control effect of Pichia caribbica against Penicillium expansum growth and patulin accumulation in apples. *Postharvest Biology and Technology* 141, 1–7.

Mahunu, G.K., Zhang, H., Yang, Q., Li, C., Zheng, X., 2015. Biological control of patulin by antagonistic yeast: A case study and possible model. *Critical Reviews in Microbiology* 42, 1–13.

Mahunu, G.K., Zhang, H., Yang, Q., Zhang, X., Li, D., Zhou, Y., 2016. Improving the biocontrol efficacy of Pichia caribbica with phytic acid against postharvest blue mold and natural decay in apples. *Biological Control* 92, 172–180.

Marín, S., Mateo, E.M., Sanchis, V., Valle-Algarra, F.M., Ramos, A.J., Jiménez, M., 2011. Patulin contamination in fruit derivatives, including baby food, from the Spanish market. *Food Chemistry* 124, 563–568.

Marin, S., Ramos, A., Cano-Sancho, G., Sanchis, V., 2013. Mycotoxins: Occurrence, toxicology, and exposure assessment. *Food and Chemical Toxicology* 60, 218–237.

Marroquín-Cardona, A., Johnson, N., Phillips, T., Hayes, A., 2014. Mycotoxins in a changing global environment–A review. *Food and Chemical Toxicology* 69, 220–230.

Matsuoka, K., 2019. Anthocyanins in Apple Fruit and Their Regulation for Health Benefits. Anthocyanins-Novel Antioxidants in Human Health and Diseases Prevention. IntechOpen.

McCallum, J., Tsao, R., Zhou, T., 2002. Factors affecting patulin production by Penicillium expansum. *Journal of Food Protection®* 65, 1937–1942.

Meng, X.-H., Qin, G.-Z., Tian, S.-P., 2010. Influences of preharvest spraying Cryptococcus laurentii combined with postharvest chitosan coating on postharvest diseases and quality of table grapes in storage. *LWT - Food Science and Technology* 43, 596–601.

Moake, M.M., Padilla-Zakour, O.I., Worobo, R.W., 2005. Comprehensive review of patulin control methods in foods. *Comprehensive Reviews in Food Science and Food Safety* 4, 8–21.

Morales, H., Marín, S., Centelles, X., Ramos, A.J., Sanchis, V., 2007a. Cold and ambient deck storage prior to processing as a critical control point for patulin accumulation. *International Journal of Food Microbiology* 116, 260–265.

Morales, H., Marín, S., Obea, L., Patiño, B., Doménech, M., Ramos, A.J., Sanchis, V., 2008a. Ecophysiological characterization of Penicillium expansum population in lleida (Spain). *International Journal of Food Microbiology* 122, 243–252.

Morales, H., Marín, S., Ramos, A.J., Sanchis, V., 2010. Influence of post-harvest technologies applied during cold storage of apples in Penicillium expansum growth and patulin accumulation: A review. *Food Control* 21, 953–962.

Morales, H., Marín, S., Rovira, A., Ramos, A.J., Sanchis, V., 2007b. Patulin accumulation in apples by Penicillium expansum during postharvest stages. *Letters in Applied Microbiology* 44, 30–35.

Morales, H., Sanchis, V., Usall, J., Ramos, A.J., Marín, S., 2008b. Effect of biocontrol agents Candida sake and Pantoea agglomerans on Penicillium expansum growth and patulin accumulation in apples. *International Journal of Food Microbiology* 122, 61–67.

Moran, C.A., 2004. Functional components of the cell wall of Saccharomyces cerevisiae: applications for yeast glucan and mannan. *Alltechs Annual Symposium*, pp. 283–296.

Moss, M., 1991. The environmental factors controlling mycotoxin formation. In Smith, J.E., and Anderson, R.A. (eds.) *Mycotoxins and Animal Foods*, pp. 37–56. CRC Press, Boca Raton, FL.

Mostafavi, H.A., Mirmajlessi, S.M., Fathollahi, H., Shahbazi, S., Mirjalili, S.M., 2013. Integrated effect of gamma radiation and biocontrol agent on quality parameters of apple fruit: An innovative commercial preservation method. *Radiation Physics and Chemistry* 91, 193–199.

Murillo-Arbizu, M., Amézqueta, S., González-Peñas, E., de Cerain, A.L., 2009. Occurrence of patulin and its dietary intake through apple juice consumption by the Spanish population. *Food Chemistry* 113, 420–423.

Neri, F., Donati, I., Veronesi, F., Mazzoni, D., Mari, M., 2010. Evaluation of Penicillium expansum isolates for aggressiveness, growth and patulin accumulation in usual and less common fruit hosts. *International Journal of Food Microbiology* 143, 109–117.

Novak, M., Vetvicka, V., 2008. β-glucans, history, and the present: Immunomodulatory aspects and mechanisms of action. *Journal of Immunotoxicology* 5, 47–57.

Ochiai, N., Fujimura, M., Oshima, M., Motoyama, T., Ichiishi, A., Yamada-Okabe, H., Yamaguchi, I., 2002. Effects of iprodione and fludioxonil on glycerol synthesis and hyphal development in Candida albicans. *Bioscience, Biotechnology, and Biochemistry* 66, 2209–2215.

Olsen, M., Lindqvist, R., Bakeeva, A., Su-lin, L.L., Sulyok, M., 2019. Distribution of mycotoxins produced by Penicillium spp. inoculated in apple jam and crème fraiche during chilled storage. *International Journal of Food Microbiology* 292, 13–20.

Organization, W.H., 2010. World Health Organization Working to overcome the global impact of neglected tropical disease. First WHO Report on Neglected Disease, World Health Organization, Geneva, Switzerland.

Oroian, M., Amariei, S., Gutt, G., 2014. Patulin in apple juices from the Romanian market. *Food Additives & Contaminants: Part B* 7, 147–150.

Park, D.L., Troxell, T.C., 2002. US perspective on mycotoxin regulatory issues. In DeVries, J.W., Trucksess, M.W., and Jackson, L.S. (eds.) *Mycotoxins and Food Safety*, pp. 277–285. Springer, Boston, MA.

Paster, N., Barkai-Golan, R., 2008. Mouldy fruits and vegetables as a source of mycotoxins: part 2. *World Mycotoxin Journal* 1, 385–396.

Paterson, R.R.M., 2007. Some fungicides and growth inhibitor/biocontrol-enhancer 2-deoxy-d-glucose increase patulin from Penicillium expansum strains in vitro. *Crop Protection* 26, 543–548.

Pitt, J., 2000. Toxigenic fungi and mycotoxins. *British Medical Bulletin* 56, 184–192.

Prusky, D., Barad, S., Luria, N., Ment, D., 2014. pH Modulation of host environment, a mechanism modulating fungal attack in postharvest pathogen interactions. In Prusky, D., and Gullino, M. (eds.) *Post-harvest Pathology*, pp. 11–25. Springer, Cham.

Prusky, D., McEvoy, J.L., Saftner, R., Conway, W.S., Jones, R., 2004. Relationship between host acidification and virulence of Penicillium spp. on apple and citrus fruit. *Phytopathology* 94, 44–51.

Puel, O., Galtier, P., Oswald, I.P., 2010. Biosynthesis and toxicological effects of patulin. *Toxins* 2, 613–631.

Qin, G., Zong, Y., Chen, Q., Hua, D., Tian, S., 2010. Inhibitory effect of boron against Botrytis cinerea on table grapes and its possible mechanisms of action. *International Journal of Food Microbiology* 138, 145–150.

Reddy, K.R., Abbas, H.K., Abel, C.A., Shier, W.T., Salleh, B., 2010. Mycotoxin Contamination of Beverages: Occurrence of Patulin in Apple Juice and Ochratoxin A in Coffee, Beer and Wine and Their Control Methods. *Toxins* 2(2), 229–261.

Reverberi, M., Fabbri, A., Zjalic, S., Ricelli, A., Punelli, F., Fanelli, C., 2005. Antioxidant enzymes stimulation in Aspergillus parasiticus by Lentinula edodes inhibits aflatoxin production. *Applied Microbiology and Biotechnology* 69, 207–215.

Reverberi, M., Zjalic, S., Ricelli, A., Punelli, F., Camera, E., Fabbri, C., Picardo, M., Fanelli, C., Fabbri, A.A., 2008. Modulation of antioxidant defense in Aspergillus parasiticus is involved in aflatoxin biosynthesis: A role for the ApyapA gene. *Eukaryotic Cell* 7, 988–1000.

Ricelli, A., Baruzzi, F., Solfrizzo, M., Morea, M., Fanizzi, F., 2007. Biotransformation of patulin by Gluconobacter oxydans. *Applied and Environmental Microbiology* 73, 785–792.

Riley, R.T., Showker, J.L., 1991. The mechanism of patulin's cytotoxicity and the antioxidant activity of indole tetramic acids. *Toxicology and Applied Pharmacology* 109, 108–126.

Romanazzi, G., Feliziani, E., 2016. Use of chitosan to control postharvest decay of temperate fruit: effectiveness and mechanisms of action. In Bautista-Baños, S., Romanazzi, G., and Jiménez-Aparicio, A. (eds.) *Chitosan in the Preservation of Agricultural Commodities*, pp. 155–177. Elsevier, Amsterdam.

Romanazzi, G., Sanzani, S.M., Bi, Y., Tian, S., Martínez, P.G., Alkan, N., 2016. Induced resistance to control postharvest decay of fruit and vegetables. *Postharvest Biology and Technology* 122, 82–94.

Rychlik, M., Kircher, F., Schusdziarra, V., Lippl, F., 2004. Absorption of the mycotoxin patulin from the rat stomach. *Food and Chemical Toxicology* 42, 729–735.

Salas, M.P., Reynoso, C.M., Céliz, G., Daz, M., Resnik, S.L., 2012. Efficacy of flavanones obtained from citrus residues to prevent patulin contamination. *Food Research International* 48, 930–934.

Sanzani, S.M., Reverberi, M., Punelli, M., Ippolito, A., Fanelli, C., 2012. Study on the role of patulin on pathogenicity and virulence of Penicillium expansum. *International Journal of Food Microbiology* 153, 323–331.

Saxena, N., Dwivedi, P.D., Ansari, K.M., Das, M., 2008. Patulin in apple juices: Incidence and likely intake in an Indian population. *Food Additives and Contaminants* 1, 140–146.

Shephard, G.S., van der Westhuizen, L., Katerere, D.R., Herbst, M., Pineiro, M., 2010. Preliminary exposure assessment of deoxynivalenol and patulin in South Africa. *Mycotoxin Research* 26, 181–185.

Smith, T., Girish, C., 2012. Prevention and control of animal feed contamination by mycotoxins and reduction of their adverse effects in livestock. *Animal Feed Contamination* 326–351.

Spadaro, D., Ciavorella, A., Dianpeng, Z., Garibaldi, A., Gullino, M.L., 2010. Effect of culture media and pH on the biomass production and biocontrol efficacy of a Metschnikowia pulcherrima strain to be used as a biofungicide for postharvest disease control. *Canadian Journal of Microbiology* 56, 128–137.

Spadaro, D., Ciavorella, A., Frati, S., Garibaldi, A., Gullino, M., 2007. Incidence and level of patulin contamination in pure and mixed apple juices marketed in Italy. *Food Control* 18, 1098–1102.

Spadaro, D., Lorè, A., Garibaldi, A., Gullino, M.L., 2013. A new strain of Metschnikowia fructicola for postharvest control of Penicillium expansum and patulin accumulation on four cultivars of apple. *Postharvest Biology and Technology* 75, 1–8.

Speijers, G.J.A., Speijers, M.H.M., 2004. Combined toxic effects of mycotoxins. *Toxicology Letters* 153, 91–98.

Spring, P., Fegan, D.F., 2005. Mycotoxins–a rising threat to aquaculture. *Feedmix* 13, 5–9.

Stevens, C., Khan, V., Lu, J., Wilson, C., Pusey, P., Kabwe, M., Igwegbe, E., Chalutz, E., Droby, S., 1998. The germicidal and hormetic effects of UV-C light on reducing brown rot disease and yeast microflora of peaches. *Crop Protection* 17, 75–84.

Stevens, C., Khan, V.A., Wilson, C.L., Lu, J.Y., Chalutz, E., Droby, S., 2005. The effect of fruit orientation of postharvest commodities following low dose ultraviolet light-C treatment on host induced resistance to decay. *Crop Protection* 24, 756–759.

Stinson, E., Osman, S., Bills, D., 1979. Water-soluble products from patulin during alcoholic fermentation of apple juice. *Journal of Food Science* 44, 788–789.

Stott, W., Bullerman, L., 1975. Patulin: A mycotoxin of potential concern in foods. *Journal of Milk and Food Technology* 38, 695–705.

Sylos, C.M.D., 1999. Incidence of patulin in fruits and fruit juices marketed in Campinas, Brazil. *Food Additives & Contaminants* 16, 71–74.

Tang, J., Liu, Y., Li, H., Wang, L., Huang, K., Chen, Z., 2015. Combining an antagonistic yeast with harpin treatment to control postharvest decay of kiwifruit. *Biological Control* 89, 61–67.

Tian, S.P., Yao, H.J., Deng, X., Xu, X.B., Qin, G.Z., Chan, Z.L., 2007. Characterization and expression of β-1, 3-glucanase genes in jujube fruit induced by the microbial biocontrol agent Cryptococcus laurentii. *Phytopathology* 97, 260–268.

Tikekar, R.V., 2009. Ultraviolet light induced degradation of patulin and ascorbic acid in apple juice.

Tolaini, V., Zjalic, S., Reverberi, M., Fanelli, C., Fabbri, A.A., Del Fiore, A., De Rossi, P., Ricelli, A., 2010. Lentinula edodes enhances the biocontrol activity of Cryptococcus laurentii against Penicillium expansum contamination and patulin production in apple fruits. *International Journal of Food Microbiology* 138, 243–249.

Tournas, V., Katsoudas, E., 2019. Effect of $CaCl_2$ and various wild yeasts from plant origin on controlling penicillium expansum postharvest decays in golden delicious apples. *Microbiology Insights* 12. doi: 10.1177/1178636119837643.

Valletrisco, M., Niola, I., Stefanelli, C., 1988. Decontamination tests on fruit juices contaminated with patulin. In Walker, R., and Quattrucci, E. (eds.) *Nutritional and Toxicological Aspects of Food Processing: Proceedings of an International Symposium Held at the Istituto Superiore di Sanita*, Rome, Italy, 14–16 April 1987. Taylor & Francis, London.

Van Egmond, H., Jonker, M., Rivka, B., 2008. Regulations and limits for mycotoxins in fruits and vegetables. *Mycotoxins in Fruits and Vegetables* 1, 45–74.

Varga, J., Kocsubé, S., Suri, K., Szigeti, G., Szekeres, A., Varga, M., Tóth, B., Bartók, T., 2010. Fumonisin contamination and fumonisin producing black Aspergilli in dried vine fruits of different origin. *International Journal of Food Microbiology* 143, 143–149.

Wang, L., Wang, Z., Yuan, Y., Cai, R., Niu, C., Yue, T., 2015. Identification of key factors involved in the biosorption of patulin by inactivated lactic acid bacteria (LAB) cells. *PloS one* 10, e0143431.

Wang, Y., Li, Y., Xu, W., Zheng, X., Zhang, X., Abdelhai, M.H., Zhao, L., Li, H., Diao, J., Zhang, H., 2018. Exploring the effect of β-glucan on the biocontrol activity of Cryptococcus podzolicus against postharvest decay of apples and the possible mechanisms involved. *Biological Control* 121, 14–22.

Wang, Y., Tang, F., Xia, J., Yu, T., Wang, J., Azhati, R., Zheng, X.D., 2011. A combination of marine yeast and food additive enhances preventive effects on postharvest decay of jujubes (Zizyphus jujuba). *Food Chemistry* 125, 835–840.

Watanabe, M., Shimizu, H., 2005. Detection of patulin in apple juices marketed in the Tohoku District, Japan. *Journal of Food Protection* 68, 610–612.

Welke, J.E., Hoeltz, M., Dottori, H.A., Noll, I.B., 2009. Effect of processing stages of apple juice concentrate on patulin levels. *Food Control* 20, 48–52.

Wisniewski, M., Droby, S., Norelli, J., Liu, J., Schena, L., 2016. Alternative management technologies for postharvest disease control: the journey from simplicity to complexity. *Postharvest Biology and Technology* 122, 3–10.

Yahia, E.M., García-Solís, P., Celis, M.E.M., 2019. Contribution of fruits and vegetables to human nutrition and health. In Yahia, E., and Carrillo-Lopez, A. (eds.) *Postharvest Physiology and Biochemistry of Fruits and Vegetables*, pp. 19–45. Elsevier, Amsterdam.

Yang, Q., Wang, H., Zhang, H., Zhang, X., Apaliya, M.T., Zheng, X., Mahunu, G.K., 2017. Effect of Yarrowia lipolytica on postharvest decay of grapes caused by Talaromyces rugulosus and the protein expression profile of T. rugulosus. *Postharvest Biology and Technology* 126, 15–22.

Yang, Q., Zhang, H., Zhang, X., Zheng, X., Qian, J., 2015. Phytic Acid Enhances Biocontrol Activity of *Rhodotorula mucilaginosa* against *Penicillium expansum* Contamination and Patulin Production in Apples. *Frontiers in Microbiology* 6, 1296.

Yiannikouris, A., Francois, J., Poughon, L., Dussap, C.-G., Bertin, G., Jeminet, G., Jouany, J.-P., 2004. Adsorption of Zearalenone by beta-D-glucans in the Saccharomyces cerevisiae cell wall. *Journal of Food Protection®* 67, 1195–1200.

Yiannikouris, A., Jouany, J.-P., Bertin, G., Lyons, T., Jacques, K., Hower, J., 2007. Counteracting mycotoxin contamination: the effectiveness of Saccharomyces cerevisiae cell wall glucans in Mycosorb® for sequestering mycotoxins. *Nutritional Biotechnology in the Feed and Food Industries: Proceedings of Alltech's 23rd Annual Symposium*. The new energy crisis: food, feed or fuel, pp. 11–19.

Youssef, K., Sanzani, S.M., Ligorio, A., Ippolito, A., Terry, L.A., 2014. Sodium carbonate and bicarbonate treatments induce resistance to postharvest green mould on citrus fruit. *Postharvest Biology and Technology* 87, 61–69.

Yu, T., Zhang, H., Li, X., Zheng, X., 2008. Biocontrol of Botrytis cinerea in apple fruit by Cryptococcus laurentii and indole-3-acetic acid. *Biological Control* 46, 171–177.

Yue, T., Guo, C., Yuan, Y., Wang, Z., Luo, Y., Wang, L., 2013. Adsorptive removal of patulin from apple juice using Ca-alginate-activated carbon beads. *Journal of Food Science* 78, T1629–T1635.

Zaied, C., Zouaoui, N., Bacha, H., and Abid, S., 2012. Natural occurrence of zearalenone in Tunisian wheat grains. *Food Control* 25(2), 773–777.

Zaied, C., Abid, S., Hlel, W., Bacha, H., 2013. Occurrence of patulin in apple-based-foods largely consumed in Tunisia. *Food Control* 31, 263–267.

Zhang, H., Chen, L., Sun, Y., Zhao, L., Zheng, X., Yang, Q., Zhang, X., 2017a. Investigating proteome and transcriptome defense response of apples induced by Yarrowia lipolytica. *Molecular Plant-Microbe Interactions* 30, 301–311.

Zhang, H., Liu, Z., Xu, B., Chen, K., Yang, Q., Zhang, Q., 2013. Burdock fructooligosaccharide enhances biocontrol of Rhodotorula mucilaginosa to postharvest decay of peaches. *Carbohydrate Polymers* 98, 366–371.

Zhang, H., Ma, L., Turner, M., Xu, H., Zheng, X., Dong, Y., Jiang, S., 2010. Salicylic acid enhances biocontrol efficacy of Rhodotorula glutinis against postharvest Rhizopus rot of strawberries and the possible mechanisms involved. *Food Chemistry* 122, 577–583.

Zhang, H., Ma, L., Wang, L., Jiang, S., Dong, Y., Zheng, X., 2008. Biocontrol of gray mold decay in peach fruit by integration of antagonistic yeast with salicylic acid and their effects on postharvest quality parameters. *Biological Control* 47, 60–65.

Zhang, X., Zhang, G., Li, P., Yang, Q., Chen, K., Zhao, L., Apaliya, M.T., Gu, X., Zhang, H., 2017b. Mechanisms of glycine betaine enhancing oxidative stress tolerance and biocontrol efficacy of Pichia caribbica against blue mold on apples. *Biological Control* 108, 55–63.

Zhang, Y., Liu, Y., Xing, F., 2009. Domestic production status of concentrated apple juice and harm of patulin on its quality. *Food Science and Technology* 34, 54–57.

Zheng, X., Yang, Q., Zhang, X., Apaliya, M.T., Ianiri, G., Zhang, H., Castoria, R., 2017. Biocontrol agents increase the specific rate of patulin production by Penicillium expansum but Decrease the disease and total patulin contamination of apples. *Frontiers in Microbiology* 8, 1240.

Zhu, R., Feussner, K., Wu, T., Yan, F., Karlovsky, P., Zheng, X., 2015. Detoxification of mycotoxin patulin by the yeast Rhodosporidium paludigenum. *Food Chemistry* 179, 1–5.

Zhu, R., Lu, L., Guo, J., Lu, H., Abudureheman, N., Yu, T., Zheng, X., 2012. Postharvest Control of green mold decay of citrus fruit using combined treatment with sodium bicarbonate and rhodosporidium paludigenum. *Food and Bioprocess Technology* 6, 1–6.

Zhu, Y., Koutchma, T., Warriner, K., Zhou, T., 2014. Reduction of patulin in apple juice products by UV light of different wavelengths in the UVC range. *Journal of Food Protection* 77, 963–971.

Zhu, Z., Zhang, X., 2016. Effect of harpin on control of postharvest decay and resistant responses of tomato fruit. *Postharvest Biology and Technology* 112, 241–246.

Zhu, Z., Zhang, Z., Qin, G., Tian, S., 2010. Effects of brassinosteroids on postharvest disease and senescence of jujube fruit in storage. *Postharvest Biology and Technology* 56, 50–55.

Zoghi, A., Khosravi-Darani, K., Sohrabvandi, S., Attar, H., Alavi, S.A., 2017. Effect of probiotics on patulin removal from synbiotic apple juice. *Journal of the Science of Food and Agriculture* 97, 2601–2609.

12 The Montreal Protocol and the Methyl Bromide Phaseout in the Soil Sector: Key Success Factors and Lessons Learned to Eliminate Synthetic Pesticide Use in Africa

Mohamed Besri
Hassan II Institute of Agronomy and Veterinary Medicine, Rabat Morocco

CONTENTS

12.1 INTRODUCTION

Molina and Rowland (1974), from the University of California, USA, found evidence that stratospheric ozone was being broken down by volatile man-made ozone depleting substances (ODSs); their prediction turned out to be of enormous environmental importance and earned them a Nobel Prize in chemistry in 1995. On September 16, 1987, governments around the world agreed on the Montreal Protocol on Substances That Deplete the Ozone Layer (chlorofluorocarbons = CFCs, hydrochlorofluorocarbons = HCFCs, hydrobromofluorocarbons = HBFCs, halons, carbon tetrachloride, and methyl chloroform) to protect human health and the environment against depletion of the stratospheric ozone layer resulting from human activities. The Montreal Protocol (MP) has been ratified by 197 parties (196 states and the European Union), making it the first universally ratified treaty in United Nations' history. Every year, September 16 is observed as the International Day for the Preservation of the Ozone Layer across the world. The celebrations are also known as World Ozone Day.

Methyl bromide (MB) was later classified as an ODS under the Copenhagen Amendment of the MP in 1992 (Andersen et al. 2007, UNEP 2019a, b).

The first use of MB as a soil fumigant occurred in France in the 1930s (MBTOC 2019). Since its discovery and implementation, MB has widely been used as a soil fumigant, in structures, stored commodities, and continues to be used for quarantine and preshipment treatments. This chemical controls a wide range of pests, including pathogens (fungi, bacteria, viruses, and nematode), insects, mites, weeds, and rodents (MBTOC 2019). The major crops using MB as a soil fumigant worldwide are vegetables (such as tomatoes, peppers, and eggplant), cucurbits (such as melon, watermelon, and squash), strawberries fruits, ornamentals, orchards (for replant disease), and nurseries (including strawberry runners and forest seedlings) (MBTOC 2019).

In this chapter, we will focus on the MP and on the MB phaseout in the soil sector in both developing (Article 5) and developed (non-Article 5) countries, on the reasons of its success and on the lessons that could be learned from this treaty and that could be applied to phase out other dangerous chemical substances used in agriculture such as synthetic pesticides, particularly in Africa.

12.2 IMPACT OF THE OZONE LAYER DEPLETION
ON MAN AND HIS ENVIRONMENT

Located in the stratosphere, the ozone layer acts as a shield to protect life on earth from the damaging effects of UV radiation type B. Intensive use of ODSs including MB has led to severe damage of the ozone layer and dramatic increases in radiation. The resultant elevated radiation levels have seriously affected human health by increasing risks of skin cancer and the incidence of eye cataracts, by suppressing the immune system. Increased UV can also lead to reduced crop yield and disruptions in the marine food chain (Hegglin et al. 2015).

A number of concerns over MB apart from ozone depletion have also led countries to impose severe restrictions on its production and use. These concerns include residues in food, toxicity to humans, and associated operator safety, public health, and detrimental effects on soil biodiversity. In some countries, pollution of surface and groundwater by MB and its bromide ion is also a concern (EPA 1992, EU 2016).

12.3 THE MONTREAL PROTOCOL AND THE METHYL BROMIDE PHASE-OUT

12.3.1 METHYL BROMIDE USES CATEGORIES IDENTIFIED BY THE PROTOCOL

The Protocol refers to four main categories of MB uses, and each is subject to different legal requirements (Table 12.1)

Two of the categories, the non-QPS fumigant uses (category 1 in Table 12.1) and laboratory and analytical uses (3), are subject to the phaseout schedules (Table 12.2), with authorized CUEs. The other two categories (2 and 4) of MB uses (QPS and feedstock used in industrial processes) are not subject to phaseout schedules but are subject to reporting requirements under the Protocol (UNEP 2019a, b).

12.3.2 METHYL BROMIDE PHASEOUT SCHEDULE

In 1997, a global phaseout schedule of this chemical was established by the Meeting of the Parties (MOP) according to which developing countries (Article 5 countries) are required to freeze consumption and production of MB by 2002, reduce its use by 20% in 2005, and complete total phaseout by 2015. In comparison, developed countries (non-Article 5 countries) must phase out most uses of MB by 2005 (UNEP 2019a) (Table 12.2). Quarantine and preshipment use of MB (QPS) was exempted from controls (UNEP 2019a, b).

12.3.3 CRITICAL USE NOMINATIONS

In 1997, the Parties to the MP recognized a justifiable need for transitional access to MB and adopted a formal decision to allow limited CUEs. Since 2005, exemptions have been allowed for "Critical Uses" in non-A5 Parties and since 2015 in A5 Parties.

Exemptions were to be granted only when the following criteria defined by the Decision IX/6 were met (UNEP 2019a):

1. Failure to provide access to MB would result in a significant market disruption.
2. There were no technically and economically feasible alternative available to an applicant that is acceptable from environmental and human health standpoints.

TABLE 12.1
Classification of Methyl Bromide Uses Under the Montreal Protocol

Methyl Bromide Uses	Status Under the Montreal Protocol
1-Non-QPS fumigant uses	Subject to production and consumption *phaseout schedules*, trade and licensing controls, and to data reporting requirements. Critical use exemptions can be authorized by the MOP for specific uses.
2-QPS fumigant uses	Exempted from reduction and phaseout schedules. Subject to data reporting requirements
3-Laboratory and analytical uses	Subject to production and consumption *phaseout schedules* except for the specific critical use exemptions. Subject to data reporting
4-Feedstock used in the manufacture of other chemicals	Exempted from phaseout schedule. Subject to data reporting requirements

Source: Ozone Secretariat, UNEP (2019a, b).
QPS, quarantine and preshipment; MOP, meeting of the parties.

TABLE 12.2

Phaseout Schedules Agreed at the Ninth Meeting of the Parties in 1997

Year	Non-Article 5 Countries	Article 5 Countries
1991	Consumption/production baseline	
1995	Freeze	
1995–1998 average		Consumption/production baseline
1999	25% reduction	
2001	50% reduction	
2002		Freeze
2003	70% reduction	Review of reductions
2005	*Phaseout with provision for CUEs*	20% reduction
2015		*Phaseout with provision for CUEs*

Source: Ozone Secretariat UNEP (2019a). Montreal Protocol Handbook.

3. The applicant had taken all feasible steps to minimize their use of MB and the associated emissions.
4. Appropriate efforts were being made to evaluate, commercialize, and register alternatives to MB for use by the applicant.

The Party submits the CUN on behalf of the country MB users to the Ozone Secretariat of the United Nations Environmental Program (UNEP), which is the administrative oversight body for the MP. The Ozone Secretariat then forwards all CUN's to the MBTOC. MBTOC provides recommendations on such nominations, for review and endorsement by the TEAP and then consideration by the Parties. Parties review MBTOC and TEAP's recommendations and may accept, reject, or modify these recommendations when making decisions on CUE requests (TEAP 2019b).

The exemption application process is extremely rigorous. When the requirements of Decision IX/6 are met, MBTOC can recommend critical uses of MB. When the requirements of Decision IX/6 are not met, MBTOC does not recommend critical uses of MB. Where some of the conditions are not fully met, MBTOC can recommend a reduced amount depending on its technical and economic evaluation or determine the CUN as "unable to assess" and request further information from the party. When the information is submitted, MBTOC is required to reassess the nomination.

12.3.4 ALTERNATIVES TO METHYL BROMIDE FOR SOIL USE

MBTOC defined "alternatives" as follows:

'any practice or treatment that can be used in place of methyl bromide. 'Existing alternatives' are those alternatives in present or past use in some regions. 'Potential alternatives' are those in the process of investigation or development.

MBTOC assumed that an alternative demonstrated in one region of the world would be technically applicable in another unless there were obvious constraints to the contrary e.g., a very different climate or pest complex.

(MBTOC 2019)

12.3.4.1 Chemical Alternatives

Soil fumigation is a widespread practice to control soilborne pathogens of many crops, e.g., tomatoes, carrots, tobacco, strawberries, and Cucurbitaceae. After the phasing out of MB, many old and new chemical alternatives including 1,3-dichloropropene, methyl isothiocyanate generators

TABLE 12.3

General Health Issues Associated to Soils Fumigants

Fumigant	Health Issues
Methyl bromide	Mutagenic potential
	Highly toxic
	Brain, kidney, respiratory toxicant
	IARC Group 3 carcinogen
1,3 D	Highly toxic
	Mutagenic potential
	Possible urinary, liver, and kidney toxicant
Dazomet	Skin and eye irritant
Metham sodium	Possible liver and urinary tract toxicant
	USEPA –probable human carcinogen
DMDS	Toxic by ingestion
	Inhalation may cause nausea, headache, and dizziness
SF	Prolonged exposure may cause pulmonary edema, nausea, and abdominal pain
	Possible kidney, CNS, blood, and bone toxicant

EU (2016)

(metam sodium/potassium and dazomet), chloropicrin, methyl iodide, and DMDS have been developed and commercially used (MBTOC 2019).

However, in many A5 and non-A5 countries, the number of chemical alternatives available for plant protection has dramatically been reduced after their reevaluation. These reevaluations have shown that some soil fumigants present a high risk of toxicity to humans and to their environment (phytotoxicity, pollution of underground water, pesticide residues, etc.). (ATSDR 1992, EU 2016, De Souza et al. 2013). General health issues reported in reference to fumigants used in the EU are summarized in Table 12.3.

12.3.4.2 Nonchemical Alternatives

Nonchemical alternatives such as substrates-resistant varieties and grafting, biofumigation, steam solarization, anaerobic soil disinfestations (ASD), biological control hot water, trap cropping, microwave radiations, and biological control continue to expand and replace MB in non-A5 and A5 countries as their technical and economical feasibility improves. Nonchemical alternatives to MB have also gained importance because of the negative health and environmental issues of most chemical alternatives (MBTOC 2019).

12.3.4.3 Integrated Pest Management

The combination of chemical and nonchemical alternatives in an Integrated Pest Management (IPM) program provides excellent results in the longer term (MBTOC 2019).

12.4 THE MONTREAL PROTOCOL AND ITS KEY IMPLEMENTATION STRATEGIES

The singular success of the MP belongs to the way its implementation was designed and particularly:

a. The creation and organization of international assessment panels
b. The identification of MB use categories and their legal requirements
c. The establishment of an MB phaseout schedules accepted by all the parties

    d. The setting up of a transitional time to adopt alternatives by allowing critical use nominations (CUNs) and exemptions (CUEs)

    e. The organization of networks of National Ozone Officers in developing countries

    f. The creation of a Multilateral Fund (MLF) to help developing countries (A5 countries) demonstrating and implementing alternatives to MB.

12.4.1   UNEP ASSESSMENT PANELS

To phase out all the ODSs, the MP has established three panels.

12.4.1.1   Technology and Economic Assessment Panel

The Technology and Economic Assessment Panel (TEAP) was established as the technology and economic advisory body to the MP Parties. TEAP provides, at the request of Parties, technical information related to the alternative technologies that have been investigated and employed to make it possible to virtually eliminate the use of ODSs (https://ozone.unep.org/science/assessment/teap).

TEAP has five Technical Option Committees (TOC): Foams TOC, Halons TOC, Medical and Chemical TOC, Refrigeration TOC, and *Methyl Bromide TOC*. The Methyl Bromide Technical Options Committee (MBTOC) was established in 1992 by the Parties to the Montreal Protocol on Substances That Deplete the Ozone Layer to identify existing and potential alternatives to MB. MBTOC provides recommendations and advices to the Parties on the technical and economic feasibility of chemical and nonchemical alternatives for controlled uses of MB. Additionally, from 2003, MBTOC has had the task of evaluating Critical Use Nominations submitted by non-Article 5 Parties to the MP and by Article 5 Parties from 2014. Currently, MBTOC has 18 members originated from 13 developing and developed countries.

12.4.1.2   Scientific Assessment Panel

The Scientific Assessment Panel (SAP) assesses the status of the depletion of the ozone layer and relevant atmospheric science issues. A report is prepared every 4years by the SAP by hundreds of top scientists from around the world (https://ozone.unep.org/science/assessment/sap.)

12.4.1.3   Environmental Effects Assessment Panel

The Environmental Effects Assessment Panel (EEAP) is one of the three Panels of experts that inform the Parties to the MP. The EEAP focuses on the effects of UV radiation on human health, terrestrial and aquatic ecosystems, air quality, and materials, as well as on the interactive effects of UV radiation and global climate change. Like the other two Panels, the EEAP produces detailed quadrennial reports every 4 years (https://ozone.unep.org/science/assessment/eeap).

12.4.2   NETWORKS OF NATIONAL OZONE OFFICERS IN DEVELOPING COUNTRIES

12.4.2.1   National Ozone Officers and Regional Networks

The strategy developed by the MP was that every one of the 147 developing countries should have a focal point headed by a National Ozone Officer (NOO) for implementation of the Protocol. All the focal points were then connected to constitute regional networkings (UNEP 2013). The world was divided into ten regions (South America, Central America, Caribbean, Europe, Anglophone Africa, Francophone Africa, Central Asia, West Asia, South East Asia, and Pacific Islands countries). Within these regions, the focal points developed their own national programs for demonstrating and implementing alternatives to MB and for sharing their experiences within the region during regional meetings. So, it was learning by sharing and learning by exchanging views. The phaseout schedule using alternatives to MB required cooperation, coordination between the countries in the region (UNEP 2013).

Incentives were developed to reward countries who have developed and implemented alternatives to MB and phased out the fumigant according to the MP schedule during international meetings to encourage them to step out of their efforts (UNEP 2013).

12.4.2.2 Networks of Ozone Officers in Africa

In Africa, there are two networks: one for the English-speaking countries (28 countries) and the other for the French-speaking countries (26 members). The Network for English-speaking Africa comprises also two Portuguese-speaking countries (Angola and Mozambique), and the French network includes one Spanish-speaking country (Guinea Equatorial) (UNEP 2013).

The main objective of these African Networks of Ozone Officers is to assist countries in the region to meet and to sustain compliance with the MP and its Amendments. At the moment, all African countries are in compliance with the MP. Some countries eliminated MB use well ahead of the 2015 deadline, which shows political will and implementation capability.

South–South cooperation has also increased within the region. Many Ozone Officers are or have been used as resource persons/consultants by Implementing Agencies to assist neighboring countries with preparations and implementation of projects for phaseout.

12.4.3 MULTILATERAL FUND

12.4.3.1 Methyl Bromide Alternatives Project

In 1990, the international community established a financial mechanism called the Multilateral Fund (MLF) to support efforts in developing countries to phase out ODSs under the MP. The Fund is financed by contributions from industrialized countries (Non-A5 countries). The MLF's activities are implemented by four international agencies: UN Environment Program (UNEP), UN Development Program (UNDP), UN Industrial Development Organization (UNIDO), and the World Bank as well as bilateral agencies of non-Article 5 countries and farmers, exporters associations, or private enterprises. The MLF supported developing countries to adopt chemical and nonchemical alternatives and to completely phaseout MB for controlled uses by January 1, 2015 (MLF 2020).

Throughout the implementation of the MP, most developing countries and particularly African countries have completely phased MB before 2015 and have demonstrated that they are able to be full partners in global efforts to protect the environment.

12.4.3.2 Lessons Learned from MLF Projects

The phaseout of MB for controlled uses in the soil sector has been completed in all A5 countries and particularly in Africa. The implementation of MLF projects has provided many useful experiences that can be summarized as follows (MBTOC 2019, TEAP 2019b):

- Technically effective alternatives to MB have been found for virtually all soilborne pathogens; however, a very small number of sectors and situations still pose challenges, for example, tomato and strawberry fruits in Argentina, strawberry runners in Australia and Canada.
- The cost and profitability of alternatives were found to be acceptable or comparable to MB in many MLF projects.
- While a number of projects have promoted alternatives that will be environmentally sustainable in the longer term (such as IPM and nonchemical approaches), some projects focused primarily on chemical alternatives. Chemical treatments, particularly fumigants, are likely to face increasing regulatory restrictions worldwide in the future.
- The involvement of an ample range of key stakeholders was essential to the success of a project.

12.5   MONTREAL PROTOCOL SUCCESSES

12.5.1   TRENDS IN METHYL BROMIDE USE FOR CUNS

First, January 2015 marked the phaseout deadline for controlled uses of MB in A5 Parties, 10years after non-A5 Parties. As of that date, controlled uses of MB are only allowed under the CUE. By the end of 2019, official reporting indicated that about 98% of the global consumption baseline (all Parties) for controlled (nonexempt) uses had been replaced with alternatives.

Since 2005, the amount of MB requested for CUNs has fallen from 18,700t to 89.57t for 2019. In 2020, MBTOC received six nominations for critical use from four Parties for use of 89.57t. This is in clear contrast with over 100 CUNs submitted in 2003 for 2005 use when the critical use process began for non-A5 Parties (TEAP 2019b) (Figure 12.1).

Complete phaseout has been achieved in many A5 countries and particularly in African countries in advance of the 2015 deadline, e.g., Kenya, Zimbabwe, Egypt, Botswana, and Morocco.

Early MB phaseout for some crops in some A5 countries has proven beneficial due to the consequent improvement in the production practices. This has increased the competitiveness of agricultural products in international markets and provided training for large numbers of growers, technical staff, and other key stakeholders, e.g., flower production in Kenya; vegetable, banana, ornamentals, and strawberry production in Morocco; and strawberry and vegetables in Lebanon.

12.5.2   RECOVERING THE OZONE LAYER

The objective of the MP is the protection of the ozone layer through control of the global production and consumption of ODSs. Projections of the future abundances of ODSs expressed as Equivalent Effective Stratospheric Chlorine (EESC) values are shown in Figure 12.2. Without the Protocol and its amendments (London, Beijing, Copenhagen), EESC values are projected to have increased significantly in the 21st century (Hegglin et al. 2015). The implication of this would have been hazardous and have been estimated to include: 20 million more cases of cancer and 130 million more cases of eye cataracts if there have been no MP. ODSs are also potent greenhouse gases, and the protocol has delivered substantial climate benefits too.

In 2003, former UN Secretary-General Kofi Annan called the MP, ratified in 1987 by all 197 UN member states, the "single most successful international agreement to date." The ozone layer is expected to return to 1980 levels between 2050 and 2070 as long as all countries continue to meet

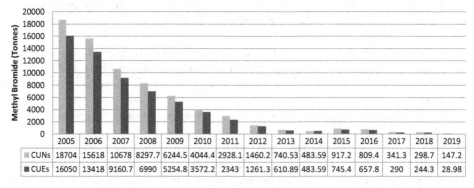

FIGURE 12.1 Amounts of methyl bromide nominated and exempted for CUE uses in nominated preplant soil and commodities sectors from 2005 to 2019 by non-A5 and A5 countries. (Ozone Secretariat Data Access Centre, accessed April 2020.)

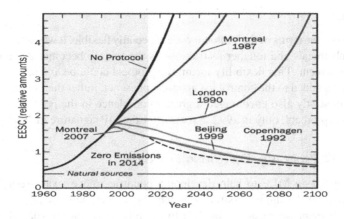

FIGURE 12.2 Effect of the Montreal Protocol long-term changes in equivalent effective stratospheric chlorine (EESC). (Hegglin et al. 2015.)

their obligations and phase out the last ODSs in the next few years and particularly MB. Phasing out ODSs has also benefited the environment more broadly, as many ODSs (TEAP 2019a) also have high global warming potential.

12.5.3 Remaining and Emerging Challenges Impacting MB Phaseout for Soil Use

The CUNs presented in 2020 by two non-A5 countries for strawberry runners production (Canada and Australia) and one Article 5 country for tomato production cited several categories of reasons for CUNs (TEAP 2019b):

- Absence of identified alternatives: e.g., resistant cultivars and rootstocks to a broad spectrum of pathogens
- Lack of chemical registration by regulatory authorities
- The cost and length of time required for obtaining registration
- Insufficient time to develop the necessary infrastructure: e.g., commercial nurseries
- Lack of training in the use of alternative and adaptation of the process to local conditions: e.g., soil less culture and biofumigation
- Unsuitability of available alternatives for local conditions: e.g., solarization
- Longer time between fumigation and planting (plant back periods) with the use of some alternatives, causing disruption to cropping programs: e.g., metam sodium, dazomet, and chloropicrin
- Available and suitable alternatives are not economically viable: e.g., propagative materials, steaming, soilless cultivation, and electric energy availability at farm level

The availability of alternatives, including those already adopted and those under development, could change in the medium to long term due to a number of issues including regulatory restrictions from environmental and health issues, increases in energy usage and application costs, etc.

12.6 LESSONS LEARNED FROM THE MONTREAL PROTOCOL THAT COULD BE APPLIED FOR THE SYNTHETIC PESTICIDE PHASEOUT IN AFRICA

Many lessons can be learned from the MP and can be applied to phase out other dangerous chemical substances such as synthetic pesticides in Africa. It is a credit to governments, industry, environmental groups, science and technical experts that such an instrument is even in existence and doing such a great job. The success of the MP is due to a mixture of reasons.

12.6.1 Flexibility

On the structure side, the terms of the Protocol were especially flexible. It was written with the ability to include more chemicals and tougher restrictions as the science becomes clearer, increasing the potency of the agreement. This flexibility meant the protocol could be amended to include stricter controls: more ODSs added to the control list and total phaseout, rather than partial phaseout, called for. Starting out modestly also encouraged a greater confidence in the process. For example, MB was included in the protocol, only in 1992, 5 years after the MP signature in 1987.

12.6.2 Financing Projects in Developing Countries

The Protocol included an MLF to help developing countries move away from MB. It provides incremental funding for developing countries to help them meet their compliance targets, to build capacity to implement phaseout activities, and to establish regional networks so they can share experiences and learn from each other

About 25% of the projects implemented and funded by the MLF were in Africa (Table 12.4).

12.6.3 Industry Cooperation

Chemical and seed companies accepted, encouraged the MB phaseout and developed new chemicals with no-ozone-depleting potential, new application techniques, new resistant varieties, and rootstocks, etc. to control most of the soil-borne pathogens. The MP also provided a stable framework that allowed industry to plan long-term research and innovation, develop new formulations with less MB and new chemicals (MBTOC 2019).

12.6.4 Farmers' Involvement

Alternatives to MB for soil disinfestations have been developed, demonstrated, accepted, and implemented in close collaboration with farmers. These alternatives have never been imposed on users (MBTOC 2019).

12.6.5 Health Reasons and Citizen Action

The inhabitants of the planet have all been made aware of the dangers of the destruction of the ozone layer on man and his environment. In many countries, demonstrations calling for a boycott of agricultural products produced with MB have been regularly organized.

TABLE 12.4
Projects Implemented by A5 Region with MLF Funding

Region	No. of Projects	Impact (ODP t)	Phased-Out (ODP t)	USD Approved
Africa	108	1,846.52	1,804.90	34,845,729
Asia and the Pacific	104	2,287.25	2,394.20	36,719,190
Europe	33	551.50	551.50	9,674,915
Latin America and Caribbean	129	3,695.60	3,710.00	54,955,552
Global	15	-	-	1,004,063
Total	**399**	8,380.87	8,460.60	137,199,449

Source: Multilateral Fund Secretariat (MLF 2020).
ODP, ozone depleting potential (tonnes).

All the governments saw more reasons to comply with the agreement than ignore it. Many international and national agencies have predicted that if nothing was done about the hole in the ozone, people would suffer millions of skin cancer deaths and millions more cases of cataracts in the coming century.

As previously reported, every year, September 16 is observed as the International Day "World Ozone Day" for the preservation of the ozone layer across the world.

12.6.6 Encouragement of Research

Some alternatives currently proposed by the MP even old (crop rotation, anaerobic soil disinfestations, resistant varieties, grafting, solarization, steaming, etc.) are still efficient in controlling soilborne pathogens, but their use was not encouraged because of the availability of MB. Other new alternatives were developed (microwaves, etc.) (MBTOC 2019).

12.6.7 Technology and Economic Assessment Panel

Another reason of the success of the MP was the establishment of expert, independent TEAP. TEAP has helped signatories reach solid and timely decisions on often-complex matters. They have given countries confidence to start their transition (MBTOC 2019).

12.6.8 Compliance Procedure

This was designed from the outset as a nonpunitive procedure. It prioritized helping wayward countries back into compliance. Developing countries work with a UN agency, the world bank, or a developed country to prepare an action plan to get them back into compliance. If necessary, resources from the MLF were made available for some short-term projects.

These are the main reasons why the MP has been so successful. These lessons could serve as a guideline for phasing out other dangerous chemicals such as synthetic pesticides.

12.7 CONCLUSION

Following MB's listing as a controlled ODS under the MP in 1992, many countries have undertaken activities to develop, transfer, and implement alternative technologies. As a result, MB use in the soil sector has almost been completely phased out in 2005 and 2015, respectively, in developed and developing countries except for critical exemptions uses.

Technical innovation, industries' and farmers' involvement, market transformation, and technology transfer were critical to the success of the MP. The implementation of the MP has shown that setting strict phaseout schedules for A5 and non-A5 countries but allowing flexibility in their implementation have promoted innovation of alternatives and speeded their transfer, acceptance, and implementation.

The MP experience demonstrated that the rules are designed to reflect the needs of farmers in both developing and developed countries and are tailored to local environmental conditions, the transition to the new technologies is easy, cost-effective, and environmentally, technically, and economically feasible (Andersen et al. 2007).

MB has been completely been phased out in Africa for soil uses. Unfortunately, many toxic synthetic pesticides to humans and to their environment are still used. About 70 synthetic highly hazardous pesticides have already been banned in the EU but, unfortunately, continue to be used in our continent. Some of these pesticides include glyphosate, Dursban, Paraquat, 1,3 D, Chloropicrin, etc. Therefore, an African agreement and an action plan are needed to substantially and sustainably reduce and then completely phase out synthetic pesticide use in the African continent.

REFERENCES

Andersen, S.O., Sarma, K.M., Taddonis, K.N. (2007). *Technology Transfer for the Ozone Layer: Lessons for Climate Change*. Earth Scan, London, Sterling VA, 418 pp.

ATSDR (Agency for Toxic Substances and Disease Registry) (1992). Toxicological Profile for Bromomethane. U.S.Public Health Service, U.S. Department of Health and Human Services, Atlanta, GA.

De Souza, A., Kedareshwar, P.S.,Sindhoora, K.V. (2013). The neurological effects of methyl bromide intoxication. *Journal of Neurological Science* 335 (1–2), 36–41.

EPA (1992). Methyl bromide (Bromoethane). https://www.epa.gov/sites/production/files/2016-09/documents/methyl-bromide.pdf.

EU (2016). European union, Guidelines on sustainable fumigation. http://ec.europa.eu/environment/life/project/Projects/index.cfm?fuseaction=home.showFile&rep=file&fil=SustUse_Fumigants_Guidelines.pdf.

Hegglin, M.I., Fahey, D.W., Mc Farland, M., Montzaka, S.A., Nash, E.R. (2015). Twenty Questions and Answers About the Ozone Layer: 2014 Update Scientific Assessment of Ozone Depletion: 2014WMO, UNEP, NOAA, NASA, EU, 81 pp.

MBTOC (2019). Report of the Methyl Bromide Technical Options Committee. 2018 Assessment. UNEP, Nairobi, Kenya, 150 pp.

MLF (2020). About the Multilateral Fund. http://www.multilateralfund.org/aboutMLF/default.aspx.

Molina, M.J., Rowland, F.S. (1974). "Stratospheric sink for chlorofluorocarbons: chlorine atom-catalysed destruction of ozone" *Nature* 249 (5460), 810–812.

TEAP (2019a). Technology and Economic Assessment Panel. June 2018 Progress Report. Ozone Secretariat, Nairobi, Kenya.

TEAP (2019b). Report of the Technology and Economic Assessment Panel, Volume 2: Evaluation of 2019 Critical Use Nominations for Methyl Bromide, Final Report, 70 pp. https://ozone.unep.org/science/assessment/teap. ISBN: 978-9966-076-76-2.

UNEP (2013). UNEP guide for national ozone officers, 160 pp.https://www.unenvironment.org/resources/report/guide-national-ozone-officers.

UNEP (2019a). Handbook for the Vienna Convention for the Protection of the Ozone Layer Twelfth edition, 83 pp. https://www.ozone.unep.org.

UNEP (2019b). Handbook for the Montreal Protocol on Substances that Deplete the Ozone Layer Thirteenth edition. https://www.ozone.unep.org.

13 Regulatory Collusion and the Illusion of Safety

Carey Gillam
U.S. Right to Know

CONTENTS

13.1 INTRODUCTION

The formation of the US Environmental Protection Agency (EPA) in 1970 heralded a new age of consumer interest in – and concern for – the condition of the air, water, and land that provide the sustenance for all life forms.

Prodded into awareness by the work of scientist and *Silent Spring* author Rachel Carson, as well as numerous other researchers and consumer advocates, Americans demanded – and received – a sweeping reform effort. President Richard Nixon established the EPA with the aim of cleaning up and protecting America's waterways, slashing toxic emissions fouling the air, and researching and reining in pollutants that might poison the populace.[1]

The agency made important strides in many areas, including deploying financial and human resources to try to clean up hazardous waste sites created by factories, abandoned mine operations, and chemical plants; reining in toxic factory emissions; relocating and assisting people whose communities suffered contamination from toxic chemicals; and working to create corporate accountability for the costs of some cleanup work, such as ocean oil spills.

Some 50 years after its founding, however, the leading US regulatory agency for protection of the environment – and by extension the health of all of us living in that environment – has proven woefully inept in achieving its overall mission. The EPA instead has accumulated a long track record of aligning with powerful industries on policies and practices that endanger, rather than protect, the public.

A close examination of EPA actions shows it to be an agency in which political will often takes precedence over proven science, leading the agency to roll back necessary protective regulations in some areas while simply ignoring or exempting dangerous products and practices in others.

A research project sponsored by Harvard University's Edmond J. Safra Center for Ethics said while the EPA has "many dedicated employees who truly believe in its mission," the agency has been "corrupted by numerous routine practices," including a "revolving door" between EPA and industry in which corporate lawyers and lobbyists gain positions of agency power; constant industry lobbying against environmental regulations; pressure from lawmakers who are beholden to donors; and meddling by the White House.[2]

This is not only a problem for the United States, but for the rest of the world as well because the EPA's reach is far broader than the boundaries of one nation. As a key regulator in one of the world's largest and most important economic engines, the agency wields deep political and cultural influence around the globe.

Under the framework of "international cooperation," for instance, the EPA operates programs in multiple countries – especially those defined as developing or "with economies in transition" – in an effort to align foreign environmental policies with those of the United States. Sub-Saharan Africa is one such country targeted by the program.[3]

The agency also collaborates on policy and public health with regulators from countries as distinct as China and Canada to "promote sustainable development, protect vulnerable populations, facilitate commerce and engage diplomatically around the world." [4]

Even in the absence of formal coordination, regulatory agencies around the world often look to the EPA for guidance in how to address dire environmental problems that sicken and kill and contribute to harmful changes in climate. But an examination of several incidences of questionable actions by the EPA makes it clear that if foreign agencies are to truly protect the public and environmental health of their nations, they must find a way to avoid the corporate influence that runs rampant through the US EPA.

13.2   THE CASE OF DIOXIN

That is no easy task considering that corporate influence upon EPA decision-making is deeply rooted, according to scientist Evaggelos Vallianatos, who worked in the EPA from 1979 to 2004. In his 2014 book, *Poison Spring,* Vallianatos makes the case that the EPA transformed from a public watchdog to a "polluter's protection agency" with public policy largely shaped by lobbyists, not science.[5]

Vallianatos cites as one example the EPA's long-standing defense of dioxins, chemical contaminants formed in pesticide manufacturing and municipal waste incineration, among industrial processes. The EPA delayed human health assessments of dioxin for years despite evidence of carcinogenicity and despite dangerous dioxin contamination of multiple communities. Critics contend that the agency failed to issue a safety standard for dioxin because of heavy pressure from the chemical and defense industries.

In the mid-1980s, research by the environmental activist group, Greenpeace, revealed collusion between the EPA and the paper bleaching industry to hide information about dioxin discharges from paper mills.[6] Leaked documents detailed an industry strategy to influence the EPA's dioxin risk assessment to help the industry avoid liability for dioxin contamination.[7]

Particular influence has been wielded on the EPA by Dow Chemical, which was found to have contributed to dioxin contamination in more than 170 locations around the United States. Dow had long insisted that there were few health effects associated with exposure to dioxins and that the company was not a major source of dioxin pollution. The EPA was so beholden to Dow that in the 1980s, agency officials changed a report to delete certain warnings to the public and details that showed Dow to be the source of dioxin contamination in Michigan waterways.

The collusion was so dire that Congress launched multiple investigations,[8] and Dow was ultimately forced to stop releasing dioxins into waterways while continuing to protest liability for contamination, even while agreeing to certain cleanup operations. A Dow lawyer who helped lead the company's legal strategy for dioxin liability was later appointed as an EPA assistant administrator.[9]

The World Health Organization says that dioxins are now persistent throughout the environment and warns that they can cause a range of health problems, including cancers, reproductive and developmental problems, and damage to the immune system.[10]

13.3   THE CASE OF CHLORPYRIFOS

EPA's allegiance to Dow also showed itself to be a factor in the agency's treatment of an insecticide called chlorpyrifos, introduced by Dow in 1965. The pesticide became widely used in agricultural settings, particularly for treating pests in corn fields. It has also been used for decades by farmers growing soybeans, fruit and nut trees, Brussels sprouts, cranberries, and cauliflower, as well as other row crops. Not surprisingly, residues of the chemical commonly have been found in food in tests conducted by government researchers. Nonagricultural uses include golf courses, turf, green houses, and utilities.

Scientists have found that both prenatal chlorpyrifos exposures and exposure of young children pose dangers, including reduced IQ, attention disorders, and delayed motor development. The American Academy for Pediatrics, which represents more than 66,000 pediatricians and pediatric surgeons, has warned that use of the chemical puts developing fetuses, infants, children, and pregnant women at great risk. And the European Food Safety Authority has stated that there is no safe exposure level of chlorpyrifos.

Dow Chemical rejected evidence of the dangers for years, but in 2020, researchers at the University of Washington reanalyzed early chlorpyrifos studies that were sponsored by Dow and submitted to the EPA and concluded that the information had been falsified. The EPA relied on the false data provided by Dow Chemical during the 1980s and 1990s to allow unsafe levels of chlorpyrifos into American homes, according to the University of Washington researchers.[11]

By the year 2000, there was so much evidence of the dangers of chlorpyrifos that the EPA reached an agreement with Dow to phase out all residential uses of the chemical, and in 2012, chlorpyrifos was banned from use around schools; however, it was not banned from agricultural use, meaning farm workers were exposed along with people eating food or drinking water contaminated with resides of the chemical. In 2015, under pressure from consumer, medical, and scientific groups, EPA scientists said that they could no longer assert there was any level of chlorpyrifos that could be considered safe in food or water.[12]

Under President Barack Obama, the EPA said it would move to ban chlorpyrifos from use in agriculture by early 2017. But after President Donald Trump took office in January 2017, Dow pressured the new administration to back away from the plan to ban the chemical. Dow also contributed $1 million to the Trump inaugural fund. Within less than a month after meeting with Dow, the EPA announced it would not proceed with the planned ban and instead would continue to study the issue.[13] In February 2020, in the face of growing calls for bans around the world, Corteva Agriscience, a successor corporation to a merger of Dow with DuPont, said it would phase out production of chlorpyrifos. The EPA has still declined to ban the chemical; however, and, as of this writing, the chemical remains legal for other companies to make and sell.

13.4   THE CASE OF GLYPHOSATE

Perhaps one of the most glaring examples of EPA and industry collusion involves the agency's backing of the weed-killing chemical glyphosate, patented in the 1970s by Monsanto and used in the company's Roundup brands. Glyphosate became the most widely used herbicide in history, thanks in no small part to the efforts of EPA leaders to back Monsanto's claims of near-complete safety, despite years of data indicating otherwise.[14]

Internal EPA records show that agency scientists were troubled as far back as 1983 when they evaluated a 2-year mouse study and found that some mice exposed to glyphosate developed tumors at statistically significant rates, while nondosed mice developed no such tumors. A February 1984 memo from EPA toxicologist William Dykstra stated: "Review of the mouse oncogenicity study indicates that glyphosate is oncogenic, producing renal tubule adenomas, a rare tumor, in a dose-related manner."[15]

The increased incidences of kidney tumors were worrisome because, while adenomas are generally benign, they have the potential to become malignant and, even in noncancerous stages, they have the potential to be harmful to other organs.

After Monsanto pushed back on the assessment, EPA statistician and toxicology branch member, Herbert Lacayo, wrote in a February 1985 memo that a "prudent person would reject the Monsanto assumption that Glyphosate dosing has no effect on kidney tumor production. Glyphosate is suspect. Monsanto's argument is unacceptable." Eight members of the EPA's toxicology branch signed a consensus review of glyphosate in March 1985 stating they were classifying glyphosate as a Category C oncogene, a substance "possibly carcinogenic to humans."[16]

Monsanto argued that the EPA scientists were wrong and pressured EPA management to overrule the staff scientists and follow Monsanto's determination as to how the data should be assessed in a manner that showed no cancer concern. The company also refused an EPA request to repeat the study.

Despite the objections of EPA staff scientists, higher-ranking EPA officials eventually agreed to classify glyphosate, not as a Category C oncogene, but as a "Group E" chemical, a classification that meant "evidence of non-carcinogenicity for humans."

The EPA stamp of approval allowed glyphosate products to be sprayed directly on food crops and helped promote not only the market for Roundup but also for crops genetically engineered by Monsanto to tolerate being sprayed directly with glyphosate herbicides. The result was an explosion of use in both Monsanto's herbicides and its glyphosate-tolerant GMO seeds that translated to billions of dollars in annual revenues for Monsanto.

Twenty years later, in 2015, the International Agency for Research on Cancer (IARC), an arm of the World Health Organization, reviewed scores of studies examining glyphosate safety and concluded – much as those early EPA scientists had – that glyphosate was a probable human carcinogen.

The IARC decision was announced on Friday, March 20, 2015 and by the following Monday morning, Monsanto regulatory affairs leader Dan Jenkins was calling and emailing EPA officials demanding they "correct" the record on glyphosate. Emails obtained through Freedom of Information requests show Jenkins submitted "talking points" to the EPA to try to contradict IARC.[17]

Several top EPA officials then helped Monsanto slow down a review of glyphosate planned by a US agency called the Agency for Toxic Substance and Disease Registry (ATSDR). The ATSDR is a US public health agency that is part of the US Department of Health and Human Services. Internal emails obtained from the EPA and Monsanto show that Monsanto was concerned that the ATSDR review would draw a conclusion similar to IARC's about glyphosate. The EPA and Monsanto worked together to delay or kill the review.

One internal Monsanto email obtained through litigation quoted EPA official Jess Rowland as saying that if he could kill the review, he should "get a medal."[18] In subsequent communications Monsanto described Rowland as an ally who would be helpful in defending glyphosate.

The ATSDR report, which initially was supposed to be published in late 2015 or early 2016, was delayed until April 2019 because of the EPA pressure. When it was finally released, as Monsanto had feared, it confirmed studies showing increased cancer risk from glyphosate exposures.

Tens of thousands of people filed lawsuits against Monsanto following the IARC classification of glyphosate, alleging that exposure to the company's herbicides caused them to develop non-Hodgkin lymphoma. Three trials led to three jury findings that the company's herbicides did cause cancer and that Monsanto had covered up the risks. [19]

German conglomerate Bayer AG bought Monsanto in 2018, and in 2020 agreed to settle the Roundup litigation for approximately $10 billion.[20]

Yet, the EPA still agreed with Monsanto that there was no evidence that its glyphosate products caused cancer. One curious twist in the saga came when a public comment was filed with the EPA in the fall of 2019 and posted to the glyphosate docket on the agency's website under the name of Patrick Breysse, the ATSDR director who had been pressured by the EPA. The comment warned about the risks of corporate influence over the EPA and pointed to studies showing cancer links to glyphosate. Yet, when the nonprofit public health research group, US Right to Know, sought to discuss the matter with Breysse, the EPA deleted his name from the comment, attributing it instead to "anonymous." The EPA then declared that Breysse did not really submit the critical comment.[21]

More evidence of deep backing for Monsanto and other chemical industry players in Washington, D.C., was seen in an internal Monsanto report that came to light during the cancer litigation over the company's Roundup herbicides.

A corporate intelligence group hired by Monsanto in 2018 "to take the temperature on current regulatory attitudes for glyphosate" reported that the White House could be counted on to defend the company's Roundup herbicides. The report from strategic intelligence and advisory firm Hakluyt reported to Monsanto that a domestic policy adviser at the White House had said: "We have Monsanto's back on pesticides regulation. We are prepared to go toe-to-toe on any disputes they may have with, for example, the EU. Monsanto need not fear any additional regulation from this administration."

In the email accompanying the report, a Hakluyt executive reported to Monsanto that the information related to issues both for the United States and for China. The report noted that the "professional" staff has "sharp" disagreement with the "political" staff in some areas, but that the concerns of some of the professional staffers would not get in the way.[22]

13.5   THE CASE OF DICAMBA

Monsanto, along with BASF and Corteva, also has been the beneficiary of friendly EPA actions with respect to an herbicide called dicamba. Farmers have been using dicamba herbicides for more than 50 years, but traditionally did not apply the herbicide during hot summer months, and rarely, if ever, over large swaths of land, due to the well-known propensity of the chemical to drift far from intended target areas. When dicamba volatizes and moves into orchards, gardens, or farm fields, it can damage and kill nontargeted plants and crops.

Despite knowledge of dicamba's volatility, the EPA approved "new uses" of dicamba starting in 2016 at the request of Monsanto, BASF, and Corteva that allowed farmers to spray the companies' dicamba products during those warm summer growing months. The companies said their new formulations of dicamba would not drift and volatize, and they wanted to encourage sales of Monsanto's dicamba-tolerant soybeans and cotton as well as sales of their new dicamba herbicides. The EPA authorized spraying the companies' dicamba formulations directly over the top of the GMO crops on millions of acres of US farmland. The EPA approval helped Monsanto sell enough of its dicamba-tolerant soybeans and cotton to cover 56 million acres in 2018, more than half of the total US acreage planted to those two crops.

Consumer and farm groups opposed the approvals and warned that widespread crop damage would result from the EPA approvals. Their warnings proved accurate as millions of acres of non-GMO crops suffered dicamba drift damage. In February 2020, a unanimous jury awarded a Missouri peach farmer $15 million in compensatory damages and $250 million in punitive damages to be paid by Monsanto owner Bayer and by BASF for dicamba damage to the farmer's property.

Just 4 months later, on June 3, 2020, the US Court of Appeals for the Ninth Circuit ruled that the EPA broke the law in approving the new dicamba herbicides. The EPA "failed entirely to acknowledge" many risks associated with its approval, the court said.[23] The court declared the products to be illegal immediately upon its ruling. Underscoring its loyalty to industry, the EPA told farmers they could continue to use the dicamba herbicides through the end of July.[24]

13.6   THE CASE OF PFAS

EPA's aloofness, inaction, and/or lax regulations have allowed for such pervasive environmental contamination that public health experts fear that some of them can never be reversed. A group of so-called "forever chemicals" – per- and polyfluoroalkyl substances (PFAS) – is a prime example. These chemicals were used for decades in a number of consumer products as well as in fire-fighting foam. PFAS have been linked to an array of adverse health effects including cancers, impaired immune system function, thyroid disease, and fertility problems. These chemicals, which do not break down over time, have been found in public water supplies, private wells, and groundwater in many cities and towns. They make their way into the human body and are now often found in human blood, urine, breast milk, and umbilical cord blood.

Lawyers and investigative journalists uncovered evidence that 3M, a major PFAS supplier, and DuPont, a manufacturer of consumer products made with PFAS, knew for decades that PFAS were dangerous to public health, but concealed that information. Even after the dangers and the corporate deception came to light in the 2000s, the EPA has been slow to strictly regulate the chemicals.[25] In 2018, the EPA secretly tried to block publication of a federal health study by the ATSDR regarding the extent of PFAS water contamination nationwide. The study was designed to show that the chemicals endanger human health at levels the EPA had been saying were safe. Emails obtained and reported by the Politico news organization revealed that the EPA feared a "public relations nightmare."[26] The report was released the month following publication of the Politico story. It estimated that safe levels of PFAS exposure were seven to ten times smaller than what the EPA had said.[27]

In another example in 2018, EPA Administrator Scott Pruitt overruled senior EPA scientists concerned about air pollution known to cause lung and heart problems in people in order for a new factory to be built in an area of Wisconsin already suffering from high levels of dangerous smog. The company pushing for the factory did not want to have to spend millions of dollars on pollution control devices in order to build the project. Internal EPA emails obtained by environmental groups showed that agency scientists objected and said the approvals were not made on scientific data, but because of political pressure from top administration officials.[28]

13.7 CONCLUSION

The EPA employs about 14,000 full-time staffers, more than half of whom are scientists, engineers, or other environmental protection specialists. History shows that these staff scientists and specialists clash often with agency administrators who are politically appointed. In the 2000s, the Government Accountability Office (GAO) found evidence that EPA scientists were forced to give drafts of their scientific assessments to the White House Office of Management and Budget for review before they could be finalized. The OMB then delayed or killed several of the chemical assessments, the GAO determined. Industry pressure played a role in weakening an EPA program called "IRIS" (Integrated Risk Information System), which was set up to catalogue scientific assessments and help the agency determine safe levels of chemical exposures. IRIS is "an important source of information on health effects that may result from exposure to chemicals in the environment," the GAO wrote in its report.[29] Investigators discovered that while agency scientists completed 32 draft chemical risk assessments from 2006 to 2007, only four were finalized because of political interference.

The GAO audit underscored concerns held by consumer and environmental advocates. One group, the Union of Concerned Scientists (UCS), issued a 2008 report titled "Interference at the EPA," that stated the following:

> EPA scientists apply their expertise to protect the public from air and water pollution, clean up hazardous waste, and study emerging threats such as global warming. Because each year brings new and potentially toxic chemicals into our homes and workplaces, because air pollution still threatens our public health, and because environmental challenges are becoming more complex and global; a strong and capable EPA is more important than ever. Yet challenges from industry lobbyists and some political leaders to the agency's decisions have too often led to the suppression and distortion of the scientific findings underlying those decisions – to the detriment of both science and the health of our nation.[30]

The EPA's fealty to industry is not a new problem by any means, nor is it a bipartisan one, persisting through both Republican and Democrat-run administrations. That said, there have been efforts by some political leaders to separate political pressure from the EPA's regulation of toxic chemicals. President Barack Obama pledged in 2009 to provide new protections for scientists working for the EPA and appointed as EPA administrator a former agency staff scientist who declared the regulation of chemicals as an area ripe for reform. Under the Obama administration, the EPA, along with the US Department of Agriculture and other agencies, was directed to create policies outlawing politically driven manipulation of scientific work produced within those agencies.[31]

Concerns about corrupt and improper influence upon and within the EPA and other federal agencies were heightened when Obama took office after the uncovering of a scandal involving the prior administration in which former Vice President Dick Cheney's staff censured congressional testimony regarding the health consequences of global warming.[32] Despite Obama's pledge, the small progress seen was almost immediately eroded when the Trump Administration took office in early 2017. Since then, the EPA has loosened regulations aimed at protecting water and air quality as well as rules for chemical evaluations. A report by the Office of Inspector General for the EPA said that 400 EPA employees surveyed in 2018 felt a manager had suppressed or otherwise interfered with the release of scientific information.[33] The Trump administration has been particularly zealous in efforts to undermine information regarding climate change and the fossil fuel industry, according to researchers and government watch dog organizations.[34]

Critics point to the fact that industry lobbyists are regularly appointed to top positions within the EPA and other federal agencies and then find lucrative jobs back with industry once they leave their government posts, cementing cozy connections that feed collusion on policies favoring corporate interests over public health interests. Within the first term of the Trump administration, no fewer than 11 former lobbyists were appointed to jobs within the EPA, including the chief administrator's position. At the time of this writing, the EPA is being led by former coal lobbyist, Andrew Wheeler.

The Natural Resources Defense Council, a nonprofit environmental advocacy organization, cites what it calls "top-down contamination" of the EPA and says the "science-backed opinions of career staff count for far less than the needs of corporate polluters." [35]

Jeff Ruch, cofounder of the watchdog group, Public Employees for Environmental Responsibility (PEER), believes the EPA is so broken that it cannot be rebuilt. He believes the US Congress must create a new system for protecting the environment and, by extension, the people who live in it. "How do you make this agency effective at protecting public and environmental health?" Ruch asks. "It goes back to the question of whether it was ever that good at it in the first place. It's our position that no, it never was."[36]

REFERENCES

1. U.S. Environmental Protection Agency, "EPA History, The Origins of the EPA," https://www.epa.gov/history/origins-epa, accessed June 23, 2020.
2. "The Economy of Influence Shaping American Public Health and the Environment." Harvard University Edmond J. Safra Center for Ethics. https://ethics.harvard.edu/economy-influence-shaping-american-public-health-and-environment, accessed July 1, 2020.
3. "Where EPA Works Around the World." U.S. Environmental Protection Agency, International Cooperation. https://www.epa.gov/international-cooperation/where-epa-works-around-world, accessed July 3, 2020.
4. Environmental Protection Agency, International Cooperation, "Collaborating Worldwide," https://www.epa.gov/international-cooperation, accessed July 3, 2020.
5. Vallianatos, E. G. and M. Jenkins, *Poison Spring: The Secret History of Pollution and the EPA*. New York: Bloomsbury Press, 2014.
6. Van Strum, C, P. Merrell, *No Margin of Safety: A Preliminary Report on Dioxin Pollution and the Need for Emergency Action in the Pulp and Paper Industry*. Washington, DC: Greenpeace, 1987.

7. Weisskopf, M. 1987. Paper Industry Campaign Defused Reaction to Dioxin Contamination. Washington Post October 25, 1987 (Washington, DC).

8. Maitland, L.,*EPA Aides Charge Superiors Forced Shift in Dow Study*, New York Times, March 19, 1983. https://www.nytimes.com/1983/03/19/us/epa-aides-charge-superiors-forced-shift-in-dow-study.html.

9. Tabuchi, H., and T. Adalbjornsson, "From Dow's 'Dioxin Lawyer' to Trump's Choice to Run Superfund," New York Times, July 28, 2018. https://www.nytimes.com/2018/07/28/climate/dow-epa-superfund.html.

10. "Dioxins and their effects on human health." World Health Organization Health Topics. October 4, 2016. https://www.who.int/news-room/fact-sheets/detail/dioxins-and-their-effects-on-human-health.

11. Sheppard, L., S. McGrew, and R. A. Fenske. "Flawed analysis of an intentional human dosing study and its impact on chlorpyrifos risk assessments," Environment International. July 2020. https://www.sciencedirect.com/science/article/pii/S0160412020318602?via%3Dihub.

12. U.S. Environmental Protection Agency. Proposed Rule. ID:EPA-HQ-OPP-2015-0653-0001. Tolerance Revocations: Chlorpyrifos. https://www.regulations.gov/document?D=EPA-HQ-OPP-2015-0653-0001.

13. Friedman, L., "E.P.A. Won't Ban Chlorpyrifos, Pesticide Tied to Children's Health Problems." New York Times, July 18, 2019. https://www.nytimes.com/2019/07/18/climate/epa-chlorpyrifos-pesticide-ban.html.

14. Gillam, C., *Whitewash – The Story of a Weed Killer, Cancer and the Corruption of Science*. Washington: Island Press, 2017.

15. U.S. Environmental Protection Agency memo, "Subject: Glyphosate; oncogenicity study in the mouse," February 10, 1984. https://usrtk.org/wp-content/uploads/2017/06/1984-mouse-is-oncogenic.pdf.

16. U.S. Environmental Protection Agency memo, "Subject: Consensus Review of Glyphosate," March 4, 1985. https://archive.epa.gov/pesticides/chemicalsearch/chemical/foia/web/pdf/103601/103601-171.pdf.

17. Gillam, C., "Collusion or Coincidence? Records Show EPA Efforts to Slow Herbicide Review Came in Coordination with Monsanto," Huffington Post, August 18, 2017. https://www.huffpost.com/entry/collusion-or-coincidence-records-show-epa-efforts_b_5994dad4e4b056a2b0ef02f1.

18. Rosenblatt, J., L. Mulvany, and P. Waldman, "EPA Official Accused of Helping Monsanto "Kill" Cancer Study," Bloomberg, March 14, 2017. https://www.bloomberg.com/news/articles/2017-03-14/monsanto-accused-of-ghost-writing-papers-on-roundup-cancer-risk.

19. Cohen, P., "$2 Billion Verdict Against Monsanto is Third to Find Roundup Caused Cancer," New York Times, May 13, 2019. https://www.nytimes.com/2019/05/13/business/monsanto-roundup-cancer-verdict.html.

20. Cohen, P., "Roundup Maker to Pay $10 Billion to Settle Cancer Suits," New York Times, June 24, 2020. https://www.nytimes.com/2020/06/24/business/roundup-settlement-lawsuits.html.

21. Gillam, C., "EPA removes name of US official from warning of glyphosate cancer links," U.S. Right to Know, July 16, 2020. https://usrtk.org/food-for-thought/epa-removes-name-of-u-s-official-from-warning-of-glyphosate-cancer-links/.

22. Alameda Superior Court for the State of California Case No.: RG17862702 Re: Pilliod v. Monsanto Company, Exhibit A. May 6, 2019. https://usrtk.org/wp-content/uploads/bsk-pdf-manager/2019/05/Monsanto-internal-emails-re-White-House-July-2018.pdf.

23. U.S. Court of Appeals for the Ninth Circuit Case No. 19-70115 Re National Family Farm Coalition, et al., v. U.S. Environmental Protection Agency, Opinion, June 3, 2020. https://usrtk.org/wp-content/uploads/2020/06/Court-decision-on-dicamba.pdf.

24. Polansek, T., "EPA says farmers can use existing supplies of controversial weed killer," Reuters, June 8, 2020. https://www.reuters.com/article/bayer-dicamba-lawsuit/u-s-says-farmers-can-use-existing-supplies-of-controversial-weed-killer-idUSL1N2DL2MS.

25. Lerner, S., "EPA Allowed Companies to Make 40 New PFAS Chemicals Despite Serious Risks," The Intercept, September, 19, 2019. https://theintercept.com/2019/09/19/epa-new-pfas-chemicals/.

26. Snider, A., "White House, EPA headed off chemical pollution study," Politico, May 14, 2018. https://www.politico.com/story/2018/05/14/emails-white-house-interfered-with-science-study-536950.

27. Agency for Toxic Substances and Disease Registry, "PFAs Toxicological Profile Key Messages," June 2018. https://www.atsdr.cdc.gov/docs/PFAS_Public_KeyMessages_June20_Final-508.pdf.

28. Gardner, T., "Emails show Trump EPA overruled career staff on Wisconsin air pollution," Reuters, May 28, 2019. https://www.reuters.com/article/us-usa-epa-smog/emails-show-trump-epa-overruled-career-staff-on-wisconsin-air-pollution-idUSKCN1SY2BP.

29. Government Accountability Office, Report to the Chairman, Committee on Environment and Public Works, U.S. Senate. March 2008. http://www.gao.gov/assets/280/273184.pdf.

30. "Interference at the EPA." The Scientific Integrity Program at the Union of Concerned Scientists. April 2008.   http://www.ucsusa.org/sites/default/files/legacy/assets/documents/scientific_integrity/interference-at-the-epa.pdf.

31. EPA Memorandum from Lisa P. Jackson to All EPA Employees; January 23, 2009. http://web.archive.org/web/20100313093014/http:/blog.epa.gov/administrator/2009/01/26/opening-memo-to-epa-employees.

32. Steven Power, "Cheney Sought to Alter Climate Discussion," The Wall Street Journal, July 9, 2008. https://www.wsj.com/articles/SB121554228130836473.

33. "Further Efforts Needed to Uphold Scientific Integrity Policy at EPA." U.S. Environmental Protection Agency Office of Inspector General, May 20, 2020. https://www.epa.gov/sites/production/files/2020-05/documents/_epaoig_20200520-20-p-0173.pdf.

34. Goldman, G.T., J.M. Carter, Y. Wang, and J.M. Larson. Perceived losses of scientific integrity under the Trump administration: A survey of federal scientists. *PLOS ONE*, April 23, 2020. https://doi.org/10.1371/journal.pone.0231929.

35. Turrentine, J., "The top-Down Contamination of the EPA," Natural Resource Defense Counsel, Natural Resource Defense Counsel, May 31, 2019. https://www.nrdc.org/onearth/top-down-contamination-epa.

36. Ruch, J., in conversation with author, July 2020.

14 The Myth of Substantial Equivalence and Safety Evaluations of Genetically Engineered Crops: A CytoSolve Systems Biology Analysis

V.A. Shiva Ayyadurai and P. Deonikar
International Center for Integrative Systems

CONTENTS

14.1 INTRODUCTION

The safety assessment of genetically modified organisms (GMOs) is a particularly contentious subject. The study, herein, provides, to the authors' knowledge, the first *systems biology* analysis for assessing the safety of GMOs. In this research, a promising computational systems biology method [1] is utilized to couple the dynamics of known perturbations caused by the genetic modification (GM) of CP4 5-enolpyruvylshikimate 3-phosphate synthase (EPSPS) in soybean, also known as Roundup Ready Soya (RRS) [2] with an integrative model of C1 metabolism and oxidative stress (two molecular systems critical to plant function), derived from previously published research [3–5]. Specifically, five biomolecules that include four enzymes: ascorbate peroxidase, catalase, glutathione (GSH) reductase, and superoxide dismutase, as well as one reactive oxygen species (ROS): hydrogen peroxide, are known to perturbed in RRS [6–8].

EPSPS is a critical enzyme that is absolutely necessary for plant survival, as the enzyme is essential to catalyze the final step in the shikimate pathway's biosynthesis of aromatic amino acids [9]. RRS contain the gene, which codes for a glyphosate-insensitive form of EPSPS, obtained from Agrobacterium sp. strain CP4 [9]. When GM occurs, the CP4 gene is incorporated into the plant genome to produce the enzyme CP4 EPSPS, which is insensitive to glyphosate, thus enabling the plant's resistance to glyphosate [10].

To date, regulatory authorities in twelve countries have approved the environmental (commercial) use of CP4 EPSPS GM in at least a total of seven plant species including *Beta vulgaris* L. (sugarbeet), *Brassica napus* L. and *Brassica rapa* L. (oilseed rape and turnip rape, respectively, although both can be referred to as canola), *Gossypium hirsutum* L. (cotton), *Medicago sativa* L. (alfalfa), *Zea mays* L. (maize), and *Glycine max* L. (soybean) [11]. The substantial advantages of such crops to be glyphosate-tolerant have resulted in rapid adoption: 94% of soybean, 91% of cotton, and 89% of corn planted in the United States, as of 2014, are glyphosate-tolerant varieties [12]. However, concerns about the potential health and environmental safety of GMOs have limited the acceptance of such seed lines and food products, particularly in Europe and Japan [10].

The results from this study suggest a substantial difference in the molecular systems of non-GMO and GMO versions of soybean, as observed in the temporal dynamics of two biomarkers, formaldehyde and glutathione, which predict metabolic disruptions in C1 metabolism. In non-GMO plants, formaldehyde, a known toxin, remains at near zero levels, as it is naturally cleared through a process of formaldehyde detoxification, a molecular system resident in all plants, bacteria and fungi [3,4]. Concomitantly, glutathione, a known anti-oxidizing agent, in non-GMO plants, is naturally replenished and remains at non-zero steady state levels, to support such system detoxification of formaldehyde [4]. However, in the GMO case of soybean, or RRS, there is a significant accumulation of formaldehyde and a concomitant depletion of glutathione, suggesting how a "small," single GM can create "large," systemic perturbations to molecular systems equilibria.

The results of this research are particularly relevant, as the United States White House on July 2, 2015, has ordered a review of rules for GM crops [11]. The computational systems biology approach, herein, and the resultant predictions, may inform regulatory agencies in their efforts for "Improving Transparency and Ensuring Continued Safety in Biotechnology," [11] to adopt a systems biology approach using a combination of in silico, computational methods used herein, and subsequent targeted experimental in vitro and in vivo designs, to develop a systems understanding of "equivalence" using biomarkers, such as formaldehyde and glutathione, which predict metabolic disruptions, as criteria for modernizing the safety assessment of GMOs, while fostering a much-needed transparent, collaborative and scientific discourse.

14.1.1 THE CURRENT DISCOURSE ON GMOS

A scientific discourse is particularly important as GMOs are a controversial topic [13–15]. Supporters of GMOs claim that due to the growing world population and shrinking resources such as arable land, there is a dire need for deploying GMOs to keep food production in pace [16-18]. This claim has been refuted by scientists and members of the sustainable agricultural community who assert that practices such as organic, biodynamic and indigenous farming methods, done within local and small farm ecosystems, can provide more than enough food to feed the world's population while avoiding risks to human and environmental health caused by GMO foods and their reliance on man-made pesticides and factory farming methods [19,20].

Proponents of GMOs assert there is no safety concern since GMOs are approved by the United States Food and Drug Administration's (FDA's) framework of substantial equivalence [21], (also referred to as "material difference" [22]), a concept used for assessing "equivalence" or "difference" of GMOs and their non-GMO counterparts, and further argue humans have been plant breeding, a form of "genetic modification," over many millennia. Anti-GMO activists counter that GMOs are unsafe since substantial equivalence is itself unscientific, and contend that targeted GM of specific genes are far different than plant breeding.

There are many aspects of this debate, which span emotional, economic, psychological, political, spiritual, and historical realms [14,23–25]. The methodology for identifying the specific and relevant criteria, used in substantial equivalence, however, appears to be the objective and tangible cause of contention [19]. There is a growing and convergent consensus that new solutions and scientific methods are needed to select the relevant criteria for substantial equivalence [26,27].

14.1.2 A SYSTEMS BIOLOGY OF GMOS

This research aims to provide a rational scientific framework, founded in principles of systems science and molecular systems biology, to discover such criteria to advance safety assessment of determining "equivalence" and/or "difference" of GMOs and their non-GMO counterparts [26,28]. Systems biology, a new discipline emerging from the post-genomic era, provides a much-needed scientific and systems-based framework to explore how genetic modifications, small or large, may affect emergent properties of whole organisms [29,30]. Computational methods emerging from systems biology are used in this study to explore critical molecular pathways and regulatory molecules, involved in critical plant function, as an approach to potentially discover key biomarkers that reflect disruptions to molecular systems equilibria and may be used as more relevant biological criteria for determining substantial equivalence.

Genetic engineering, like the splitting of the atom, is a significant human technological achievement. However, modern safety assessment methods for GMO's have lagged. Systems biology methods, such as CytoSolve, the computational systems biology approach used in this study, emerging out of research from the Massachusetts Institute of Technology (M.I.T.), one of the leading institutions that pioneered critical advancements in genetic engineering [31], now provides a systematic methodology to integrate known molecular pathway knowledge and to mathematically model such information to understand the complexity of biological organisms, towards assessing safety of a GMO with its non-GMO counterpart. Moreover, systems biology approaches can aid in biosafety questions of GMOs in scientific risk assessment for decision-making [32].

The computational systems biology methods employed in this effort provide a framework for not only making predictions but also for informing intelligent in vivo and in vitro experimental designs to verify the predictions observed in this research. This framework aims to foster open, transparent and collaborative scientific discourse, now possible through a combination of in silico, computational methods used herein, and subsequent targeted experimental in vitro and in vivo designs to develop a systems understanding of "equivalence" of GMOs and their non-GMO counterparts, not only in soybean but also across all plants.

14.1.3   Need for Standards for In Vitro and In Vivo Testing of GMOs and Non-GMO

The results from the in silico computational systems biology methods used in the study, suggest a substantial and material difference in a soybean GMO versus its non-GMO counterpart; in particular, the analysis suggests that GMOs in plants lead to significant changes in concentrations of the two biomarkers: formaldehyde and glutathione, suggesting deleterious biological impacts. The next logical phase of this study would be to conduct in vitro and in vivo experiments to verify these predictions.

However, such in vitro and in vivo testing are untenable for at least two reasons. First, it is difficult to acquire source material in an objective and independent manner, while maintaining compliance with existing legal constraints on the use of such GMO source material. Second, given the current environment of debate and controversy, any isolated experiment, done by either proponents of GMOs or those against GMOs, will be vigorously contested, since there are no agreed upon industry standards for conducting such testing to compare a GMO with its non-GMO counterpart.

Given the significant difference in levels of the two biomarkers of formaldehyde and glutathione, across GMO and non-GMO soybean, predicted from the in silico computational results of this research, and given the need to perform such in vitro and in vivo experiments, it becomes imperative to develop such objective industry standards to perform the necessary in vitro and in vivo testing, to advance the current discourse, beyond debate and controversy.

The timely White House initiative for "Improving Transparency and Ensuring Continued Safety in Biotechnology" for reviewing rules concerning safety of biotechnology products [25] perhaps provides a unique opportunity for conducting such discourse to develop the much-needed industry standards for conducting objective in vitro and in vivo testing of a GMO and its non-GMO counterpart.

14.2   SUBSTANTIAL EQUIVALENCE

The concept of substantial equivalence, first appearing in a legislative amendment enacted by President Gerald Ford on May 28, 1976, enabled the FDA to compare the "equivalence" of newly developed medical devices with its traditional counterpart [21] (now also referred to as "material difference" [22]). Substantial equivalence is based on the concept that existing products can serve as a benchmark for assessing the safety of new products. Using this concept, if a new product is found to be "equivalent" to its traditional counterpart, it is deemed safe. The intention of using this concept was toward fast-tracking the manufacturing of low- and moderate-risk medical devices to the market without the requirement of rigorous safety testing [28].

With the advent of GMO crops, substantial equivalence was horizontally adopted from the medical systems and devices industry to the realm of agricultural systems and foods to become the mainstay for safety assessment of GMOs [27]. The Organization for Economic Co-operation and Development (OECD), one of the organizations along with the WHO and FAO focused on providing guidelines for safety assessment of foods derived from biotechnology, first mentioned the term "substantial equivalence" in the context of food safety in 1996 [27]. In 2001, the FDA issued a draft guidance for voluntary labeling of foods developed using bioengineering [22]. The draft guidance reiterates the FDA's 1992 requirements for labeling products that have a significant change in nutrient content, contain a proven allergen, or have a "material difference" from the conventional counterpart. Using this concept, meta-level criteria were identified for substantial equivalence and/or material difference, relative to foods, in terms of: use, nutritional value, composition, nutritional effects, metabolism, and level of undesirable substances [27]. In this application of substantial equivalence, foods derived from GMOs are considered "substantially equivalent" to their non-GMO and traditional counterparts if there are no intended or unintended alterations in the composition, if there are no adverse effects on dietary value, and if they pose no harm to the consumer and environment [27].

The first instance of the application of the substantial equivalence concept to a GMO was for the safety assessment of Flavr Savr tomato in 1994, when it was proven that the GM tomato was equivalent to the wild type, in terms of molecular and chemical composition [27]. The newly introduced traits were further studied and certified by the FDA for safety [27]. Another report on the establishment of compositional equivalence includes details on GM corn and soybean [33]. In this case, the investigators put forth that the compositional variation between GM varieties and their conventional counterparts is encompassed within the natural variability of the crop, and only when there is a difference of $> \pm 20\%$, additional analyses are required [33].

There is, however, significant disagreement concerning the application of substantial equivalence for safety assessment of GMOs even among the scientific community [34]. Anti-GMO activists argue that substantial equivalence is unscientific and outdated since it was originally developed in the 1970s for medical devices, which are not comparable to the complexity of biological systems. They also argue that by using substantial equivalence, industries could try to "have it both ways" [35] by stating that GM foods are novel in certain respects, which allow them to be patented, and in the same breath by using substantial equivalence, they can prove they are "not so novel" and "equivalent" to their non-GMO counterparts, thereby allowing them to mollify safety concerns as they pose no risks to health or environment [35].

The concept has also been criticized as "pseudo-scientific" as it provides excuses for not conducting toxicological tests and prevents further scientific research into the possible risks of GMO-derived foods [34]. Per the current policy, as long as the GM food industries do not market GM foods with an alarmingly different chemical composition from those of foods already on the market, their new GM products are permitted without any safety or toxicological tests [36]. A counter to such views puts forth the position that substantial equivalence can be seen as, "… merely a regulatory shorthand for defining those new foods that do not raise safety issues that require special, intensive, case by case scrutiny," and the principle itself is not intended to be a scientific formulation [35]. In addition, proponents of GMOs and supporters of substantial equivalence mention that critics of the concept have ignored many other quality assurance procedures that plant seeds undergo before sale [35].

What is clear is that the criteria used for establishing substantial equivalence needs to be assessed more closely [26], since criteria used to assess such equivalence may not be sufficient and refined to measure the effects of "small" genetic modifications on the complex and potentially "large" systemic changes in the end food product [34–36]. Moreover, since the methodology used in substantial equivalence plays a significant role in influencing the labeling of GM foods, if a GM food is determined to be substantially equivalent to its non-GM counterpart, based on using a unidimensional criteria such as *nutritional standpoint alone*, then there is no reason why the two sorts of foods should be distinctly labeled, and the freedom of choice among buyers is confused and limited [26]. Others argue that substantial equivalence needs to be adapted to situations where the composition of GMOs has been deliberately altered for novel traits [36]. This argument is based on the rationale that if the wrong cluster of properties or criteria is selected for comparison of GM and non-GM foods, the establishment of equivalence could be influenced [26].

14.2.1 Beyond "Substantial Equivalence": The Need for a Systems-Based Approach

Within the debate, there appears to be an emerging and directional convergence, including from the scientific community, starting in 2000, that perhaps other solutions and newer scientific methods are necessary for identifying criteria and methods for advancing the use of substantial equivalence in assessing safety of GMOs [27,37–40], as best exemplified in this statement:

> Establishing substantial equivalence is not a safety assessment in itself, but is a pragmatic tool to ana-
> lyze the safety of a new food. It goes without saying that in the testing of new foods, use has to be made
> of the latest scientific methods.

[27]

The field of systems theory and systems science may provide a foundational and beneficial perspective toward defining the characteristics of those "latest scientific methods." In systems science, the definition of a "system" originates from generalized systems theory (GST), which arose out of several disciplines, including biology, mathematics, philosophy, and the social sciences [41]. GST came into prominence in the 1950s [41]. Von Bertalanffy began thinking of GST in the 1930s; however, his ideas were not popular at the time and did not receive widespread attention until much later [42]. The aim of GST was to be a "...unifying theoretical construct for all of the sciences" [42].

One broad definition of GST is, "...a set of related definitions, assumptions, and propositions which deal with reality as an integrated hierarchy of organizations of matter and energy" [39]. Another definition is, "...a collection of general concepts, principles, tools, problems, methods, and techniques associated with systems" [40]. From the context of GST, a generalized definition of a system emerges as: "An arrangement of certain components so interrelated as to form a whole" [40].

Since the mid-1960s, systems science and *systems thinking* have evolved into a definitive discipline that is based on a holistic, systems-based approach, which recognizes that systems cannot be understood by taking them apart and studying just their parts. The systems approach arose in contrast to the *reductionism* of the Newtonian method. In reductionism, a system or object is broken down into individual parts. To understand the system, the behavior of each part is studied individually, without considering the interactions among the set of parts [42].

Such a reductionist approach, which focuses on the parts versus the interconnections, while valuable in understanding the individual part, is unable to account for complex and emergent behavior, denoted as *emergent properties*, which manifest as the parts of the system interact together [43]. From these interactions, new properties of the systems "emerge"—properties that cannot be predicted from the properties of any individual part. Biological organisms and food, from a systems theory perspective, are themselves complex systems of interacting systems of interconnected molecular pathways.

Modifications to any component or to the interconnection of a food's molecular system, given its complexity, will likely yield a new system with concomitant variations to its system dynamics and properties that can be wide-ranging depending on the nature of such modifications [43].

14.2.2 The "Old" Biology: A Lesson in Misplaced Criteria of "Substantial Equivalence"

The field of biology is fundamentally an experimental science. Biologists execute many experiments to understand genes, proteins, protein–protein interactions. An example of perhaps the largest experiment in biology is the Human Genome Project (HGP), which began in 1990 and was completed in 2005. The HGP, when it began, was predicated on the hypothesis that what made humans *different or nonequivalent* to a nematode (or worm) was the number of genes. In some sense, the HGP used the criteria of number of genes to determine the "substantial equivalence" of complexity across organisms—the theory being that the number of genes equated to complexity.

Originally, it was estimated that a human had approximately 100,000 genes [44]. The HGP concluded that humans have only 20,000–25,000 genes, far less than what was originally theorized [45], and near the same number of genes as a worm, the nematode *Caenorhabditis elegans*, of approximately 19,000 genes [46]. The genome of the starlet sea anemone, *Nematostella vectensis*, a delicate, few-inch-long animal in the form of a transparent, multi-tentacled tube has approximately 18,000 genes [47].

The HGP revealed that whether human or a worm (or sea anemone), they all have a similar number of genes, but a great difference in complexity of function as whole organisms. This contradiction led biologists to conclude that perhaps the number of genes in the genome is not connected or the basis of "equivalence" with the complexity of an organism and that much of an organism's complexity can be ascribed to regulation of existing genes by other substances (such as proteins) rather than to novel genes [47].

What ironically emerged, therefore, from the HGP is that nature of being human is *not* predicated on the number of genes but rather by the complex interconnection of molecular interactions across the nucleus, cytoplasm, and organelles. Being human is an "emergent property" of those specific interactions. Systems of interconnections, across myriad systems of molecular pathways, determine the difference between a human and a worm, not the number of genes. Therefore, equivalence in biological systems cannot rely on particular arbitrary criteria such as the number of genes but must be determined through a different and non-reductionist approach.

14.3 SYSTEMS BIOLOGY

Systems biology emerges from where the HGP ends and provides such a non-reductionist approach to understand the complexity of biological systems. Reductionist thinking and the central dogma theory of Watson and Crick [48] had emphasized that genes alone are what make us who we are [49]. Systems biology rose in response to this reductionism and focuses attention not on just on one part, such as the genome, but on the complex interaction of systems of systems across genes, proteins, and complex molecular pathways, which are all influenced by an epigenetic layer [50] affected by both endogenous and exogenous systems including nutrition, environment, and perhaps, even thoughts [30].

While systems biology, as a field, is only a decade old, building systems-level understanding of biology is not a new phenomenon. Over 5,000 years ago, many traditional systems of medicine including Siddha, Unani, Ayurveda, and Traditional Chinese Medicine (TCM) proposed systems approaches to describing the whole human physiome [29,51]. During modern times, starting in 1930s, with the concepts of homeostasis [52], allostasis, and biological cybernetics [53], attempts were made to understand biology from a systems level using the modern language of physics and control systems engineering.

Systems biology is now developing a system-level understanding by connecting knowledge at the molecular level to higher-level biological functions [54]. Previous attempts at system-level approaches to biology were primarily focused on the description and analysis of biological systems, limited to the physiological level. Since these approaches had little to no knowledge of how molecular interactions were linked to biological functions—a systems-based biology of connecting molecular interactions to biological functions was not previously possible [54].

Modern systems biology, as a new field of biology, offers the opportunity, as never before, to link the behaviors of molecules to the characteristics of biological systems. This new field is enabling a description of systems of systems (SoS) of cells, tissues, organs, and human beings within a consistent framework governed by the basic principles of physics [54]. This framework, therefore, provides a much needed scientific foundation in the current GMO debate to identify those "latest scientific methods" to evaluate "equivalence" of complex systems, such as GMOs and their non-GMO counterparts.

14.3.1 Computational Systems Biology: Modeling Molecular Systems

A grand challenge of modern systems biology is to develop tools that enable the analysis and modeling of complex cellular functions, including the whole cell, by considering molecular pathways as being the elemental modules of complex cellular functions. Biological systems are thought to have large number of parts almost all of which are related in complex ways [55]. Functionality emerges as the result of interactions between many proteins relating to each other in multiple cascades and in interaction with the cellular environment. By computing these interactions, it can be used to determine the logic of healthy and diseased states [56]. One way to model the whole cell is through a bottom-up reconstruction. Such bottom-up reconstruction, for example, of the human metabolic network, was done primarily through a manual process of integrating databases and pathway models [57].

It is possible, for example, to regard signaling networks as systems that decode complex inputs in time, space, and chemistry into combinatorial output patterns of signaling activity [58]. By treating molecular pathways as modules, our minds can still deal with the complexity [59]. In this way, accurate experimentation and detailed modeling of network behavior in terms of molecular properties can reinforce each other [60]. The goal then becomes that of linking kinetic models on small parts to build larger models to form detailed kinetic models of larger chunks of molecular pathways, such as metabolism, for example, and ultimately of the entire living cell [61].

The value of integrating systems of molecular pathways is to demonstrate that the integrated networks show emergent properties that the individual pathways do not possess, such as extended signal duration, activation of feedback loops, thresholds for biological effects, or a multitude of signal outputs [62]. In this sense, a cell can be seen as an adaptive autonomous agent or as a society of such agents, where each can exhibit a particular behavior depending on its cognitive capabilities.

Unique mathematical frameworks will be needed to obtain an integrated perspective on these complex systems, which operate over wide length and timescales. These may involve a multilayered, hierarchical approach, wherein the overall signaling network, at one layer, is modeled in terms of effective "circuit" or "algorithm" modules [30], and then at other layers, each module is correspondingly modeled with more detailed incorporation of its actual underlying biochemical/biophysical molecular interactions [63]. The mammalian cell may be considered as a central signaling network connected to various cellular machines that are responsible for phenotypic functions. Cellular machines such as transcriptional, translational, motility, and secretory machinery can be represented as sets of interacting components that form functional local networks [64].

As biology begins to move into the post-genomic era, a key emerging question is how to approach the understanding of how complex molecular pathways function as dynamical systems. Prominent examples include multimolecular protein "machines," intracellular signal transduction cascades, and cell–cell communication mechanisms. As the proportion of identified systems involved in any of these molecular pathways continues to increase, in certain instances already asymptotically, the daunting challenge of developing useful models—both mathematical as well as conceptual—for how they work, is drawing increased interest [65].

The scientific methods that emerge from such computational systems biology may provide more resilient and sophisticated tools to explore the effects of modifications to the myriad systems of interconnected molecular pathways inherent in organisms and food itself. Such approaches are likely more relevant and meaningful in assessing "equivalence" of GMO and non-GMO foods rather than reliance on unidimensional criteria such as nutritional value, composition, and nutritional effects based on methods adopted for relatively simpler engineering systems such as medical devices nearly 40 years ago.

14.3.2 CytoSolve®: A Framework Modeling the Whole Cell and Complex Molecular Systems

One aspect of the grand challenge of systems biology is to create a platform to model the whole cell as well as complex molecular systems. This challenge led to the development of CytoSolve, starting in 2003 at the Massachusetts of Technology (M.I.T.) [66]. The development of CytoSolve recognized the siloed nature of biology, where biologists work in isolated and domain-specific groups, to investigate, understand, and document particular molecular pathways. CytoSolve aggregates existing peer-reviewed scientific literature and mines this literature to extract molecular pathways of biological processes. The platform abstracts complex cellular functions as a plurality of such molecular pathways, each of which can be treated as individual model, as illustrated in Figure 14.1.

The CytoSolve platform computationally integrates the individual molecular pathway models, each of which may span multiple spatial and temporal scales, across compartments, cell types, and biological domains [1,59] to provide a computational architecture, as shown in Figure 14.2, for coupling individual molecular pathway models dynamically without the need to create a monolithic model. This approach allows for an inherent scalability to build models of complex biological phenomena, not afforded by other known methods, since the approach obviates the need to create one large monolithic model [1], which can neither be modularly scaled nor maintained, given the dynamic nature of biological research.

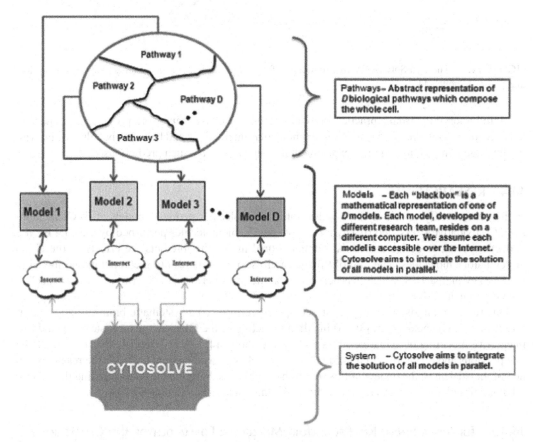

FIGURE 14.1 CytoSolve provides a framework for integrating systems of systems of molecular pathway models [44].

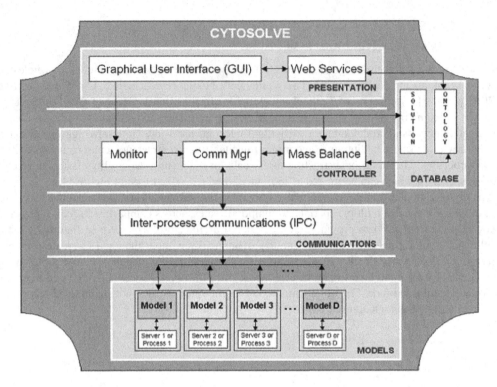

FIGURE 14.2   The CytoSolve software architecture framework for integrating systems of systems of molecular pathway models [44].

More importantly, from a practical standpoint, the CytoSolve framework provides a mechanism not only for making predictions of complex molecular interactions and behavior but also for informing intelligent in vivo and in vitro experimental designs to verify such predictions.

14.4   RESEARCH AIM

In this research, CytoSolve is employed to integrate molecular pathway models of (i) C1 metabolism, (ii) oxidative stress, and (iii) dynamics of specific biomolecules perturbed by the CP4 EPSPS GM of soybean, to derive a modular computational model that predicts the effects of the GMO on regulatory molecules. The aim of this research is to use this resulting computational model to identify key biomarkers, which may serve as more definitive criteria to determine "equivalence" of GMOs and non-GMOs.

This research builds on three recent and specific efforts: (i) a systematic bioinformatics literature review of C1 metabolism [3], (ii) in silico modeling of the C1 metabolism system [4], and (iii) integrative modeling of oxidative stress and C1 metabolism [5]. A brief review of the results from that previous and recent research [4,5] is provided below as a contextual basis for the research aim herein. In particular, the review of this research provides the baseline for understanding the temporal dynamics of key biomarkers associated with plant metabolism in non-GMOs.

14.4.1   IDENTIFICATION OF KEY REGULATORY MOLECULES: FORMALDEHYDE AND GLUTATHIONE

The previous work in in silico modeling of C1 metabolism [3–5] provides a cogent systems biology framework, which demonstrates that formaldehyde (HCHO) and GSH are two important regulatory molecules involved in the control systems of oxidative stress and C1 Metabolism. The temporal dynamics of formaldehyde and GSH, in the non-GMO case, will be provided in Section 14.3.2, as a baseline for comparing with the predictions for the GMO case in Section 14.5.

14.4.2 Review of In Silico Modeling of C1 Metabolism

Recent efforts, using systems biology approaches in the field of plant sciences, have resulted in a comprehensive computational model of C1 metabolism [4]. C1 metabolism is one of the most important biological processes in living systems responsible for providing one-carbon units for proteins, nucleic acids, methylated compounds, and other biomolecules. The C1 metabolism system is mostly found in plants, bacteria, fungi, and mammals [67,68]. A wide variety of important biomolecules are synthesized in C1 metabolism such as methionine, formylmethionine-tRNA, pantothenate, thymidylate, adenosine, and serine. More importantly, the C1 metabolism process provides the one-carbon units essential for DNA methylation, which controls plant growth and development, with a particular involvement in regulation of gene expression and DNA replication [69].

Simulation results from the in silico model of C1 metabolism [4] provide new insights and predictions of temporal changes to formaldehyde, sarcosine, and GSH. The integrative model of C1 metabolism predicts that in normal, non-GMO plants, formaldehyde is evanescently produced and rapidly detoxified between ~1.5 and ~2 days [4] as shown in Figure 14.3a. In non-GMO plants,

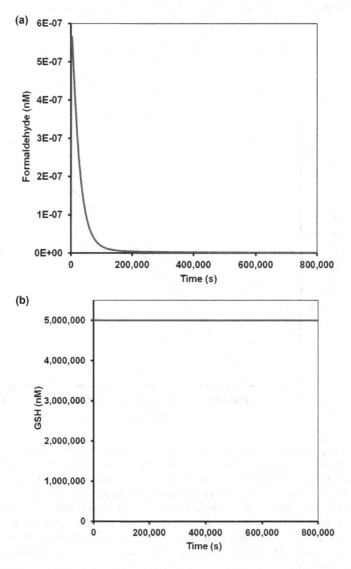

FIGURE 14.3 (a) Temporal dynamics of formaldehyde in non-GMO plants. (b) Temporal dynamics of glutathione in non-GMO plants.

GSH levels are minimally affected and maintain a steady state of 5,000,000 nM [4] as shown in Figure 14.3b (the log-scale version of this figure is in Figure 14.B of Appendix 14.3B).

Finally, in that research, it was predicted that sarcosine is fully consumed during C1 metabolism [55]. Parameter sensitivity analysis of the C1 metabolism model revealed that variations in kGSH-HCHO, the binding rate constant of GSH and formaldehyde (HCHO), affect formaldehyde concentration in normal plants [55]. The sensitivity analysis demonstrated that even an order of magnitude variation in this parameter still results in complete formaldehyde detoxification.

14.4.3 REVIEW OF INTEGRATED MODEL OF OXIDATIVE STRESS WITH C1 METABOLISM

Another related and recent work, concerning C1 metabolism [5], the study employed computational systems biology approaches to explore how dysregulation to C1 metabolism may result from the influence of oxidative stress on C1 metabolism. The simulation results from the in silico modeling of integration of oxidative stress [5] with the fully integrative model of C1 metabolism [4] suggested that in non-GMO plants, oxidative stress causes accumulation of formaldehyde [5] as shown in Figure 14.4a and depletion of GSH [5] as shown in Figure 14.4b (the log-scale version of this figure is in Figure 14.B2 of Appendix 14.B).

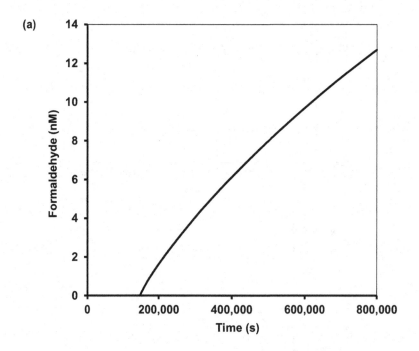

FIGURE 14.4A Temporal dynamics of formaldehyde in non-GMO plants undergoing oxidative stress. Parameter sensitivity analysis, in that study, relative to variations of the binding rate constant of formaldehyde (HCHO) and GSH, kGSH-HCHO, demonstrated that formaldehyde accumulation, as well as GSH depletion, remained [5]. Similarly, relative to variations of the rate of formation of sarcosine from glycine, VMTG, parameter sensitivity analysis demonstrated that formaldehyde accumulation, as well as GSH depletion, remained [5]. Finally, relative to variations of the rate of superoxide production, kO_2^-, parameter sensitivity analysis demonstrated that formaldehyde accumulation, as well as GSH depletion, remained [5]. In summary, there is a consistent accumulation of formaldehyde and depletion of GSH in non-GMO plants, when they undergo oxidative stress.

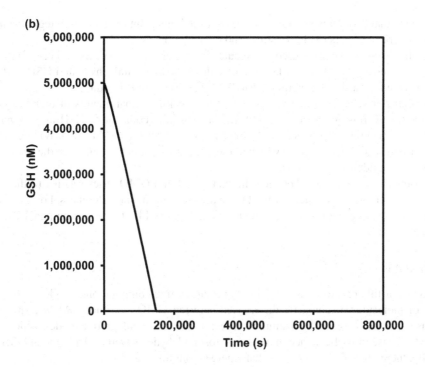

FIGURE 14.4B   Temporal dynamics of glutathione in non-GMO plants undergoing oxidative stress.

14.5   METHODS

In this effort, the CytoSolve® Collaboratory™ [3–5] is used to develop an in silico computational model to understand the effects to C1 metabolism as a result of GMO of soybean [3–5].

First, a systematic bioinformatics literature review is conducted to discover any molecular mechanisms affected by GMO of soybean. Literature collection from an informatics standpoint is executed to ensure high recall to acquire the *initial set*. Based on the research question of "What effect does genetic modification have on C1 metabolism via oxidative stress?" 22 search criteria were developed and are listed in *Supplementary Materials'* Appendix A. Online databases including PubMed and Google Scholar were searched using the search criteria. An initial set was produced as a result of 22 parallel independent searches. The initial set was searched by constraining the search criteria within the Titles or Abstracts to GM, oxidative stress, and C1 metabolism in plants to acquire the *relevant set*.

The papers from relevant set are reviewed by domain experts to determine the *study set* paper, from the relevant set, containing molecular pathway information such as:

1. cellular compartments containing species and reactions
2. kinetics parameters, oxidative stress pathways
3. fold changes in relevant enzymes and key molecular species concentrations.

In this detection process, priority is given to those articles that are the most recent and that contained information and/or studies on oxidative stress and soybean. The final step of this literature review is to discover the dynamics of molecular interactions induced by GMO in soybean.

Second, any dynamics of molecular interactions, induced by GM, identified from the literature review, are incorporated to expand the *systems architecture* for oxidative stress and C1 metabolism, developed in earlier work [5].

Third, the updated systems architecture is used as the blueprint to create an integrative model of how GM of soybean affects oxidative stress and C1 metabolism.

Fourth, the resultant model is used to execute simulations to observe the effects of GM of soybean on the homeostasis of key regulatory molecules such as formaldehyde and GSH. All simulations are executed for a simulation time period of 800,000 seconds (~9 days).

Fifth, parameter sensitivity analysis is performed on kinetic parameters of (i) rate of formaldehyde production from methanol (VCAT), (ii) binding rate constant of GSH and formaldehyde (kGSH-HCHO), (iii) rate of production of sarcosine from glycine (VGMT), and (iv) rate of production of superoxide (kO_2^-) to estimate which of these kinetic parameters influence the key regulatory molecules of formaldehyde and GSH.

Sixth, comparisons are made between the non-GMO and GMO cases relative to the temporal dynamics of key biomarkers to determine if there are any significant differences. For the non-GMO case, data exists from previous work, as reviewed in Figures 14.3 (panel A and panel B) and 14.4 (panel A and panel B).

14.6 RESULTS

The outcomes of this research are twofold: (i) an integrative computational model that allows for the study of molecular mechanistic differences between GMOs and non-GMOs using GMO of soybean as a first use case, and (ii) simulation results using this integrative model, which suggests that in the GM case of soybean, accumulation of formaldehyde as well as depletion of GSH occurs. Specifically, there are six sets of results that emerge from this study.

The first set of results is a systematic literature review of effect of GM on C1 metabolism, described in Section 14.6.1. The second set of results is an integrated computational systems architecture of C1 metabolism, oxidative stress, and GM, in Section 14.6.2, which reveals the interfaces of the interactions between GM, oxidative stress system, and C1 metabolism. The third and fourth sets of results are simulation outputs from the integration of GM and oxidative stress system separately with methionine biosynthesis and formaldehyde detoxification, in Sections 14.6.3.1 and 14.6.3.2, respectively.

The fifth set of results is the simulation output from the integration of GM and oxidative stress system with the *entire* model of C1 metabolism, in Section 14.6.4. Finally, the sixth set of results is the parameter sensitivity analysis, in Section 14.6.5, which provides a detailed understanding of which parameters are most sensitive to variations in the integrative computation of GM, oxidative stress, and C1 metabolism.

14.6.1 SYSTEMATIC BIOINFORMATICS LITERATURE REVIEW OF GMO CROPS AND MOLECULAR PATHWAYS

A systematic bioinformatics literature review is conducted for identification of molecular pathways involved in GMO crops, similar to the method used to identify the key molecular pathways of C1 metabolism [3]. Based on the framing of the research question and the application of the search criteria, in Appendix 14.10.A, through a parallel strategy, the literature collection of an initial set of 107 papers is identified from online databases such as PubMed and Google Scholar. The final results of the systematic review are summarized in Figure 14.5, which identified four critical mechanisms that are affected by the GM of soybean, as well as five biomolecules that included four enzymes and one ROS.

14.6.1.1 Identification of Altered Molecular Interactions from GM Soybean

The 107 papers of the initial set from the systematic bioinformatics literature review yielded important insights, in particular, on the molecular interactions of GM of soybean relative to their effects on oxidative stress. For example, comparative studies of GM of CP4 EPSPS in soybean with non-GM

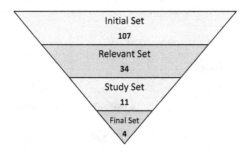

FIGURE 14.5 Systematic review results. There are 107 scientific papers (initial set), which met the search criteria. Of those, 34 papers (relevant set) appeared to be relevant based on the title and abstract. Upon further review, 11 papers (study set) were chosen as the study set upon which this systematic review is based. With this study set, four critical mechanisms (final set) in the oxidative stress pathways were identified that were affected by GM.

plants reported differences in metal uptake ability and content, thereby making the plant susceptible to oxidative stress [7,8].

The GM of CP4 EPSPS in soybean, in particular, has been found to be involved in four molecular mechanisms, as shown in Figure 14.6a and b, which upregulate four key enzymes such as superoxide dismutase, catalase, ascorbate peroxidase, and glutathione reductase and also affect hydrogen peroxide, a ROS across two important molecular systems in oxidative stress [70]: ascorbate–GSH pathway and ROS synthesis pathway. The specific kinetics, relative to the dynamics of these five biomolecules and their molecular interactions, are derived from the literature and provided in the *Supplementary Materials* in Table 14.S1 along with the references.

In addition, GM of CP4 EPSPS in soybean has been estimated to contain higher levels of hydrogen peroxide and malondialdehyde (MDA), indicative of lipid peroxidation [6]. During stressful conditions, such as a drought, it is well known that plants respond to such stressful conditions by altering their gene expression. In the case of GM of soybean with CP4 EPSPS, it has been reported that key enzymes such as catalase, involved in combating oxidative stress, are upregulated in comparison to their non-GM counterparts, indicating changes in cellular redox state [6]. Catalase has a feature of functioning in two modes: (i) the catalytic mode: catalyzing the direct decomposition of hydrogen peroxide, or (ii) the peroxidatic mode: the utilization of hydrogen peroxide to oxidize organic substrates such as methanol, yielding formaldehyde [67,70]. Upregulation of catalase enzyme due to oxidative stress could be a factor in increasing formaldehyde production through its peroxidatic activity.

Substantial literature exists that GSH is an important antioxidizing agent that serves to maintain cellular redox homeostasis. Although in healthy cells, most of the GSH exists in its reduced state, the oxidative stress condition is characterized by the presence of higher amounts of the oxidized form, GSSG. In certain cases, upregulation of GSH reductase, the enzyme catalyzing the conversion of GSSG to GSH to maintain glutathione homeostasis is insufficient to counter the GSH consumption [71].

The literature review reveals that the action of sarcosine oxidase on sarcosine is also known to generate formaldehyde [67]. Formaldehyde is a toxic compound produced during plant C1 metabolism. The main sources of formaldehyde in plants are dissociation of 5,10-methylene-THF and oxidation of methanol [72]. GSH-dependent formaldehyde dehydrogenase is the major enzyme involved in the detoxification of formaldehyde. It acts on a nonenzymatically formed adduct of GSH and formaldehyde [72], indicating the crucial role of GSH in the process.

Finally, relative to formaldehyde detoxification, previous research reveals that GSH-dependent formaldehyde dehydrogenase also acts to detoxify products of lipid peroxidation generated during oxidative stress conditions [73], thereby acting as a competing process to its usual function of formaldehyde detoxification. In addition, results from in silico modeling studies of C1 metabolism and oxidative stress [4,5] conclude that oxidative stress perturbs formaldehyde detoxification in C1 metabolism.

(a)

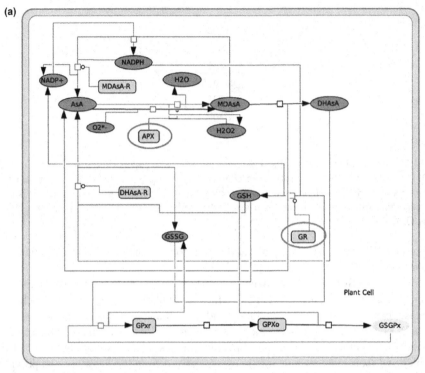

(b)

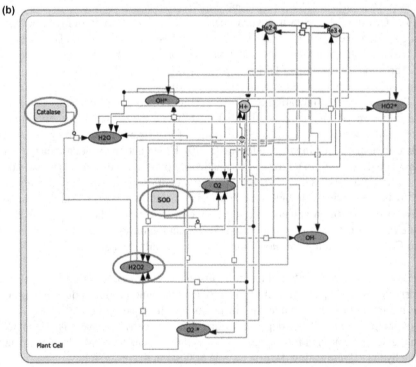

FIGURE 14.6 (a) Ascorbate–glutathione pathway. Molecular interactions of two enzymes, ascorbate peroxidase (APX) and glutathione reductase (GR), circled in red, which are upregulated by GM of soybean in the ascorbate–glutathione pathway [16]. (b) reactive oxygen species synthesis pathway. Molecular interactions of two enzymes, ascorbate peroxidase (APX) and glutathione reductase (GR), circled in red, which are upregulated by GM of soybean in the ascorbate–glutathione pathway [16].

14.6.2 SYSTEMS ARCHITECTURE OF GM, OXIDATIVE STRESS, AND C1 METABOLISM

The literature review of the GM in soybean and its effects on molecular pathways, in Section 14.6.1, provides valuable information on the interface of the GMO with oxidative stress. Earlier work on systematic review and modeling of C1 metabolism and oxidative stress revealed the systems architecture of interfaces between these two molecular systems for C1 metabolism [4,5].

In Figure 14.7, an integrative molecular systems architecture is presented by coupling the dynamics of the GMO molecular interaction in the heretofore known literature, accessible and aggregated by the authors, with the systems architecture of oxidative stress and C1 metabolism derived in earlier work [5].

In Figure 14.7, GM, based on the literature review, interacts with oxidative stress pathways by interfacing with ROS synthesis and ascorbate–GSH pathways. In addition, the oxidative stress pathways interact, as reported in previous research [6–8], with C1 metabolism by interfacing through the ascorbate–GSH pathway, which interfaces with both methionine biosynthesis and formaldehyde detoxification pathways of C1 metabolism. These interfaces will be relevant in developing and testing the in silico modeling of the GMO's effects on C1 metabolism.

14.6.3 INTERACTION OF GM SOYBEAN WITH OXIDATIVE STRESS AND INDIVIDUAL MOLECULAR PATHWAYS OF C1 METABOLISM

The integrative systems architecture of oxidative stress and C1 metabolism, in Figure 14.7, reveals that genetic modification interfaces with oxidative stress pathways, which in turn, as shown previously [5], interface with methionine biosynthesis and formaldehyde detoxification pathways of C1 metabolism.

Relative to the interface between genetic modifications and oxidative stress pathways, genetic modification increases the production of ROS and upregulates the enzymes catalase, ascorbate peroxidase (APX), glutathione reductase (GR), and superoxide dismutase (SOD) in the oxidative stress pathways [74].

Relative to the interface of oxidative stress with the methionine biosynthesis pathway, hydrogen peroxide (H_2O_2), a product of oxidative stress, is used to oxidize glyoxylate, in the methionine biosynthesis pathway, to create formate [75].

Relative to the interface of oxidative stress with the formaldehyde detoxification pathway, GSH, a main substrate for the antioxidant activity of GR in the oxidative pathway [73,74] binds with formaldehyde, which is the first step in clearing formaldehyde [76]. Additionally, catalase, an antioxidant enzyme from the oxidative stress pathway, catalyzes the conversion of methanol to formaldehyde [70].

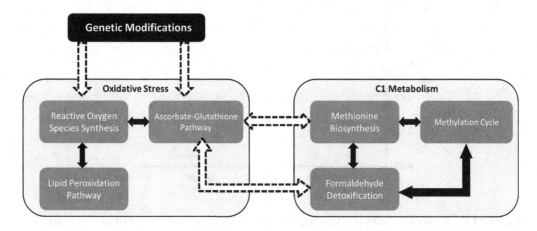

FIGURE 14.7 Systems architecture of GM of soybean on oxidative stress pathway and C1 metabolism.

14.6.3.1 Simulation Results from In Silico Modeling of GM Soybean and Oxidative Stress with *Only* the Methionine Biosynthesis System of C1 Metabolism

Herein, results from in silico modeling of the dynamics of GM of soybean and oxidative stress, with only the methionine biosynthesis system of C1 metabolism, are presented. The results obtained from this integrative model reveal the temporal dynamics of formaldehyde as shown in Figure 14.8. GSH is not observed since it is not a part of methionine biosynthesis system. The simulations are executed for a simulation time period of 800,000 seconds (~9 days).

The concentrations of formaldehyde, as shown in Figure 14.8, increase rapidly and reach a steady state at a concentration level of ~0.06 nM. This result is consistent since methionine biosynthesis is a source of formaldehyde, and the production of formaldehyde will be not affected by any of the by-products of oxidative stress. For example, hydrogen peroxide (H_2O_2), which is a product of oxidative stress, has no effect on the formation of formaldehyde in the methionine biosynthesis model, though H_2O_2 does affect oxidation of glyoxylate to formate [77].

14.6.3.2 Simulation Results from In Silico Modeling of GM of Soybean and Oxidative Stress with Only Formaldehyde Detoxification System of C1 Metabolism

Herein, results from in silico modeling of the dynamics of GM of soybean and oxidative stress with only the formaldehyde detoxification system of C1 metabolism are presented. The results obtained from this integrative model reveal the temporal dynamics of formaldehyde and GSH, as shown in Figure 14.9a and b, respectively. The simulations were executed for a simulation time period of 800,000 seconds (~9 days). The results show an *accumulation* of formaldehyde concentrations (Figure 14.9a) and *depletion* of GSH concentrations (Figure 14.9b) in the presence of oxidative stress with the GMO.

In Figure 14.9a, the simulation results indicate that the formaldehyde concentration increases after a simulation period of ~180,000 seconds (~2 days) and reaches a level of ~0.25 nM at 800,000 seconds (~9 days). This result can likely be explained by understanding the dynamics of interactions between oxidative stress due to the GMO and formaldehyde detoxification. In the presence of oxidative stress due to the GMO, the synthesis of formaldehyde is increased and formaldehyde

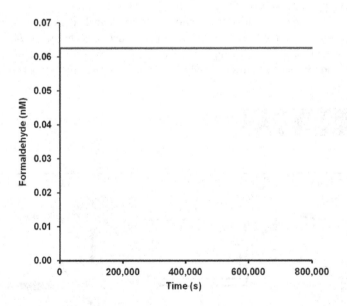

FIGURE 14.8 Simulation results of GM of soybean and oxidative stress on formaldehyde concentration in in methionine biosynthesis model.

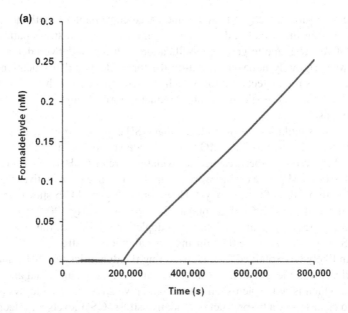

FIGURE 14.9A   Simulation results of GM of soybean and oxidative stress on formaldehyde concentration in formaldehyde detoxification pathway.

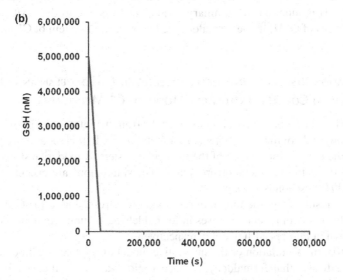

FIGURE 14.9B   Simulation results of GM of soybean and oxidative stress on glutathione (GSH) concentration in formaldehyde detoxification model.

detoxification is lowered leading to accumulation of formaldehyde; however, there is a temporal delay in which the increases in formaldehyde concentrations become apparent.

This temporal delay of increased formaldehyde concentrations is likely because of the time evolution of *two synergistic phenomena*. The first phenomenon results from the accelerated consumption and depletion of GSH, which results in reduced detoxification of formaldehyde. This is because there is a competitive need for GSH to be used: (i) to clear H_2O_2, a by-product of oxidative stress, *and* (ii) to enable the detoxification of formaldehyde. The second phenomenon results from the increased production of formaldehyde from the increased conversion of methanol to formaldehyde by catalase, which is an important and integral enzyme of oxidative stress pathway [70,76]. The coupling of oxidative stress due to genetic modification with formaldehyde detoxification exposes catalase, which was originally nonexistent within the formaldehyde detoxification pathway alone. In summary, oxidative stress affects the formaldehyde detoxification pathway by increasing form-aldehyde concentrations, synergistically through reducing formaldehyde clearance and increasing formaldehyde synthesis.

In Figure 14.9b, the simulation results indicate that GSH concentration varies significantly in the presence of oxidative stress due to the GMO (the log-scale version of this figure is in Figure 14.B3 of Appendix 14.12.B). Without the presence of oxidative stress, GSH levels remain at the steady state value of 5,000,000 nM [4]. In earlier work [5], in the presence of oxidative stress, *without GM*, GSH is depleted within ~180,000 s (~2 days). However, as Figure 14.7b shows, in the presence of oxidative stress, induced by GM, GSH is completely depleted, nearly 300% faster, within ~50,000 seconds (~0.5 days). The significant acceleration in depletion of GSH, in the GMO case, is because of increased ROS synthesis, far more than during normal oxidative stress.

This result can likely be explained by understanding the dynamics of GSH's dual role in oxida-tive stress as well as formaldehyde detoxification. In this simulation, an initial and finite amount of GSH is provided, which is not replenished. In the oxidative stress molecular system alone, where GSH is needed to clear H_2O_2, a by-product of oxidative stress, GSH levels will decrease over time.

In the formaldehyde detoxification system, where glutathione is needed to clear and detoxify formaldehyde accumulation, GSH is used and replenished in a cycle with a temporal periodicity. The simulation reveals that for a finite and initial amount of GSH, the integration of oxidative stress with formaldehyde detoxification will eventually lead to depletion of GSH, notwithstanding any new sources of GSH production. In summary, oxidative stress due to the GMO significantly per-turbs the homeostasis of GSH, in the formaldehyde detoxification system of C1 metabolism.

14.6.4   SIMULATION RESULTS OF IN SILICO MODELING OF GM SOYBEAN AND OXIDATIVE STRESS WITH COMPLETE INTEGRATIVE MODEL OF C1 METABOLISM

The previous Section 14.7.3 provided simulation results from the integration of oxidative stress with only the formaldehyde detoxification system of C1 metabolism. In this section, we present the simu-lation results of the holistic integration of the molecular system of oxidative stress induced by GM of soybean, with the entire C1 metabolism system. The simulations are executed for a simulation time period of 800,000 seconds (~9 days).

The simulation results from the integration of oxidative stress, due to the GMO, with the com-plete C1 metabolism system reveal increases in formaldehyde accumulation and concomitant GSH depletion, as shown in Figures 14.10a and b, respectively.

In Figure 14.10a, the simulation of the integrative model of the GMO with oxidative stress and C1 metabolism indicates that formaldehyde concentration varies significantly in the presence of oxidative stress induced by the GMO. Without the presence of oxidative stress, formaldehyde does not accumulate in the C1 metabolism system [4]. In the presence of oxidative stress induced by the GMO, formaldehyde accumulates in the C1 metabolism system, starting at ~50,000 seconds (~0.5 days), and continues accumulating nonlinearly to ~30 nM in 800,000 seconds (~9 day).

This simulation result is consistent with the previous integration of oxidative stress induced by the GMO with formaldehyde detoxification alone, as shown in Figure 14.9a. There are two key differences, however, in the temporal accumulation of formaldehyde in the integrative model of oxidative stress induced by the GMO and C1 metabolism (Figure 14.10a) versus the interaction of oxidative stress induced by the GMO with formaldehyde detoxification alone (Figure 14.9a). The first difference is that in the integrative model of oxidative stress due to genetic modification and C1 metabolism, formaldehyde accumulation begins nearly four times sooner at ~50,000 seconds

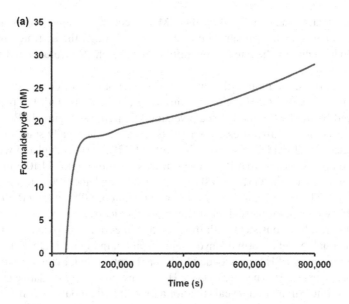

FIGURE 14.10A Simulation results of GM of soybean and oxidative stress on formaldehyde concentration in C1 metabolism model.

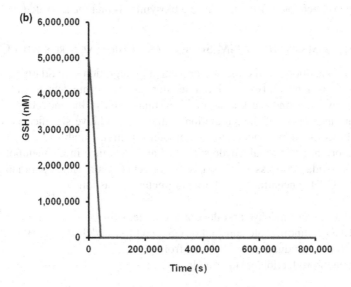

FIGURE 14.10B Simulation results of GM of soybean and oxidative stress on glutathione (GSH) concentration in integrative C1 metabolism model.

(~0.5 days) versus at ~180,000 seconds (~2 days). The second difference is that at 800,000 seconds (~9 days), the formaldehyde concentration in the integrative model of oxidative stress and C1 metabolism is ~120 times higher at ~30 nM versus at ~0.25 nM.

These results indicate that in the integrative model of oxidative stress induced by the GMO and C1 metabolism, formaldehyde accumulation occurs faster and achieves a significantly higher level during the same temporal period. This is likely due to the fact that in the C1 metabolism model, not only are the effects on the formaldehyde detoxification being considered, but also its coupled effects with methionine biosynthesis and the activated methyl cycle. The activated methyl cycle contributes to formaldehyde production from sarcosine [78].

In summary, oxidative stress induced by the GMO affects C1 metabolism by not only increasing the rate and quantity of formaldehyde concentrations through the activated methyl cycle, for example, but also by reducing the rate and quantity of formaldehyde clearance in the formaldehyde detoxification pathway.

In Figure 14.10b, the simulation of the integrative model of the GMO with oxidative stress and C1 metabolism indicates that GSH concentrations vary significantly with the presence of oxidative stress induced by the GMO (the log-scale version this figure is Figure 14.B4 of Appendix B). Without the presence of oxidative stress, GSH levels remain at the steady state value of 5,000,000 nM [4]. In the presence of oxidative stress, without the GMO, GSH is depleted within ~180,000 s (~2 days) based on previous research [5]. However, as shown in Figure 14.10b, in the presence of oxidative stress induced by the GMOs, GSH is completely depleted 300% faster, within ~50,000 seconds (~0.5 days). The significant acceleration in depletion of GSH in the GMO case is because of increased ROS synthesis, far more than during normal oxidative stress.

This simulation result is consistent with the previous integration of oxidative stress induced by the GMO in individual model of formaldehyde detoxification, in Figure 14.9b. In this simulation, an initial and finite amount of GSH is provided, which is not replenished. In the oxidative stress molecular system alone, where GSH is needed to clear H_2O_2, a by-product of oxidative stress, GSH levels will decrease over time. Simulation reveals that for a finite initial amount of GSH, the integration of oxidative stress with C1 metabolism will eventually lead to depletion of GSH, notwithstanding any new sources of GSH production. The close similarity of this result (Figure 14.10b) with the previous result (Figure 14.9b) is because GSH directly affects and couples oxidative stress and formaldehyde detoxification and is decoupled from methionine biosynthesis and the activated methyl cycle.

14.6.5 Parameter Sensitivity of GM Soybean and Oxidative Stress with C1 Metabolism

The results from simulations of the molecular systems integration of oxidative stress and the C1 metabolism provide insights on two key biomolecular species: formaldehyde and GSH. The integrity of literature reviewed and the kinetic rate constants used in the modeling is critical for the interpretation and usefulness of the simulation results. The relative significance of these critical parameters can be assessed by conducting a parameter sensitivity analysis.

Given the importance of formaldehyde synthesis and clearance in C1 metabolism, and central role of GSH in the oxidative stress homeostasis, the effects of four critical parameters are tested on formaldehyde and GSH concentrations. These parameters are as follows:

1. VCAT—Rate of formaldehyde production from methanol
2. kGSH-HCHO—Binding rate constant of GSH and formaldehyde (HCHO)
3. VGMT—Rate of production of sarcosine from glycine
4. $kO_2{}^-$—Rate of production of superoxide.

Four sets of results emerge from the parameter sensitivity analysis of the four parameters itemized above. The first parameter that is analyzed is VCAT. VCAT is varied from 22 to 100 nM s^{-1}, and the resulting formaldehyde and GSH concentrations are observed for the integrated oxidative stress induced by GM of soybean and C1 metabolism model in Figure 14.11a and b, respectively (the log-scale version of Figure 14.11b is in Figure 14.B5 of Appendix 14.12.B).

These results indicate that both formaldehyde and GSH concentrations are not sensitive to changes in VCAT for the integrated oxidative stress induced by the GMO and C1 metabolism model. In all cases, formaldehyde accumulates to the same levels, and GSH (GSH) is fully depleted.

The second parameter that is varied is kGSH-HCHO. kGSH-HCHO is varied from 0.000864 to 0.00864 nM^{-1} s^{-1}, and the resulting formaldehyde and GSH concentrations are observed for the integrated oxidative stress induced by the GMO and C1 metabolism model in Figures 14.12a and b, respectively, (the log-scale version of Figure 14.12b is in Figure 14.B6 of Appendix 14.12.B).

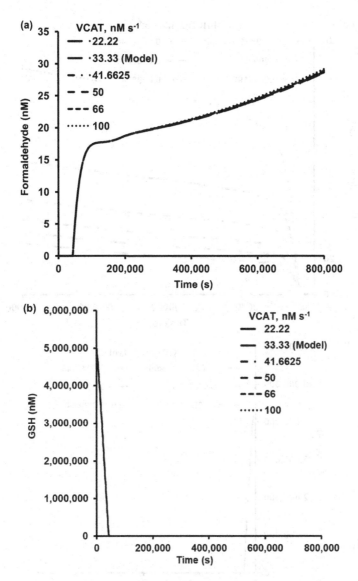

FIGURE 14.11   (a) Parameter sensitivity analysis of VCAT on formaldehyde in the integrated GM soybean and oxidative stress model with C1 metabolism. (b) Parameter sensitivity analysis of VCAT on glutathione (GSH) in the integrated GM of soybean and oxidative stress model with C1 metabolism.

These results indicate that kGSH-HCHO is sensitive and affects formaldehyde concentrations. As Figure 14.12a illustrates, an order of magnitude variation kGSH-HCHO results in a nonlinear variation at ~800,000 seconds (~9 days) of formaldehyde concentrations by approximately six times. These results, relative to GSH, in Figure 14.12b, however, indicate that kGSH-HCHO is insensitive and does not affect GSH concentrations. In all cases, formaldehyde accumulates though to varying levels, and GSH is fully depleted.

The third parameter that is varied is VMTG. VMTG is varied from 20 to 87 nM s⁻¹, and the resulting formaldehyde and GSH concentrations are observed for the integrated oxidative stress and C1 metabolism model in Figure 14.13a and b, respectively, (the log-scale version of Figure 14.13b is in Figure 14.B7 of Appendix 14.12.B).

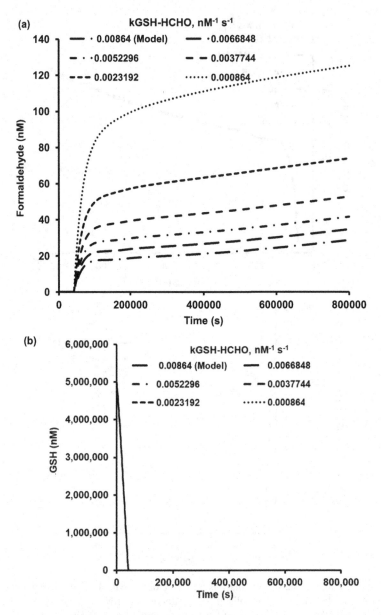

FIGURE 14.12 (a) Parameter sensitivity analysis of kGSH-HCHO on formaldehyde in the integrated GM soybean and oxidative stress model with C1 metabolism. (b) Parameter sensitivity analysis of kGSH-HCHO on glutathione (GSH) in the integrated GM soybean and oxidative stress model with C1 metabolism.

These results indicate that VMTG is sensitive and affects formaldehyde concentrations. As Figure 14.13a illustrates, a four times variation of VMTG results in a nonlinear variation at ~800,000 seconds (~9 days) of formaldehyde concentrations by approximately six times. In all cases, formaldehyde is shown to accumulate consistently and is never depleted. These results, relative to GSH, as shown in Figure 14.13b; however, indicate that VMTG is insensitive and does not affect GSH concentrations. In all cases, formaldehyde accumulates though to varying levels, and GSH is fully depleted.

The fourth parameter that is varied is $kO_2{}^-$. $kO_2{}^-$ is varied from 20 to 100 nM s^{-1}, and the resulting formaldehyde and GSH concentrations are observed for the integrated oxidative stress and C1

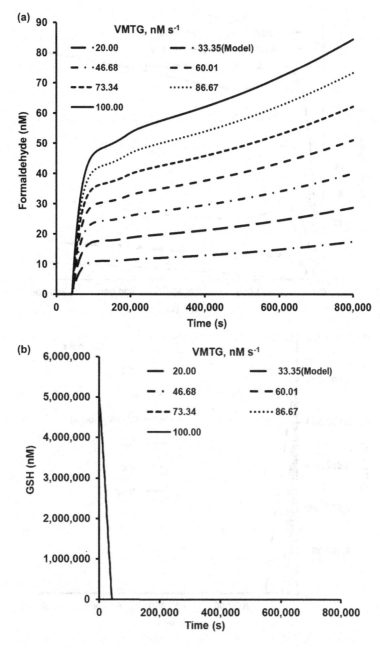

FIGURE 14.13 (a) Parameter sensitivity analysis of VMTG on formaldehyde in the integrated GM of soybean and oxidative stress model with C1 metabolism. (b) Parameter sensitivity analysis of VMTG on glutathione (GSH) in the integrated GM soybean and oxidative stress model with C1 metabolism.

metabolism model in Figures 14.14a and b, respectively (the log-scale version of Figure 14.14b is in Figure 14.B8 of Appendix 14.12.B).

These results indicate that kO_2^- is sensitive and affects formaldehyde concentrations. As Figure 14.14a illustrates, a five times variation in kO_2^- results in a nonlinear variation, at ~800,000 seconds (~9 days), of formaldehyde concentrations by ~ten times. These results indicate that kO_2^- is highly sensitive and affects GSH concentrations. As Figure 14.14b illustrates, a five times variation in kO_2^- results in an acceleration of GSH depletion by a factor of 6. In all cases, formaldehyde

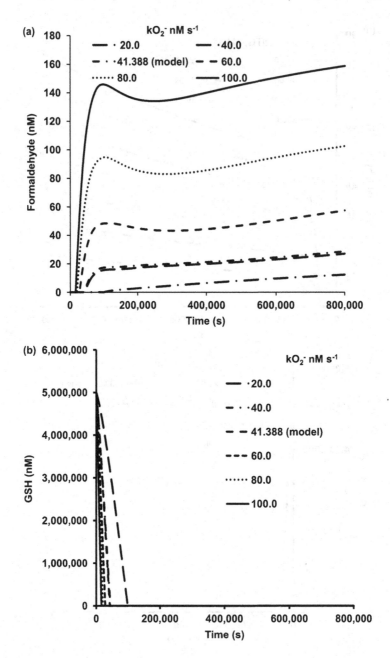

FIGURE 14.14 (a) Parameter sensitivity analysis of kO_2^- on formaldehyde in the integrated GM of soybean and oxidative stress model with C1 metabolism. (b) Parameter sensitivity analysis of kO_2^- on glutathione (GSH) in the integrated GM of soybean and oxidative stress model with C1 metabolism.

consistently accumulates to varying degrees, and GSH, concomitantly, is fully depleted, though temporally accelerated to varying levels.

14.7   DISCUSSION AND CONCLUSIONS

There are many important outcomes and conclusions that are derived from this research on what appears, to the authors' knowledge, to be the first systems biology of GMOs.

First, a scalable and modular in silico computational systems biology framework for understanding GMOs now exists, for not only making predictions but also for informing intelligent in vivo and in vitro experimental designs to verify the predictions observed in this research, using GM of soybean as a use case.

Second, this work, along with the previous work [4,5], now provides a methodology for comparative analysis of the dynamics of two key biomarkers, formaldehyde and GSH (and can be expanded to any number of biomarkers) between the non-GMO and GMO counterpart. In Figure 14.15, the comparative results of the temporal dynamics of formaldehyde in the non-GMO (Figure 14.15a) and GMO (Figure 14.15b) are shown. In Figure 14.16, the comparative results of the temporal dynamics of GSH in the non-GMO (Figure 14.16a) and GMO (Figure 14.16b) are shown.

Without GM, as shown in Figure 14.15a, formaldehyde begins at a peak of ~6×10^{-7} nM and is detoxified within 200,000 seconds (~2.5 days) and does not accumulate. With the GMO, as shown in Figure 14.15b, formaldehyde is not detoxified and begins to accumulate at 50,000 seconds (~0.5 days) and reaches a peak of ~30 nM at 800,000 seconds (~9 days). As discussed, the GMO induces

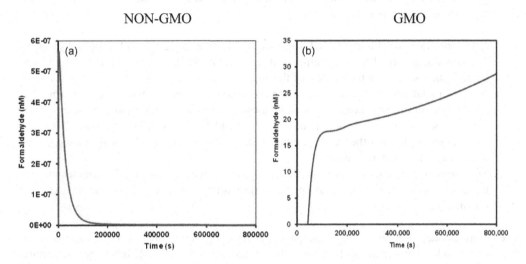

FIGURE 14.15 The temporal dynamics of formaldehyde (HCHO) concentration levels in non-GMO [54] and with the GMO. (a) is the non-GMO, and (b) is the GMO.

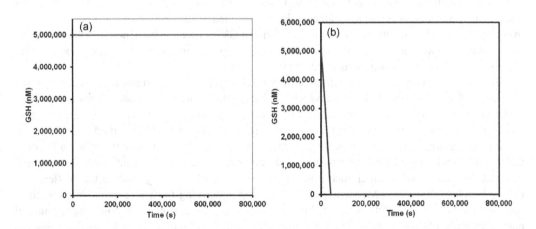

FIGURE 14.16 The temporal dynamics of glutathione (GSH) concentration levels in non-GMO [54] and with the GMO. (a) is the non-GMO, and (b) is the GMO.

oxidative stress, which forces formaldehyde accumulation, and more importantly, unlike normal non-GMO-induced oxidative stress, as shown in Figure 14.4a, the formaldehyde levels produced by the oxidative stress induced by the GMO are different in two ways: (i) concentration levels of formaldehyde, at 800,000 seconds (~9 days), are slightly over two times higher in the GMO case than in the non-GMO case with oxidative stress, and (ii) accumulation in the GMO begins approximately three times sooner case than the non-GMO case with oxidative stress.

Without GM, as shown in Figure 14.16a, GSH remains at a steady-state level of 5,000,000 nM and is maintained for 800,000 seconds (~9 days). With the GMO, as shown in Figure 14.16b, glutathione is completely depleted within 50,000 seconds (~0.5 days). As discussed, the GMO induces oxidative stress that leads to GSH depletion. More importantly, unlike normal non-GMO-induced oxidative stress, as shown in Figure 14.4b, the GSH depletion caused by the oxidative stress induced by the GMO occurs approximately four times faster.

Third, the authors recognize that mathematical modeling, in general, is highly dependent on many variables and assumptions. The models developed and integrated herein are based on literature aggregated, by the authors, from the known and accessible scientific literature. In this study, the critical assumptions are as follows:

1. All the reactions used in the models discussed occur in a single cell and at the cell surface.
2. The cell was assumed to be a well-mixed reactor with uniform concentration of a given biomolecular species in the volume of the cell.
3. All the simulations were performed over a continuous time period without considering the effect of environmental factors such as solar cycle, temperature, and soil condition.
4. The results predict the temporal dynamics of formaldehyde and GSH in the non-GMO and GMO soybean, but not the exact quantitative behavior as may be observed in subsequent in vitro and in vivo experiments.
5. Most importantly, the results from this study are dependent on kinetic parameters and initial conditions of biomolecular species, information of which is based on the existing scientific literature.

The computational systems biology framework, afforded by CytoSolve, recognizes these limitations and provides a promising methodology that is scalable and accessible in an open, transparent, and collaborative manner, to continually integrate and update: (i) new findings from publicly available literature, (ii) corrections to errors in previous research, and (iii) proprietary information, to dynamically reevaluate, expand and refine the models, and publish new predictions.

Fourth, because the system is modular and allows for a "plug-and-play" type methodology for integrating molecular systems, as evidenced by the systematic integration of C1 metabolism with oxidative stress and then with molecular mechanisms of GM soybean, which induce oxidative stress, it is possible to continue to expand and refine this investigation by integrating other biomolecular phenomena and execute simulations under different conditions.

Fifth, the simulation shows that without GM, formaldehyde is produced and is detoxified to near-zero levels. This allows one to understand why, in aggregate, normal, non-GMO plants always have a certain "background level" of formaldehyde, since at any point in time, there is a certain, very low, nonzero amount of formaldehyde, at that particular time, that is yet to be detoxified.

Sixth, concerning whether GMOs are "substantially equivalent" or "materially different" to non-GMOs, this work has focused the discussion on the criteria used for substantial equivalence and material difference and whether such criteria are sensible and can truly capture the "difference" or "equivalence" between GMOs and non-GMOs. The underlying meta-level parameters, such as nutritional value, composition, nutritional effects, and metabolism, used in determining substantial equivalence are philosophically derived from performance parameters used for medical devices and hardware systems, which may not meet the needs for assessing the equivalence of biological organisms.

Seventh, as the HGP demonstrated, attempting to establish substantial equivalence, based on meta-level parameters, such as the number of genes, to establish equivalence or differences between the complexity of two organisms is scientifically unsound. Had science used the number of genes as the criteria to determine the equivalence of organisms, then one would conclude that human beings are "equivalent" to a worm. The reality is humans are different from a worm, regardless of whether they have the same number of genes, because small differences in how genes interact, what proteins they upregulate, the interactions at the epigenetic level, etc., give rise to emergent properties and system dynamics, which ultimately define the difference between a human and a worm and not the number of genes. Even a 0.01% difference in genetic matter can be "substantial" to making a big difference, depending on where such differences lie in the genome and what effects such difference has in the complex interactions of molecular pathways.

Eighth, systems biology and the latest scientific methods, such as CytoSolve, now provide science with a capability and framework, though not perfect or complete, to acknowledge this complexity, and based on the known science, to integrate documented molecular pathway phenomena to predict, or at least know the bounds, of the range of effects that GMOs may have, and more importantly, to discover key regulatory mechanisms and critical molecules, which affect the regulation of these complex processes. Such methods can also be personalized to understand the specific behavior of particular GMOs by incorporating their unique biochemistries to discover other relevant and more specific biomarkers, beyond the two identified from this effort.

Ninth, in this study, through an important use case of GM in soybean, it has been demonstrated how a GMO can induce oxidative stress mechanisms, which then affect important molecular systems such as C1 metabolism, a pathway that is central to the functioning of all plants, bacteria, and fungi. This effort has resulted in the identification of two regulatory molecules, or biomarkers, GSH and formaldehyde, whose levels vary significantly, in the GMO and non-GMO case.

Tenth, this approach may provide a generalized method for discovering molecular mechanistic criteria such as the concentration levels of formaldehyde (HCHO) and GSH, found in this use case, which could perhaps, more systematically and rationally, address the equivalence or nonequivalence of GMOs and their non-GMO counterparts.

Eleventh, more recently, there is a growing confusion, even being promoted by eminent media outlets [17,79] that GMOs are a "natural" process in "genetic modifications" resulting from human involvement in plant breeding. Such arguments point to plant breeding, done by indigenous people over millennia, as an example of this natural process. This research demonstrates that a GM is more akin to a targeted single-molecule drug therapy, which can perturb molecular systems equilibria leading to side effects that can be enormous, uncertain, and far-reaching. Even in the case of drug development, which is by no means perfect, the FDA requires significant testing and clinical trials, spanning upward of 15 years from the time of compound identification, to final approval. Moreover, plant breeding, unlike GM, was done over larger time spans and did involve "genetic modifications," but such modifications were not just a single gene, but likely a choreography of "modifications" that resulted over time where many subsystems and genes were induced and modulated, by many external factors, thus, regulating side effects. At a minimum, given the potentially far-reaching effects of a single GM, as this research suggests for the GMO soybean, at least a process, similar to the FDA clinical trials for single compound drugs, seems rational for GMO safety assessment and approval.

Twelfth, one important question that emerges from this effort is whether in vitro and in vivo testing should have been performed to verify the predictions. However, such testing is beyond the scope of this project for two reasons. First, it is difficult to acquire source material in an objective and independent manner, while ensuring legal compliance. Second, and more importantly, given the current environment, any isolated experiment, done by either proponents of GMOs or those against GMOs, will be vigorously contested since there are no agreed-upon industry standards for conducting such testing to compare a GMO with its non-GMO counterpart.

The substantial and material difference in the levels of biomarkers of formaldehyde and glutathione, across non-GMO and GMO soybean, predicted from this in silico systems biology analysis, however, demands such in vitro and in vivo testing. In order for the results of such testing to be broadly accepted by the scientific community, it is necessary to develop objective *industry standards* that define the exact protocols, processes, and procedures, on how such in vitro and in vivo testing is to be performed. The timely White House initiative for *Improving Transparency and Ensuring Continued Safety in Biotechnology* to review rules concerning safety of biotechnology products [25] perhaps provides an unique opportunity for building consensus to develop these much-needed industry standards to conduct such in vitro and in vivo testing.

In conclusion, systems science may provide the path forward in moving beyond the current debate and controversy, constrained by a reliance on reductionist approaches, to a new paradigm of systems biology that enables a systems understanding of "equivalence" an/or "difference" between a GMO and its non-GMO counterpart.

14.8   FUTURE DIRECTION

The methodology and results of this effort provide many areas of future research. There are four areas, in particular, that are relevant undertakings as logical and immediate next steps.

First, is to test the hypothesis that glyphosate action in endocrine disruption may likely be mediated through its upregulation of formaldehyde. The current framework is poised to conduct such in silico analysis. Second, since C1 metabolism is present in gut bacteria of animals, it is possible to predict formaldehyde accumulation and GSH depletion in the microbiome and its subsequent effect on various disease models affecting the health of the host.

Third, to address the logical question of why a GMO product survives and appears to maintain a phenotypic homeostasis, in spite of the deleterious biological impacts predicted from this research, future research can be conducted to demonstrate that it is likely that the GMO is in a perturbed state and has "adjusted" to an unnatural allostasis, a result of a significant disruption from its normal homeostasis.

Fourth, since little is known concerning the mechanism of methylation and how it affects the regulation of genes, future work can incorporate emerging research toward understanding how changes in the numbers of methyl groups modify methylation processes and how genes are targeted for methylation, affecting epigenetic phenomena in plants.

14.9   SUPPLEMENTARY MATERIALS

TABLE 14.S1

List of Parameters Used in In Silico Models of Oxidative Stress

Kinetic Parameter	Description	Reference
kO_2^-	Rate constant for superoxide production	[1]
kmO_2^-	Michaelis–Menten constant for superoxide production	[1]
kFe^3	Rate constant for the conversion of superoxide to oxygen with simultaneous reduction of Fe^{3+} to Fe^{2+}	[2]
kH_2O_2	Rate constant for the production of hydrogen peroxide and oxygen from superoxide and H^+ (nonenzymatic)	[3]
kSOD	Rate constant for superoxide dismutase producing hydrogen peroxide from superoxide	[3]
KmH_2O_2	Michaelis–Menten constant for catalase-induced conversion of H_2O_2 to H_2O	[4]
Kcata	Rate constant for catalase-induced conversion of H_2O_2 to H_2O	[4]

(Continued)

TABLE 14.S1 (*Continued*)

List of Parameters Used in In Silico Models of Oxidative Stress

Kinetic Parameter	Description	Reference
kFe1	Fenton reaction rate constant (hydrogen peroxide forming hydroxyl radical and anion with simultaneous conversion of Fe^{2+} to Fe^{3+})	[2]
kinitLR	Rate constant for lipid peroxidation reaction by hydroxyl radicals, forming lipid radicals	[5]
kLPO	Rate constant for the oxidation of lipid radicals	[5]
kLR1	Rate constant for the formation of L* and LOOH from LH and LOO*	[5]
kLRFe1	Rate constant for Fe^{2+}-induced formation of LO* from LOOH	[6]
kLRFe2	Rate constant for Fe^{3+}-induced formation of LOO* from LOOH	[6]
kfrLOO	Rate constant for LOO* fragmentation to alkane radical and aldehyde product	[7]
kFe4	Rate constant for OH*-induced formation of HO_2* from H_2O_2	[2]
kFe5	Rate constant for Fe^{3+}-induced formation of HO* from H_2O_2	[2]
kFe8	Rate constant for H_2O_2 formation from HO_2*	[2]
kFe9	Rate constant for the conversion of HO_2* and H_2O_2 to H_2O and OH*	[2]
kFe6	Rate constant of $Fe2+$induced conversion of OH* to OH$^-$	[2]
kFe7	Rate constant for the conversion of OH* and HO_2* to H_2O and O_2	[2]
kdH2O	Dissociation rate of H_2O to H+ and OH-	[8]
KH2O	Association rate of H+ and OH- to H_2O	[8]
kAPX	Rate constant for APX-induced conversion of Ascorbate to MDA	[3]
KAPX	Michaelis–Menten constant for APX-induced conversion of ASC to MDA	[3]
KAPXH	Michaelis–Menten constant for APX-induced conversion of H_2O_2 to H_2O	[3]
k_ASCH$_2$O$_2$	Rate constant for ASC and H_2O_2	[3]
k_ASCO$_2$	Rate constant for superoxide reacting with ascorbate	[3]
kMDAR	Rate constant for molecular MDAR activity	[3]
KMDARM	Michaelis–Menten constant of MDAR for MDA	[3]
KMDARN	Michaelis–Menten constant of MDAR for NADPH	[3]
k_MDAMDA	Apparent rate constant of MDA	[3]
kDAR	Rate constant for molecular DAR activity	[3]
KDAR	Michaelis–Menten constant of DAR for DHA	[3]
KDARG	Michaelis–Menten constant of DAR for GSH	[3]
k_DHAGSH	Apparent rate constant of GSH and DHA	[3]
kGPxr	Rate constant of reduced GPx with H_2O_2	[9]
kGPxo	Rate constant of oxidized GPx with GSH to form intermediate GSGPx	[9]
kGSSG	Rate constant of GSGPx with GSH to recycle reduced Gpx	[9]
kGR	Rate constant for molecular GR activity	[3]
KGR	Michaelis–Menten constant of GR for GSSG	[3]
KGRN	Michaelis–Menten constant of GR for NADPH	[3]
kNAP	Rate constant for the conversion of NADP to NADPH	[3]

14.10 APPENDIX A: LIST OF KEYWORDS

1. Genetic modification oxidative stress signaling pathways
2. Agriculture AND genetic modification
3. Impact of GM CP4 EPSP-induced oxidative stress
4. Population AND food security AND genetic modification
5. Kinetics of iron uptake in plants

6. Hydrogen peroxide and glutathione
7. Hydrogen peroxide and glutathione peroxidase in plants
8. Superoxide production AND photosynthesis
9. Perhydroxyl radical AND oxidative stress in plants
10. Fenton reaction AND oxidative stress in plants
11. Factors affecting formaldehyde dehydrogenase activity AND oxidative stress
12. Formaldehyde dehydrogenase acting on lipid peroxide
13. Lipid peroxide as substrate for formaldehyde dehydrogenase
14. ROS AND catalase expression in plants
15. Competitive inhibitors of formaldehyde dehydrogenase AND plant
16. Requirement of GSH for formaldehyde dehydrogenase activity
17. Glutathione depletion and formaldehyde dehydrogenase
18. Hydrogen peroxide levels AND ascorbate glutathione cycle
19. Iron AND oxidative stress AND glutathione level AND plants
20. EPSP synthase AND photosynthetic electron transport chain
21. Effect of illumination on glutathione reductase activity in chloroplasts
22. Sarcosine oxidase activity versus sarcosine concentration

14.11   SUPPLEMENTARY MATERIALS REFERENCES

1. Asada, K. (1999) The water-water cycle in chloroplasts: Scavenging of active oxygens and dissipation of excess photons. *Annu. Rev. Plant Physiol. Plant Mol. Biol.*, **50**, 601–639.
2. Henle, E.S., Luo, Y., and Linn, S. (1996) Fe 2 +, Fe 3 +, and oxygen react with DNA-derived radicals formed during. *Biochemistry.***35**, 12212–12219.
3. Polle, A. (2001) Dissecting the superoxide dismutase-ascorbate- glutathione-pathway in chloroplasts by metabolic modeling. Computer simulations as a step towards flux analysis. *Plant Physiol.*, **126**, 445–462.
4. Havir, E. and McHale, N. (1989) Enhanced-peroxidatic activity in specific catalase isozymes of tobacco, barley, and maize. *Plant Physiol.*, **91**, 812–815.
5. Antunes, F., Salvador, A., Marinho, H.S., Alves, R., and Pinto, R.E. (1996) Lipid peroxidation in mitochondrial inner membranes. I. An integrative kinetic model. *Free Radic. Biol. Med.*, **21**, 917–943.
6. Xue, C., Chou, C.-S., Kao, C.-Y., Sen, C.K., and Friedman, A. (2013) propagation of cutaneous thermal injury: A mathematical model. *Wound Repair Regen.*, **20**, 114–122.
7. Pratt, D.A., Tallman, K.A., and Porter, N.A. (2011) Free radical oxidation of polyunsaturated lipids: new mechanistic insights and the development of peroxyl radical clocks. *Acc Chem Res.*, **44**, 458–467.
8. Stillinger, F.H. (1978) Proton transfer reactions and kinetics in water. *Theor. Chem.*, **3,** 177–234.
9. Buettner, G.R., Ng, C.F., Wang, M., Rodgers, V.G.J., and Schafer, F.Q. (2006) A new paradigm: Manganese superoxide dismutase influences the production of H_2O_2 in cells and thereby their biological state. *Free Radic. Biol. Med.*, **41**, 1338–1350.

14.12 APPENDIX B: LOG-SCALE FIGURE FOR GLUTATHIONE (GSH) TEMPORAL DYNAMICS

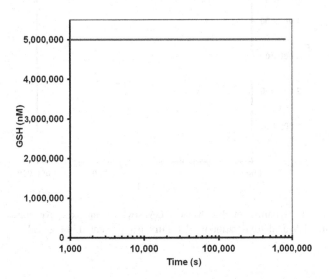

FIGURE 14.B1   Temporal dynamics of glutathione in non-GMO plants. Time is represented in log-scale.

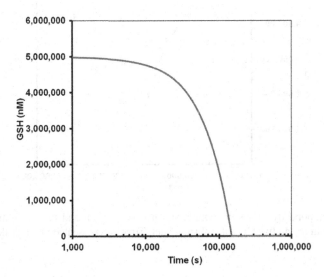

FIGURE 14.B2   Temporal dynamics of glutathione in non-GMO plants undergoing oxidative stress. Time is represented in log-scale.

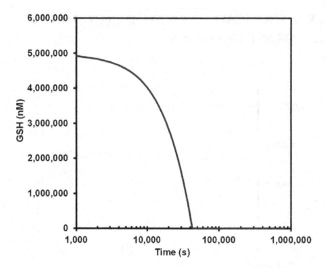

FIGURE 14.B3   Temporal dynamics of glutathione in GM soybean and oxidative stress on glutathione (GSH) concentration in formaldehyde detoxification model. Time is represented in log-scale.

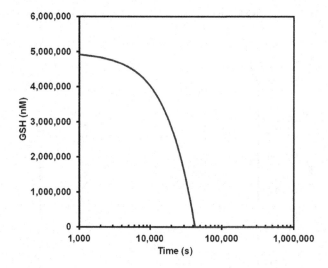

FIGURE 14.B4   Temporal dynamics of glutathione in GM soybean and oxidative stress on glutathione (GSH) concentration in integrative C1 metabolism model. Time is represented in log-scale.

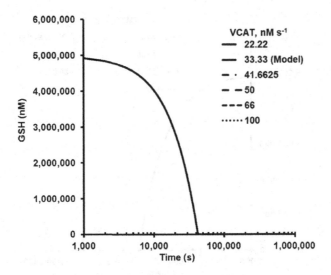

FIGURE 14.B5 Parameter sensitivity analysis of VCAT on glutathione (GSH) in the integrated GM soybean and oxidative stress model with C1 metabolism. Time is represented in log-scale.

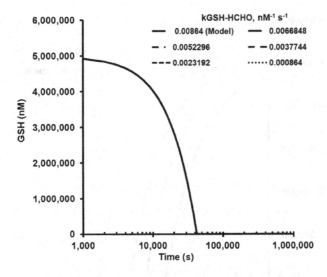

FIGURE 14.B6 Parameter sensitivity analysis of kGSH-HCHO on glutathione (GSH) in the integrated GM soybean and oxidative stress model with C1 metabolism. Time is represented in log-scale.

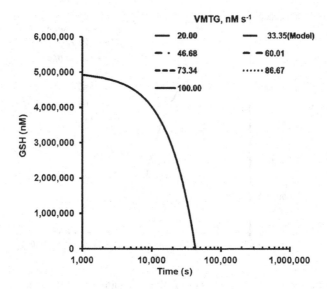

FIGURE 14.B7   Parameter sensitivity analysis of VMTG on glutathione (GSH) in the integrated GM soybean and oxidative stress model with C1 metabolism. Time is represented in log-scale.

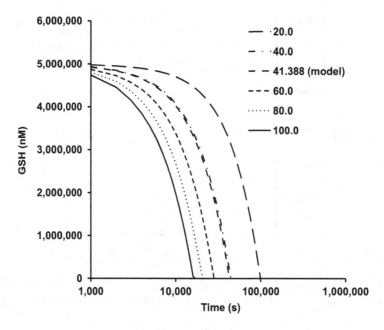

FIGURE 14.B8   Parameter sensitivity analysis of kO_2^- on glutathione (GSH) in the integrated GM soybean and oxidative stress model with C1 metabolism. Time is represented in log-scale.

ACKNOWLEDGMENTS

We thank Drs. Michael Hansen, John Fagan, and Ray Seidler for their feedback.

REFERENCES

1. Ayyadurai, V.A. and Dewey, C.F. (2011) CytoSolve: A scalable computational method for dynamic integration of multiple molecular pathway models. *Cell Mol Bioeng*, **4**(1), 28–45. 10.1007/s12195-010-0143-x.
2. Kodama, T., et al. (2011) Qualitative PCR method for roundup ready soybean: interlaboratory study. *J AOAC Int*, **94**(1), 224–31.
3. Deonikar, P., et al. (2015) Discovery of key molecular pathways of C1 metabolism and formaldehyde detoxification in maize through a systematic bioinformatics literature review. *Agric Sci*, **6**(5), 571–85
4. Kothandaram, S., Deonikar, P., Mohan, M., Venugopal, V., and Ayyadurai, V.A.S. (2015) In silico modeling of C1 metabolism. *Am J Plant Sci*, **6**(9), 1444–65.
5. Mohan, M., Kothandaram, S., Deonikar, P., Venugopal, V., and Ayyadurai, V.A.S. (2015) Integrative modeling of oxidative stress and C1 metabolism reveals upregulation of formaldehyde and down regulation of glutathione. *Am J Plant Sci*, **6**(9), 1527–42.
6. Arruda, S.C., Barbosa, H.S., Azevedo, R.A., and Arruda, M.A. (2013) Comparative studies focusing on transgenic through cp4EPSPS gene and non-transgenic soybean plants: An analysis of protein species and enzymes. *J Proteomics*, **93**, 107–16. 10.1016/j.jprot.2013.05.039.
7. Barbosa, H.S., Arruda, S.C., Azevedo, R.A., and Arruda, M.A. (2012) New insights on proteomics of transgenic soybean seeds: Evaluation of differential expressions of enzymes and proteins. *Anal Bioanal Chem*, **402**(1), 299–314. 10.1007/s00216-011-5409-1.
8. Mataveli, L.R., Pohl, P., Mounicou, S., Arruda, M.A., and Szpunar, J. (2010) A comparative study of element concentrations and binding in transgenic and non-transgenic soybean seeds. *Metallomics*, **2**(12), 800–5.
9. Sammons, R.D. and Gaines, T.A. (2014) Glyphosate resistance: state of knowledge. *Pest Manag Sci*, **70**(9), 1367–77.
10. Funke, T., Han, H., Healy-Fried, M.L., Fischer, M., and Schonbrunn, E. (2006) Molecular basis for the herbicide resistance of roundup ready crops. *Proc Natl Acad Sci USA*, **103**(35), 13010-5. 10.1073/pnas.0603638103.
11. Roberts, A.F., *A Review of the Environmental Safety of the CP4 EPSPS Protein*. 2010, Center for Environmental Risk Assessment, ILSI Research Foundation: Washington D.C.
12. Fernandez-Cornejo, J. *Recent Trends in GE Adoption*. 2014 (cited 2015 July 2); Available from: http://www.ers.usda.gov/data-products/adoption-of-genetically-engineered-crops-in-the-us/recent-trends-in-ge-adoption.aspx.
13. de Vendomois, J.S., et al. (2010) Debate on GMOs health risks after statistical findings in regulatory tests. *Int J Biol Sci*, **6**(6), 590–8.
14. Devos, Y., Sanvido, O., Tait, J., and Raybould, A. (2014) Towards a more open debate about values in decision-making on agricultural biotechnology. *Transgenic Res*, **23**(6), 933–43. 10.1007/s11248-013-9754-z.
15. McHughen, A. (2013) GM crops and foods: What do consumers want to know? *GM Crops Food*, **4**(3), 172–82. 10.4161/gmcr.26532.
16. Ricroch, A.E., and Hénard-Damave, M.C. (2015) Next biotech plants: New traits, crops, developers and technologies for addressing global challenges. *Crit Rev Biotechnol*, **36**, 1–16. **[Epub ahead of print]**
17. Rottman, D. (2013) Why we will need genetically modified foods.
18. Aldemita, R.R., Reaño, I.M.E., Renando, O.S., and Hautea, R.A. (2015) Trends in global approvals of biotech crops (1992–2014). *GM Crops Food Biotech. Agric Food Chain*, doi: 10.1080/21645698.2015.1056972.
19. Domingo, J.L. and Gine Bordonaba, J. (2011) A literature review on the safety assessment of genetically modified plants. *Environ Int*, **37**(4), 734–42. 10.1016/j.envint.2011.01.003.
20. Paoletti, C.F., Flamm, E., Yan, W., Meek, S., Renckens, S., Fellous, M., and Kuiper, H. (2008) GMO risk assessment around the world: Some examples. *Trends Food Sci Technol*, **19**, S70–78.
21. Goldberger, B.A. (2001) The evolution of substantial equivalence in FDA's premarket review of medical devices. *Food Drug Law J*, **56**(3), 317–37.
22. FAD Administration, Draft guidance for industry: Voluntary labeling indicating whether foods have or have not been developed using bioengineering: Notice of availability. 2001. Federal Register.
23. Marris, C. (2001) Public views on GMOs: deconstructing the myths. Stakeholders in the GMO debate often describe public opinion as irrational. But do they really understand the public? *EMBO Rep*, **2**(7), 545–8. 10.1093/embo-reports/kve142.
24. Phillips, T. (2008) Genetically modified organisms (GMOs): Transgenic crops and recombinant DNA technology. *Nature Educ*, **1**(1), 213.

25. Holdren, J.P., Shelanski, H., Vetter, D., and Glodfuss, C. (2015) *Improving Transparency and Ensuring Continued Safety in Biotechnology.* July 2, 2015 (cited 2015 July 4). Available from: https://www.white-house.gov/blog/2015/07/02/improving-transparency-and-ensuring-continued-safety-biotechnology.

26. Borghini, A., Substantial Equivalence. In *Encyclopedia of Food and Agricultural Ethics*, D.M.K. Paul, B. Thompson, (Eds.) 2014, Springer Science+Business Media, Dordrecht, pp. 1–6.

27. Schauzu, M. (2000) The concept of substantial equivalence in safety assessment of foods derived from genetically modified organisms. *Ag. Biotech. Net*, **2**, 1–4.

28. Lennox, K. (2014) Substantially unequivalent: Reforming FDA regulation of medical devices. University of Illinois Law, pp. 1363–1400.

29. Patwardhan, B., Warude, D., Pushpangadan, P., and Bhatt, N. (2005) Ayurveda and traditional Chinese medicine: a comparative overview. *Evid Based Complement Alternat Med*, **2**(4), 465–73. doi:10.1093/ecam/neh140.

30. Ayyadurai, V.A.S. (2014) The control systems engineering foundation of traditional Indian medicine: The rosetta stone for siddha and ayurveda. *Int J Syst Syst Eng*, **5**(2). doi: 10.1504/IJSSE.2014.064836.

31. Mio Technology, *Research.* (cited 2015 July 3). Available from: http://be.mit.edu/research.

32. Tepfer, M., Jacquemond, M., and Garcia-Arenal, F. (2015) A critical evaluation of whether recombination in virus-resistant transgenic plants will lead to the emergence of novel viral diseases. *New Phytol*, doi:10.1111/nph.13358.

33. Harrigan, G.G., et al. (2010) Natural variation in crop composition and the impact of transgenesis. *Nat Biotechnol*, **28**(5), 402–4. doi:10.1038/nbt0510-402.

34. Millstone, E., Brunner, E., and Mayer, S. (1999) Beyond 'substantial equivalence'. *Nature*, **401**(6753), 525–6. doi:10.1038/44006.

35. Miller, H.I. (1999) Substantial equivalence: Its uses and abuses. *Nat Biotechnol*, **17**(11), 1042–3. doi:10.1038/14987.

36. Halford, N.G., et al. (2014) Safety assessment of genetically modified plants with deliberately altered composition. *Plant Biotechnol J*, **12**(6), 651–4. doi:10.1111/pbi.12194.

37. Simo, C., Ibanez, C., Valdes, A., Cifuentes, A., and Garcia-Canas, V. (2014) Metabolomics of genetically modified crops. *Int J Mol Sci*, **15**(10), 18941–66. doi:10.3390/ijms151018941.

38. Vahl, C.I., and Kang, Q. (2015) Equivalence criteria for the safety evaluation of a genetically modified crop: A statistical perspective. *J Agric Sci*. doi:10.1017/S0021859615000271.

39. Miller, J.G., *Living Systems.* 1978, New York: McGraw-Hill.

40. Klir, G.J., *Trends in General Systems Theory.* 1972, New York: Wiley-Interscience.

41. Boulding, K.E., *The Organization Revolution: A Study in The Ethics of Economic Organization.* 1953, New York: Harper.

42. von Bertalanffy, L., *General System Theory.* 1968, New York: George Braziller, Inc.

43. Ideker, T., et al. (2001) Integrated genomic and proteomic analyses of a systematically perturbed metabolic network. *Science*, **292**(5518), 929–34. doi:10.1126/science.292.5518.929.

44. Schuler, G.D., et al. (1996) A gene map of the human genome. *Science*, **274**(5287), 540–6.

45. Pennisi, E. (2003) Human genome. A low number wins the GeneSweep Pool. *Science*, **300**(5625), 1484. doi:10.1126/science.300.5625.1484b.

46. Hodgkin, J. (2001) What does a worm want with 20,000 genes? *Genome Biol*, **2**(11), Comment2008.1.

47. Putnam, N.H., et al. (2007) Sea anemone genome reveals ancestral eumetazoan gene repertoire and genomic organization. *Science*, **317**(5834), 86–94. doi:10.1126/science.1139158.

48. Watson, J.D. and Crick, F.H. (1953) Molecular structure of nucleic acids; A structure for deoxyribose nucleic acid. *Nature*, **171**(4356), 737–8.

49. Schaffner, K.F. (1969) The Watson-Crick model and reductionism. *Brit J Phil Sci*, **20** 325–348.

50. Hood, L., Heath, J.R., Phelps, M.E., and Lin, B. (2004) Systems biology and new technologies enable predictive and preventative medicine. *Science*, **306**(5696), 640–3. doi:10.1126/science.1104635.

51. Subbarayappa, B.V. (1997) Siddha medicine: An overview. *Lancet*, **350**(9094), 1841–4.

52. Cannon, W.B., *The Wisdom of the Body.* 1933, New York: Norton.

53. Wiener, N., *Cybernetics or Control and Communication in the Animal Machine.* 1948, Cambridge: The MIT Press.

54. Kitano, H., *Foundations of Systems Biology.* 2001, Cambridge: The MIT Press.

55. Keller, E.F. (2007) A clash of two cultures. *Nature*, **445**(7128), 603. doi:10.1038/445603a.

56. Noble, D. (2006) Systems biology and the heart. *Biosystems*, **83**(2–3), 75–80. doi:10.1016/j.biosystems.2005.05.013.

57. Duarte, N.C., et al. (2007) Global reconstruction of the human metabolic network based on genomic and bibliomic data. *Proc Natl Acad Sci USA*, **104**(6), 1777–82. doi:10.1073/pnas.0610772104.

58. Bhalla, U.S. (2003) Understanding complex signaling networks through models and metaphors. *Prog Biophys Mol Biol*, **81**(1), 45–65.
59. Ayyadurai, V.A.S. (2011) Services-based systems architecture for modeling the whole cell: A distributed collaborative engineering systems approach. *Commun Med Care Compunetics*, **1**, 115–168.
60. Hornberg, J.J., Bruggeman, F.J., Westerhoff, H.V., and Lankelma, J. (2006) Cancer: A Systems Biology disease. *Biosystems*, **83**(2–3), 81–90. doi:10.1016/j.biosystems.2005.05.014.
61. Snoep, J.L., Bruggeman, F., Olivier, B.G., and Westerhoff, H.V. (2006) Towards building the silicon cell: a modular approach. *Biosystems*, **83**(2–3), 207–16. doi:10.1016/j.biosystems.2005.07.006.
62. Klipp, E. and Liebermeister, W. (2006) Mathematical modeling of intracellular signaling pathways. *BMC Neurosci*, **7** (Suppl 1), S10. doi:10.1186/1471-2202-7-S1-S10.
63. Asthagiri, A.R. and Lauffenburger, D.A. (2000) Bioengineering models of cell signaling. *Annu Rev Biomed Eng*, **2**, 31–53. doi:10.1146/annurev.bioeng.2.1.31.
64. Ma'ayan, A., et al. (2005) Formation of regulatory patterns during signal propagation in a Mammalian cellular network. *Science*, **309**(5737), 1078–83. doi:10.1126/science.1108876.
65. Lauffenburger, D.A. (2000) Cell signaling pathways as control modules: Complexity for simplicity? *Proc Natl Acad Sci USA*, **97**(10), 5031–3.
66. Shiva, V.A., *Scalable computational architecture for integrating biological pathway models*. 2007, Massachusetts Institute of Technology, Biological Engineering Division. p. 2 v. (xvii, 303 leaves).
67. Hanson, A.D. and Roje, S. (2001) One-carbon metabolism in higher plants. *Annu Rev Plant Physiol Plant Mol Biol*, **52**, 119–137. doi:10.1146/annurev.arplant.52.1.119.
68. Hanson, A.D., Gage, D.A., and Shachar-Hill, Y. (2000) Plant one-carbon metabolism and its engineering. *Trends Plant Sci*, **5**(5), 206–13.
69. Vanyushin, B.F. (2006) DNA methylation in plants. *Curr Top Microbiol Immunol*, **301**, 67–122.
70. Havir, E.A. and McHale, N.A. (1989) Enhanced-peroxidatic activity in specific catalase isozymes of tobacco, barley, and maize. *Plant Physiol*, **91**(3), 812–5.
71. Jozefczak, M., Remans, T., Vangronsveld, J., and Cuypers, A. (2012) Glutathione is a key player in metal-induced oxidative stress defenses. *Int J Mol Sci*, **13**(3), 3145–75. doi:10.3390/ijms13033145.
72. Achkor, H., et al. (2003) Enhanced formaldehyde detoxification by overexpression of glutathione-dependent formaldehyde dehydrogenase from Arabidopsis. *Plant Physiol*, **132**(4), 2248–55.
73. Wippermann, U., et al. (1999) Maize glutathione-dependent formaldehyde dehydrogenase: Protein sequence and catalytic properties. *Planta*, **208**(1), 12–8.
74. Chew, O., Whelan, J., and Millar, A.H. (2003) Molecular definition of the ascorbate-glutathione cycle in Arabidopsis mitochondria reveals dual targeting of antioxidant defenses in plants. *J Biol Chem*, **278**(47), 46869–77. doi:10.1074/jbc.M307525200.
75. Martinez, M.C., et al. (1996) Arabidopsis formaldehyde dehydrogenase. Molecular properties of plant class III alcohol dehydrogenase provide further insights into the origins, structure and function of plant class p and liver class I alcohol dehydrogenases. *Eur J Biochem*, **241**(3), 849–57.
76. Nijhout, H.F., Reed, M.C., Budu, P., and Ulrich, C.M. (2004) A mathematical model of the folate cycle: new insights into folate homeostasis. *J Biol Chem*, **279**(53), 55008–16. doi:10.1074/jbc.M410818200.
77. Yokota, A., Kitaoka, S., Miura, K., and Wadano, A. (1985) Reactivity of glyoxylate with hydrogen perioxide and simulation of the glycolate pathway of C3 plants and Euglena. *Planta*, **165**(1), 59–67. doi:10.1007/BF00392212.
78. Goyer, A., et al. (2004) Characterization and metabolic function of a peroxisomal sarcosine and pipecolate oxidase from Arabidopsis. *J Biol Chem*, **279**(17), 16947–53. 10.1074/jbc.M400071200.
79. Brody, J.E., *Fears, Not Facts, Support G.M.O.-Free Food, in Well*. 2015, New York Times: New York.

Index

Note: **Bold** page numbers refer to tables and *italic* page numbers refer to figures.

Printed in the United States
by Baker & Taylor Publisher Services